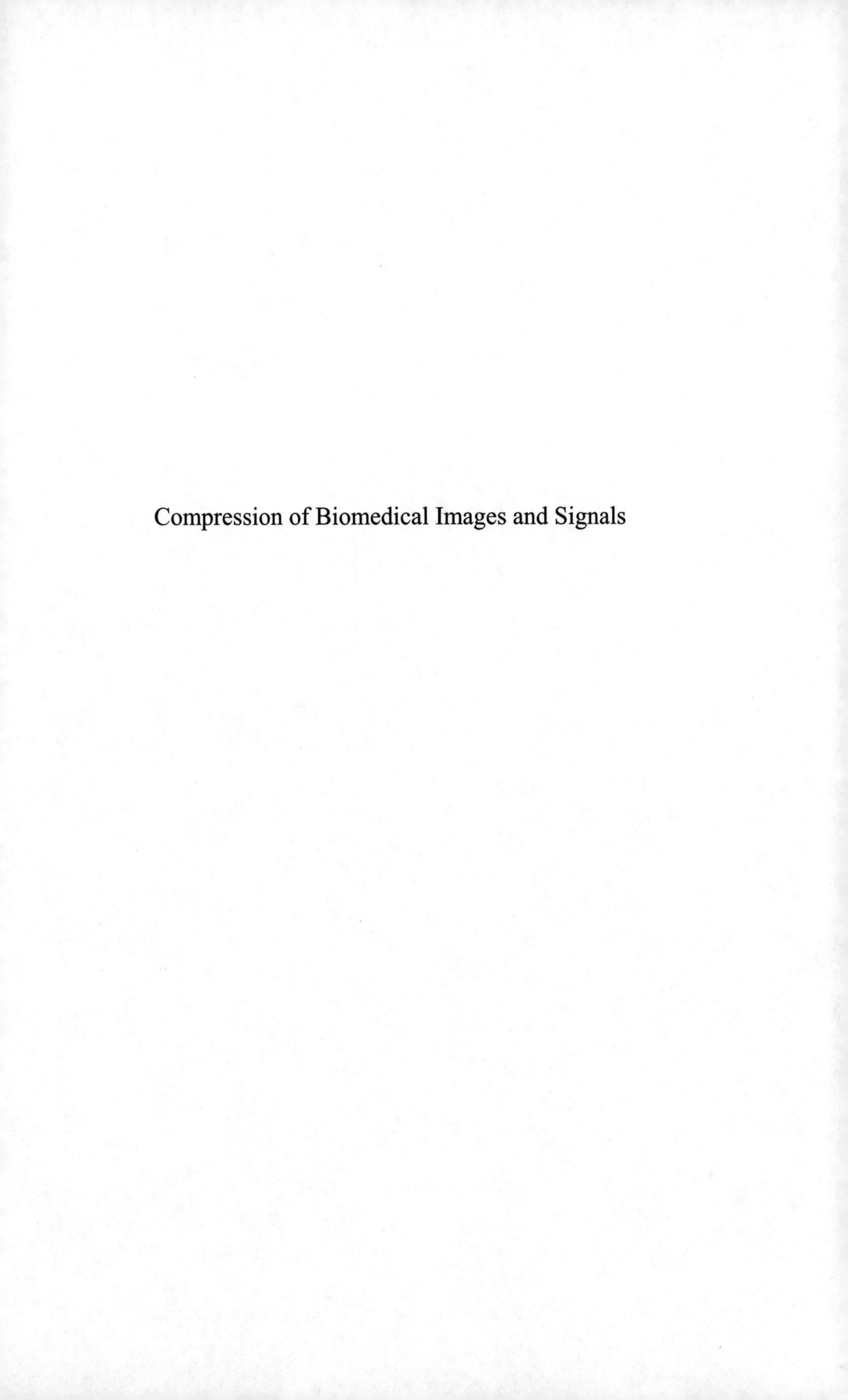

Compression of Biomedical Images and Signals

Compression of Biomedical Images and Signals

Edited by

Amine Naït-Ali

Christine Cavaro-Ménard

First published in France in 2007 by Hermes Science/Lavoisier entitled "Compression des images et des signaux médicaux"
First published in Great Britain and the United States in 2008 by ISTE Ltd and John Wiley & Sons, Inc.

ISTE Ltd
6 Fitzroy Square
London W1T 5DX
UK

www.iste.co.uk

John Wiley & Sons, Inc.
111 River Street
Hoboken, NJ 07030
USA

www.wiley.com

Library of Congress Cataloging-in-Publication Data

Compression des images et des signaux médicaux. English.
Compression of biomedical images and signals / edited by Amine Naït-Ali, Christine Cavaro-Menard.
p. ; cm.
Includes bibliographical references and index.
Translated from French.
ISBN 978-1-84821-028-8
1. Diagnosis--Data processing. 2. Data compression (Computer science) 3. Medical informatics. I. Naït-Ali, Amine. II. Cavaro-Menard, Christine. III. Title.
[DNLM: 1. Data Compression. 2. Diagnostic Imaging. WN 26.5 C7355c 2008a]
RC78.7.D35C63813 2008
616.07'50285--dc22

2008003130

British Library Cataloguing-in-Publication Data
A CIP record for this book is available from the British Library
ISBN: 978-1-84821-028-8

Printed and bound in Great Britain by Antony Rowe Ltd, Chippenham, Wiltshire.

Table of Contents

Preface

Although we might not be aware of this, compression methods are used on a daily basis to store or transmit data. We can find examples of this by looking at our computers (which compress large folders with a simple click of the mouse), our mobile phones (which integrate Codecs), our digital and video cameras (including post-compression recording on flash memory or others), our CD and MP3 players (which are capable of storing hundreds or thousands of songs), our High Definition digital televisions (using the MPEG-2/MPEG-4 compression standards) and our DVD players (which allow us to visualize data in various formats, such as the MPEG-4 format).

Consideration of these can lead us to ask the following question: how does this apply to the medical field?

Although some of the thousands of observations made by physicians are still recorded on paper using radiological film, much of the data acquired (signals, images) are now digital. In order to properly manage the huge amount of medical information, it is essential to exploit all of this digital data efficiently.

It is obvious that most doctors, wherever they are located, would appreciate efficient and fast access to the medical information pertaining to their patient. For instance, suppose that the doctor uses some type of mobile imaging system (for instance, an ultrasound system) for the purpose of analysis. As a consequence, the main clinical observations can be transmitted to a medical center for a preliminary check-up. Of course, in this case, secure data might be transmitted by telephone line or simply through the Internet. In fact, this acquisition/transmission protocol can be established so that the patient could be directed efficiently to the most appropriate clinical service in order to pursue the medical examination further.

Written by Amine NAÏT-ALI and Christine CAVARO-MÉNARD.

For such changes to take place, data compression will be necessary both for the transmission as well as for the storage of all medical information. In fact, many authors have been interested in the medical compression field, and numerous techniques have been dedicated to this purpose. However, as the title *Compression of Biomedical Images and Signals* suggests, we have aimed to work collectively on this topic while giving detailed consideration to the use of recent technology in medicine, focusing particularly on compression.

This book will address questions such as the following: should bioelectric or physiological signals be compressed as audio signals? Should we compress a medical image as if it had been acquired by a simple camera? What about three-dimensional images? In other words, should we directly apply common compression methods to medical data? Should we compress the images with or without losing any information at all? Is there a compression method specific to medical data? In order to answer questions on such a sensitive and delicate topic, we have gathered the skills of over 20 researchers from all corners of France and from various medical and scientific communities including: the signal and image community and the medical community. Such a topic cannot simply be seen from the perspective of a single community, in the sense that one community cannot provide objective judgment on the topic whilst at the same time being involved in its activity. Moreover, a multi-disciplinary reflection is enriching and produces more fruitful work. We therefore hope that this piece of work will serve as a starting point for all young researchers in scientific and medical communities wishing to engage in this particular field. It should thus be used as additional reading to any specialized course module at a Masters level (in science or in medicine).

This book is organized into 11 chapters and structured in the following way.

Chapter 1 describes how important the role of compression is in the medical field. It is built on the observations and points of view of medical experts in images and signals. Their experiences as doctors working in imaging poles have helped us outline the function of medical information compression. It is important to note however that the views upon which our argument is based are specifically relevant to the current state of technological developments (2006) and that innovations in this field are recognized and significant.

Chapter 2 deals with the state of compression methods, and more generally the different compression norms. Some of them can be used to compress medical data while others cannot. Throughout the following nine chapters we will be making constant references to this particular chapter, most notably when comparing the different methods applied to medical data.

Chapter 3 is an introduction to the subsequent chapters. It outlines important features of medical signals and images that are used throughout the discussion in the rest of the book and in various descriptions of certain specific compression methods.

Chapter 4 describes the role of compression norms applied to medical images. This chapter will introduce standardization committees present in the field of medical information exchange as well as the DICOM standard which encompasses almost all medical images. This standard is undergoing constant improvement and incorporates a variety of different compression methods.

Strong compressions with a high risk of information loss are not used in clinical routine for the simple reason that such possible degradations may thwart the medical diagnosis. Chapter 5 outlines the different approaches commonly used to evaluate the quality of reconstructed medical images following lossy compression.

Chapter 6 specifically concerns the compression of physiological signals. Specific attention will be given to electrocardiogram (ECG) compression.

Chapter 7 reviews the different techniques applied (and often adapted) to medical images. It will look at lossless, lossy and progressive compression methods.

Chapter 8 will look into the compression methods of image sequences, represented as videos (2D+t) or as a non-geometrical volume (3D). The use and popularity of this type of imaging is growing rapidly.

Chapter 9 deals more particularly with geometrical (3D) and (3D+t) compression methods. These techniques are particularly interesting today as they have become the main subjects of various studies and practices on organs such as the heart and lungs. This chapter will conclude with a look at potential prospects and opportunities for the use of such methods.

The security aspects of medical imageries will be looked at in Chapter 10. This chapter will also address encrypting techniques.

The final chapter, Chapter 11, looks at wireless transmission of medical images as well as the potential problems that may arise linked to transmission channels. Various solutions will then be suggested as a possible answer to such problems.

Various medical images used as illustrations throughout this work have been taken from the MeDEISA[1] database *Medical Database for the Evaluation of Image and Signal Processing*, created in 2006. This evolving database can be accessed freely through the Internet and gathers a number of images obtained by different acquisition methods (based on recent acquisition systems). Researchers are encouraged to use this database in order to evaluate their own algorithms.

We would like to thank everyone who has participated in the creation of this work. Special thanks go to Christian Olivier and William Puech for their precious help with planning the structure of the book. We would also like to thank Marie Lamy and Helen Bird for the translation and Sophie Fuggle and Amitava Chattejee for their corrections. Thank you all.

---

1 Accessible at http://www.medeisa.net.

Chapter 1

Relevance of Biomedical Data Compression

1.1. Introduction

Medical information, composed of clinical data, images and other physiological signals, has become an essential part of a patient's care, whether during screening, the diagnostic stage or the treatment phase. Data in the form of images and signals form part of each patient's medical file, and as such have to be stored and often transmitted from one place to another.

Over the past 30 years, information technology (IT) has facilitated the development of digital medical imaging. This development has mainly concerned Computed Tomography (CT), Magnetic Resonance Imaging (MRI), the different digital radiological processes for vascular, cardiovascular and contrast imaging, mammography, diagnostic ultrasound imaging, nuclear medical imaging with Single Photon Emission Computed Tomography (SPECT) and Positron Emission Tomography (PET). All these processes (which will be examined in Chapter 3) are producing ever-increasing quantities of images. The same is true for optical imaging: video-endoscopies, microscopy, etc.

The development of this digital imaging creates the obvious problem of the transmission of the images within healthcare centers, and from one establishment to another, as well as the problem of storage and archival. Compression techniques can therefore be extremely useful when we consider the large quantities of data in question.

Chapter written by Jean-Yves TANGUY, Pierre JALLET, Christel LE BOZEC and Guy FRIJA.

Ten years ago, physicians were hostile towards the compression of data. The risk of losing a piece of diagnostic information does not sit well with medical ethics. Failing to identify a life-threatening illness in its early stages due to lost information is unthinkable, given the importance of early diagnosis in such cases. The evolution of digital imaging, retrieval systems and Picture Archiving and Communication Systems (PACS), alongside compression systems, has resulted in changing attitudes, and compression is now accepted and even desired by medical experts.

In this chapter, we will begin by presenting the IT systems which enable the safe archival and communication of medical data, their usefulness and their limitations (section 1.2). Next, with the help of three examples, we will look at the increase – which has been considerable over the past 30 years – in digital data collected in health centers (section 1.3). The problem of the archival and communication of data will then be examined in section 1.4 in relation to clinical practice and legal issues. Each of these areas of comment and debate will help to establish the advantages of compressing medical data, which is the key objective of this chapter. The concerns of the medical community regarding compression, and the ways to tackle these objections, are discussed in section 1.6. The conclusion aims to present possibilities for the foreseeable future, as enabled by compression.

1.2. The management of digital data using PACS

A PACS is composed of an archival system, a quantity (variable in size) of examinations available in real-time from a storage space reachable at high-speed, a system allowing these data to be accessed by those carrying out the examinations, and also a system for the communication of examination results, including images, within a healthcare center and also externally. This communication is generally carried out by a server, on demand, with Internet technology as its basis.

The European countries where this equipment is most prolific are Austria, Norway and Sweden [FOO 01]. In North America and Scandinavia, some establishments are already at the stage where they are re-equipping themselves with these systems.

1.2.1. *Usefulness of PACS*

There are many reasons to support a system for the archival and communication of medical data.

The quality of analyses made can be significantly improved compared to the quality achieved by data stored on film [REI 02]. With PACS, clinicians and radiologists have easy access to data from previous examinations (e.g. images and results), which leads to more reliable hypotheses thanks to the ability to compare a patient's symptoms over time with the known progression of an illness. The time it takes to access and file images is reduced, which allows medical professionals to devote more of their energy to studying the images. Evaluating the progression of a disease – so crucial for chronic illnesses such as cancer – is also made easier.

The time an analysis takes can, thus, be significantly reduced. Some tasks are simply made redundant, such as making telephone calls in order to pass on results, the steps previously needed to display data or check the quality of films, or the searching within medical records to retrieve previous results and to display them alongside the latest data. All of these steps are carried out automatically by a PACS. This leads to two possible benefits: the freeing-up of time to be spent on other tasks or increased productivity. Due to this, it has been observed that productivity increases, leading to a return on the investment in a PACS within three and a half years and real savings to be made from this point onwards [CHA 02]. In this way, the time it takes for the clinician responsible for the patient to receive the examination results is greatly reduced.

A PACS also allows teaching materials to be created more quickly and efficiently than a system that works using copies of films [TRU 05].

1.2.2. *The limitations of installing a PACS*

The main reasons cited for an unwillingness to install a PACS have been: a lack of sufficiently powerful machines for the management of large volumes of medical data, the space required to house such equipment, the time taken to transfer data and the extremely high cost of installation in medical centers. Over time, progress made in the IT field has improved the ratios between the cost and the power of machines on the one hand, and the cost and the storage and transmission capabilities on the other.

Today, the emergence of new techniques such as multi-sectioning scanners and virtual slides in anatomic pathology, and the development of existing techniques such as high resolution digital radiographs, and 3D MRI, have resulted in a significant increase in data quantities, at the very moment when PACS finally seemed to present a feasible ratio of cost to technological advantages.

Nevertheless, sending information outside of the PACS via a low-rate connection remains fraught with problems. The electronic transmission of the results from a biological examination, composed of text and digital data, can be easily carried out, whereas sending image results outside of hospitals or imaging centers is more difficult for numerous reasons. First of all, the DICOM 3 images have to be converted to a commonly-used format, or alternatively a multi-platform image display program must be included with the documents. When dealing with confidential information, security is an issue, as is extranet access. Finally, sending images electronically over a low-rate connection is problematic due to the sheer volume of data, which leads to a very slow transmission speed. However, the need for medical images in the field of telemedicine is great. It can prevent, for example, the need to move a patient from one hospital to another, if a decision can be made based on results in the form of images and signals sent from one physician to another [HAZ 04]. It can also minimize the number of radiological screenings which a patient undergoes, thereby avoiding exposing the patient to excessive quantities of radiation.

1.3. The increasing quantities of digital data

In order to carry out a quantitative analysis of the digital data produced in health centers, three representative fields – each a source of digital data – have been studied: radiology, anatomic pathology and cardiology including the ElectroCardioGram (ECG). The specifics of these fields and others (such as MRI and diagnostic ultrasound) will be presented in Chapter 3.

1.3.1. *An example from radiology*

In radiology, the most obvious example used to demonstrate the increase in the quantity of data collected is that of computed tomography. At the beginning of the 1990s, a scan of the thorax was typically composed of 25 contiguous slices (512 x 512 x 16 bits each), with a thickness of 10 mm, after the injection of a contrast media. The time necessary to acquire and reconstruct a slice on a machine commonly used at the time (CGR CE10000) was approximately 30 seconds. The emergence at the end of the decade of the continuous rotation technique and the spiral computed tomography scanner was incredible progress: scanners can now capture a slice per second. Single-slice devices, thus, collect a quantity of information allowing for the reconstruction of a series of slices of 5 mm, overlapping at 3 mm intervals i.e. 80 slices. At the beginning of the 21st century, multi-slice scanners appeared. For the same clinical condition, today's scan on a model running on 16 channels can carry 600 overlapping slices of 1 mm each, with a matrix of 768 x 768. The quantity of electronic data produced from the same examination, thus, has risen in a few years from 12.5 to 40, and then to 675 MB. As we can see in the trend curve given in Figure 1.1, this increase is nothing short of exponential.

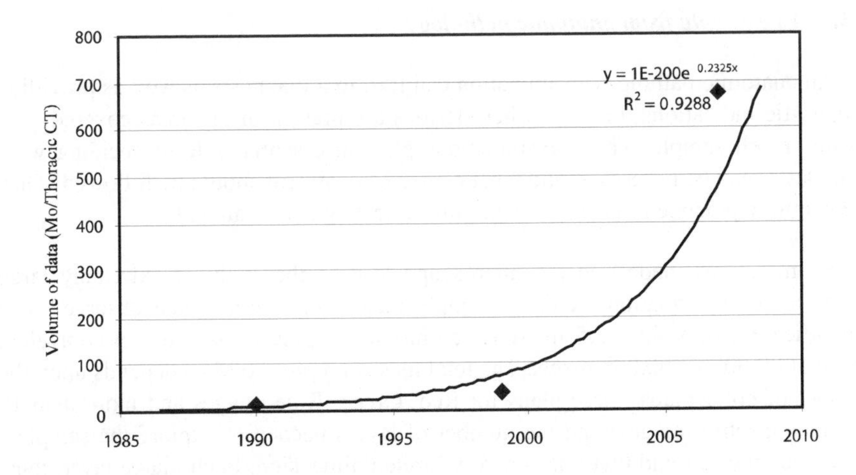

Figure 1.1. *The evolving quantities of image data produced by a thoracic CT examination. The trend curve is given with its equation and the correlation coefficient* R^2

The quantity of data produced by the various sources of medical imaging is thus ever-increasing due to the parallel progress being made in IT and in capture techniques. The number of slices to be studied after each examination is growing at the same rate for every modality. At the same time, the quality of images has been enhanced both in terms of contrast as well as spatial resolution for most modalities. The diagnoses made based on these examinations are therefore becoming more accurate. The price to pay for this improvement is that medical imaging services are carrying out a far greater number of examinations than was previously the case.

In two studies carried out in the radiology departments of 23 US medical centers from 1996 to 2003, using an average workload estimate in a Relative Value Unit form (RVU), and considering both the time required and the difficulty of each procedure, Lu and Arenson [ARE 01] [LU 05] reported that in five years, the number of examinations had increased by 17% Full-Time Equivalent. This increase goes hand-in-hand with a greater average workload for each examination, as shown by a 32% increase in RVU from 1998 to 2003, and as much as 55% when compared with 1996. Indeed, the RVU average per examination increases by 13%, reflecting the developments in slide imaging and in interventional radiology, which have led to a greater complexity in the standard radiological procedure. We can attribute part of this evolution to the need for *a posteriori* use of image treatment software, in order to display the image data in a format accessible to clinicians. If we are to compensate for this extra time spent, we need to reduce the amount of time physicians spend physically organizing the images, whether these images are on film or CD-R, which is where the role of a PACS comes in.

1.3.2. *An example from anatomic pathology*

An anatomic pathology examination can lead to a diagnosis as well as providing prognostic indications, in cases where lesions are present in the areas covered by a tissue or cell sample. These examinations play an essential role in deciding what other tests may be necessary, and what course of treatment should be followed. They are common practice in the process of clinical testing and treatment.

Examinations made under a microscope lead to the study of extremely large quantities of information. A tissue sample measuring 5 cm^2 when observed at a magnification of x 40 (0.25 microns resolution per pixel), takes up an equivalent space of 80,000 x 100,000 pixels. The total mass of a piece of data depends upon the number of color plans (three plans for Red, Green, Blue images and more than 10 for multispectral images) and the number of layers needed to explore the sample's depth, bearing in mind that this is a very limited dimension. Each image layer, thus, takes up 8 or 16 GB, depending on whether it is coded in 8 or 16 bits. This raises the question of how to store and transmit such large quantities of data [WEI 05]. Until recently, there was no effective, practical and repeatable solution for digitizing such material.

Today, the recent development of techniques allowing the quick digitization of whole slides (between one and 20 minutes per slide) (Figure 1.2) allowing for the creation of "virtual slides", alongside the development of viewing systems which are equally effective *in situ* as over a network, have allowed clinicians to increase their productivity through the management of the workflow in the laboratory. It has become possible to relocate the task of examining slides onto the laboratory's network, to carry out quality controls easily, to keep detailed and up-to-date documentation, and to make use of IT for the retrieval of specific elements and for quantification [KAY 06] [GIL 06].

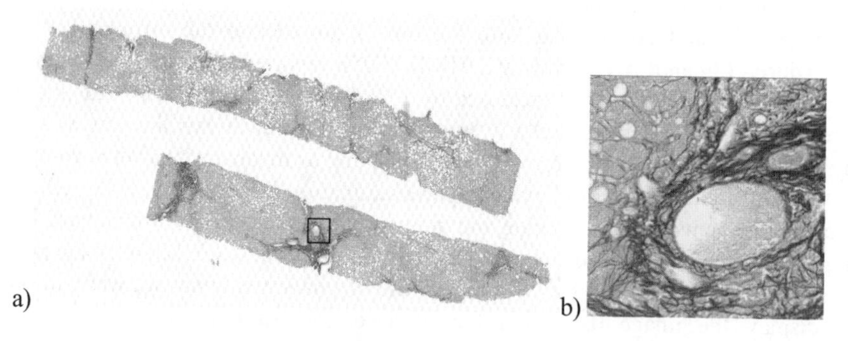

Figure 1.2. *Whole slide image showing a liver sample: a) whole slide image (14,000 x 19,000 pixels coded in 3 x 8 bits); b) detail from the image (256 x 256) equivalent to a x20 zoom*

Certain studies have already claimed that diagnoses made based on digital imaging are reliable [HEL 05], but it is only with the most recent technological developments leading to the production of whole slide images which give a closer reproduction of anatomic pathological material that digital imaging can begin to play a more significant role in the diagnostic process [KAY 06] [GIL 06]. Thus, the speed at which scanning can be performed, and even more importantly, storage volumes, will be key questions in the future.

In the field of anatomic pathology, therefore, we have moved beyond the analog era and into the era of a proliferation of digital data.

1.3.3. *An example from cardiology with ECG*

As current clinical practice shows, studying a signal produced by an ECG on paper is a practice which is beginning to disappear; being replaced by digital displays. Moreover, in some cases the cardiac data needs to be stored on Holter[1] devices or similar systems, in order for physicians to acquire long recordings (e.g. 24 hours) at the patient's home. The aim of this technique is to observe and diagnose problems which are not constant, and so may not be observed on a shorter recording. The digital information gathered (with recent models of the Holter device) is stored via flash memory.

If we consider, for example, an ECG signal sampled at 180 samples per second (180 Hz) and we suppose that each sample is coded on 12 bits, a simple calculation will show us that 24 hours worth of information will amount to 22.8 MB. This figure increases at a rapid rate in accordance with, on the one hand, the number of channels and, on the other, the length of the recording. Table 1.1 displays the quantity of data (rounded values) gathered during recordings of 24 to 96 hours, using 1, 3 or 12 channels.

	1 channel	**3 channels**	**12 channels**
24 hours	22 MB	68 MB	269 MB
48 hours	45 MB	135 MB	538 MB
72 hours	67 MB	202 MB	806 MB
96 hours	90 MB	269 MB	1 GB

Table 1.1. *The quantities of data (rounded values) stored depending upon the length of a recording and the number of channels*

1 Invented by Norman Holter in the late 1940s, this portable system is used to record a patient's cardiac activity over a long period of time.

It is also important to remember that some computer applications work with mapping systems, which gather 64 or even 256 simultaneous recordings, leading to an increase of the same magnitude in the quantity of data acquired.

1.3.4. *Increases in the number of explorative examinations*

Along with these developments in the fields of images and signals, the number of examinations carried out upon patients is growing constantly, particularly in disciplines such as cancer research, where the staging of an illness and the follow-up treatment can lead to large amounts of images. Thankfully, the capacities of storage and transmission systems have increased in recent years, yet they remain limited: to 10 GB per DVD. The challenge now is to acknowledge the need for patients' medical files, including related images, to be stored, while finding a way to deal with the huge quantities of digital data which are currently being gathered.

1.4. Legal and practical matters

Legislation on the archival of medical images manages to be both clear and ambiguous. For public healthcare centers – in France for example – the law insists upon the storage of a patient's medical file for at least 20 years, and more often than not 70 years (in the domains of paediatrics, neurology, stomatology, chronic illnesses, etc.). In the case of hereditary diseases, the files must be stored indefinitely. A patient's medical file can include diagnoses, notes, test results, radiographs and electrocardiograms. This means that images are part of the material which goes to make up a medical file. The recent project of the Personal Health Record, shared by healthcare professionals and stored by selected hosts, promises to result in recommendations on the archival of images [ZOR 06].

The issue of compressing images with a potential loss of detail has not yet been tackled. On a day-to-day basis, medical professionals adhere as closely as possible – bearing in mind practical, technical and financial requirements and limitations – to the recommendations drawn up by academic bodies. They refer particularly to the American College of Radiology's report on teleradiology [AME 05]:

– the compression process must be carried out under the supervision of a qualified physician, and must not result in any significant reduction in the diagnostic usefulness of data;

– the hardware used for the transmission of data must meet the DICOM standard, and be up-to-date;

– the display method must allow the images to be viewed as they will be habitually needed, including windowing, and density calculations for CT images;

– patient confidentiality must be retained.

Currently at the majority of imaging centers, numerous examinations – particularly CT scans – are transferred via CD-R, alongside a selection of key images which are chosen to be printed out. Without a PACS, digital images which have not been archived are typically lost after a few days or weeks, depending on the hard-drive capacities of the machines or the capabilities of the image treatment centers concerned. Examination results which are used in teaching are often stored at teaching hospitals, on physical media such as optomagnetic discs or compact discs. In fact, the legislation serving to protect individuals who undergo biomedical studies (the French Huriet Law) specifically requires that examination results be stored, although no details are given on the format in which these archives should be created. In such cases, imaging centers store the digital data accordingly.

At this point we feel it necessary to highlight the limitations of archival methods, which are heavily dependent upon physical media. Managing these items is no easy task, and they are not always reliable, which results all too frequently in the loss of data. Furthermore, the fact that technology is always advancing leads to problems, since new systems for reading data frequently emerge, which can turn archives into "data graveyards" which cannot be accessed.

1.5. The role of data compression

PACS can resolve, at least in part, the problems of storage and communication, but the ever-growing quantities of information needing to be managed also have to be taken into consideration. Compression, by reducing the volume of data needed to display an image or other signal, seems, then, to offer an effective solution which would allow the introduction of a PACS. Compression presents a less costly alternative to the repeated updating and increasing of storage capacities and lines of communication. It would be possible, for example, to use a compression technique in order to maximize the quantity of data available for quick access online in a health center, thereby avoiding too great and too costly an investment in storage space [AVR 00]. This would improve the performance of the establishment's image distribution systems [BER 04].

Compression is also vital in telemedicine, whether it be for images or other signals, daily practice or for research and teaching. In teaching, compression will allow for the easier creation of more complete banks of images and other reference information required for medical training, and transferred via a digital medium (CD-ROM or the Internet) [LUN 04] [ZAP 02]. In clinical practice, the exchange of images between medical teams occurs every day, in order to compare or examine certain results in detail, or to draw-up images for use as reference tools. In the field of research, the sharing of digital data will revolutionize certain practices, allowing, for example, the analysis of preliminary examination results from a distance, which may prevent unnecessary journeys. Similarly, it will be possible to obtain quick second opinions from national or international experts working in a certain field. Until recently, telemedicine had not extended beyond small, often local, networks of varying levels of experience and expertise. Compression, by reducing the amount of data to be stored and communicated, allows for faster and less expensive transmission, and thus makes it possible to use telemedicine on a daily basis.

For legal and ethical reasons, lossless (or reversible) compression techniques are preferable because they can produce an exact reproduction of the original image. Such are the techniques which are currently present in PACS (as described in Chapter 4). However, lossless systems are only of limited usefulness due to their compression rate (between 3:1 and 6:1 depending on the image involved) [KIV 98], and thus do not present a long-term solution. Only lossy (or irreversible) compression techniques, i.e. those involving a permanent loss of data, allow for more significant compression rates. However, as the American College of Radiology points out, compression should only be carried out if it results in no loss of diagnostic information [AME 05]. The compression-decompression process must avoid, at all costs, creating any distortion which may lead to a change in the qualitative and diagnostic interpretation of the images involved.

1.6. Diagnostic quality

How are we to measure diagnostic quality? The answer is that the pathological condition is what determines the information which must be retained in any given medical data. This information may be large in volume, but not contrast greatly with the surrounding tissue, or perhaps small, linear or punctiform details are needed; varying only very slightly if at all from the original noise or resolution gathered by the initial technique. In fact, both of these categories of information may be required within one image, for diagnostic purposes. Diagnostic quality is therefore heavily dependent upon the protocol of both the respective gathering technique and the pathological condition concerned.

1.6.1. *Evaluation*

The evaluation of compression techniques is therefore a difficult task. The data gathering techniques vary, and the images produced by each are different; whether in spatial resolution, contrast or type and quantity of noise. For this reason, we often refer to studies evaluating the legibility of the diagnosis by a radiologist. Receiver Operating Characteristic (ROC) curves are widely used, but such studies are laborious and difficult to organize [PRZ 04]. In practice, it is no easy task to assemble a selection of examinations representing an accurate sample of different pathological conditions and/or a sample which allows for the analysis of each type of data, in order to make significant comparisons. The loss of detail which accompanies a lossy compression technique is more apparent for certain types of lesions, as Ko showed in the case of different categories of pulmonary nodules [KO 03] [KO 05]. We examine this area in depth in Chapter 5.

1.6.2. *Reticence*

Reversible compression techniques are of limited usefulness. Lossy compression techniques are a worry for physicians. This is because they cannot accept the possibility of losing any parts of an image which are "useful" for diagnosis [RIT 05]. In fact, the compression algorithms in common use lead quite quickly to a visible loss of quality, when applied at high frequencies. It is therefore essential, in order to retain the visual quality and diagnostic usefulness of an image, to limit the compression rates; for example around 10:1 for JPEG images [SLO 03]. The progress first of home computing, and then of the Internet, have presented physicians with limitations in image compression, when it comes to the legibility of diagnostic information.

The fear of destroying evidence with the threat of legal action makes the idea of lossy compression very unattractive. However, the current practice, which relies upon printed films, which set the gray level of each pixel, also greatly reduces the amount of information available. The number of slices produced by certain examinations is so great that not all can be printed, so radiology teams produce a selection of relevant images and a series of images reconstructed from the whole, through modifications in the format, or averaging techniques. If the current practice, recommended by academic bodies, could work towards a compromise – with the law agreeing that the necessary protocol had been adhered to – then we can envisage a mutually-agreed solution. For example, we could imagine using compression techniques on a large part of the data acquired in an examination, but avoiding any loss for the images judged the most important. On an international level, current thinking among the academic bodies in medical imaging is directed towards the sophisticated application of compression methods: tailored to each image gathering technique, and even each pathological condition involved.

Ideas constantly evolve with the emergence of promising new compression techniques. Opinions are thus gradually changing, as a result of numerous studies assessing the efficiency of methods of compression by wavelets [ERI 02] [SUN 02]

[PEN 05]. The current situation however is still rather complicated and we must therefore remain cautious. Detailed studies on a variety of pathologies must be carried out so as to determine the compression thresholds that are not to be exceeded for each examination technique [ERI 02].

1.7. Conclusion

When we consider the large quantities of digital data which go into making up a patient's medical file, the usefulness of compression is quite clear. Therefore, the implementation of data compression methods allows for numerous benefits, including:

– "new generation" PACS: extremely user-friendly, allows for longer-term storage on a quick-access storage platform, before transferral to a slower-access archival system;

– the medical record of each patient stored on an individual memory card, which could soon hold all their images and other clinical data.

Current lossless (i.e. reversible) compression methods are of limited usefulness, and only lossy (i.e. irreversible) will allow us to achieve very significant compression rates. The compression techniques used in medical imaging should not only allow for high compression rates, but more importantly they should retain the diagnostic usefulness of the original image. It seems essential, then, that the losses brought about by any given compression technique should be evaluated before the implementation of such compression within a storage and communication system. Moreover, these methods should include a way of tackling the demands raised by interoperability and durability in the healthcare sector. There is still progress to be made, therefore, if a compression system perfectly-suited to the medical needs of tomorrow is to be developed.

1.8. Bibliography

[AME 05] AMERICAN COLLEGE OF RADIOLOGY (ACR), "ACR technical standard for teleradiology", *ACR Practice Guideline*, http://www.acr.org/, p. 801-810, October 2005.

[ARE 01] ARENSON R.L., LU Y., ELLIOTT S.C., JOVAIS C., AVRIN D.E., "Measuring the academic radiologist's clinical productivity: survey results for subspecialty sections", *Academic Radiology*, vol. 8, no. 6, p. 524-532, June 2001.

[AVR 00] AVRIN D.E., ANDRIOLE K.P., YIN L., GOULD R., ARENSON R.L., "Simulation of disaster recovery of a picture archiving and communications system using off-site hierarchal storage management", *Journal of Digital Imaging*, vol. 13, no. 2 Suppl. 1, p. 168-170, May 2000.

[BER 04] BERGH B., PIETSCH M., SCHLAEFKE A., GARCIA I., VOGL T.J., "Upload capacity and time-to-display of an image Web system during simultaneous up- and download processes", *European Radiology*, vol. 14, no. 3, p. 526-533, March 2004.

[CHA 02] CHAN L., TRAMBERT M., KYWI A., HARTZMAN S., "PACS in private practice - effect on profits and productivity", *Journal of Digital Imaging*, vol. 15, Suppl. 1, p. 131-136, March 2002.

[ERI 02] ERICKSON B.J., "Irreversible compression of medical images", *Journal of Digital Imaging*, vol. 15, no. 1, p. 5-14, March 2002.

[FOO 01] FOORD K., "Year 2000: status of picture archiving and digital imaging in European hospitals", *European Radiology,* vol. 11, no. 3, p. 513-524, February 2001.

[GIL 06] GILBERTSON J.R., HO J., ANTHONY L., JUKIC D.M., YAGI Y., PARWANI A.V., "Primary histologic diagnosis using automated whole slide imaging: a validation study", *BMC Clinical Pathology*, vol. 6, April 2006.

[HAZ 04] HAZEBROUCQ V., FERY-LEMONNIER E., "The value of teleradiology in the management of neuroradiologic emergencies", *Journal of Neuroradiology*, vol. 31, no. 4, p. 334-339, September 2004.

[HEL 05] HELIN H., LUNDIN M., LUNDIN J., MARTIKAINEN P., TAMMELA T., HELIN H., VAN DER KWAST T., ISOLA J., "Web-based virtual microscopy in teaching and standardizing Gleason grading", *Human Pathology*, vol. 36, no. 4, p. 381-386, April 2005.

[KAY 06] KAYSER K., RADZISZOWSKI D., BZDYL P., SOMMER R., KAYSER G., "Towards an automated virtual slide screening: theoretical considerations and practical experiences of automated tissue-based virtual diagnosis to be implemented in the Internet", *Diagnostic Pathology*, vol. 1, June 2006.

[KIV 98] KIVIJÄRVI J., OJALA T., KAUKORANTA T., KUBA A., NYUL L., NEVALAINEN O., "A comparison of lossless compression methods for medical images", *Computerized Medical Imaging and Graphics*, vol. 22, no. 4, p. 323-339, August 1998.

[KO 03] KO J.P., RUSINEK H., NAIDICH D.P., MCGUINNESS G., RUBINOWITZ A.N., LEITMAN B.S., MARTINO J.M., "Wavelet compression of low-dose chest CT data: effect on lung nodule detection", *Radiology*, vol. 228, no. 1, p. 70-75, July 2003.

[KO 05] KO J.P., CHANG J., BOMSZTYK E., BABB J.S., NAIDICH D.P., RUSINEK H., "Effect of CT image compression on computer-assisted lung nodule volume measurement", *Radiology*, vol. 237, no. 1, p. 83-88, October 2005.

[LU 05] LU Y., ARENSON R.L., "The academic radiologist's clinical productivity: an update", *Academic Radiology*, vol. 12, no. 9, p. 1211-1223, September 2005.

[LUN 04] LUNDIN M., LUNDIN J., HELIN H., ISOLA J., "A digital atlas of breast histopathology: an application of web based virtual microscopy", *Journal of Clinical Pathology*, vol. 57, no. 12, p. 1288-1291, December 2004.

[PEN 05] PENEDO M., SOUTO M., TAHOCES P.G., CARREIRA J.M., VILLALON J., PORTO G., SEOANE C., VIDAL J.J., BERDAUM K.S., CHAKRABORTY D.P., FAJARDO L.L., "Free-response receiver operating characteristic evaluation of lossy JPEG2000 and object-based set partitioning in hierarchical trees compression of digitized mammograms", *Radiology*, vol. 237, no. 2, p. 450-457, November 2005.

[PRZ 04] PRZELASKOWSKI A., "Vector quality measure of lossy compressed medical images", *Computers in Biology and Medicine*, vol. 34, no. 3, p. 193-207, April 2004.

[REI 02] REINER B.I., SIEGEL E.L., HOOPER F.J., "Accuracy of interpretation of CT scans: comparing PACS monitor displays and hard-copy images", *AJR American Journal of Roentgenology*, vol. 179, no. 6, p. 1407-1410, December 2002.

[RIT 99] RITENOUR E.R., "Lossy compression should not be used in certain imaging applications such as chest radiography. For the proposition", *Medical Physics*, vol. 26, no. 9, p. 1773-1774, September 1999.

[SLO 03] SLONE R.M., MUKA E., PILGRAM T.K., "Irreversible JPEG compression of digital chest radiographs for primary interpretation: assessment of visually lossless threshold", *Radiology*, vol. 228, no. 2, p. 425-429, August 2003.

[SUN 02] SUN M.M., KIM H.J, YOO S.K., CHOI B.W., NAM J.E., KIM H.S., LEE J.H., YOO H.S., "Clinical evaluation of compression ratios using JPEG2000 on computed radiography chest images", *Journal of Digital Imaging*, vol. 15, no. 2, p. 78-83, June 2002.

[TRU 05] TRUMM C., DUGAS M., WIRTH S., TREITL M., LICKE A., KUTTNER B., PANDER E., CLEVERT D.A., GLASER C., REISER M., "Digital teaching archive. Concept, implementation, and experiences in a university setting", *Radiology*, vol. 45, no. 8, p. 724-734, August 2005.

[WEI 05] WEINSTEIN R.S., "Innovations in medical imaging and virtual microscopy", *Humam Pathology*, vol. 36, no. 4, p. 317-319, April 2005.

[ZAP 02] ZAPLETAL E., LE BOZEC C., GUINEBRETIÈRE J.M., JAULENT M.C., HÉMET J., MARTIN E., "TRIDEM: production of consensual cases in pathology using Teleslide over Internet", *6th European Congress of Telepathology*, Heraklion, September 2002.

[ZOR 06] ZORN C., "The place of personal health record in information. Stage assessment: The contents, management and the access to Personal Health Record (PHR)", *Oncology*, vol. 8 (supp. 12), p. HS113-HS117, December 2006.

Chapter 2

State of the Art of Compression Methods

2.1. Introduction

The development of new techniques in the fields of IT and communications has a great impact on our daily lives. In parallel to the constant evolution of information systems, the increase of bandwidths allows us to access and share huge quantities of data proposed by innovative services and uses. Exciting scientific problems arise concerning multimedia content, network architecture and protocols, services and uses, information-sharing and security issues. In this context, data compression remains an essential step both for transmission and for archiving.

Since the 1980s, a wide community of researchers has been working on compression techniques. Their work has led to significant advances: the broadcasting of digital television at home using a reduced bandwidth ADSL; the archival of high quality digital images on the reduced memory of a digital camera; the storage of hours of music in MP3 format on a flash drive player. To give a well known example, the JPEG standard for the compression of still images is the result of the efforts of a large scientific community between 1987 and 1993, when the standard was set. The work that led to the creation of this standard was instigated even earlier, with the proposal of a discrete cosine transform in 1974 by [AHM 74].

The collaboration of the international research community has continued with developments of quite interesting techniques for the compression of video, audio and 3D files. All these compression methods attempt to find an optimal compromise: minimize the bitrates whilst retaining the maximum visual or audio

---

Chapter written by Atilla BASKURT.

quality. In parallel to these developments, the community of researchers working on network protocols has proposed specific protocols for multimedia. These protocols, called RTP and RTCP, allow for the real-time transmission of data with a guaranteed quality of service. The Internet is a very good example of the convergence of data, image and video applications on networks.

Among the numerous compression techniques suggested in the literature, some aim for a perfect reconstruction of the original data. These methods are described as lossless. However, such techniques lead to relatively small compression ratios and are used in some delicate application domains such as for medical or satellite images, as well as the compression of computer files. Examples are entropic coding such as Huffman coding, arithmetic coding or LZW coding (the encoding of computer files such as ZIP, PDF, GIF, PNG, etc.). The general aim of these coding techniques is to get as close as possible to the real entropy of a given image. To learn more about these lossless methods, see section 2.3.

When an application requires limited bitrates, we use methods which enable a supervised loss of information (a loss often so small that it cannot be detected by the human eye). These so-called "lossy" methods combine high compression ratios with an acceptable visual quality (a rate of 8-10 for the JPEG standard and 20-30 for the JPEG 2000 standard). These losses can take the form of blocking effects, reduced color quality, blurriness, or step effect around the contours, oscillations in the transition areas, etc. We can see why the levels of loss and/or distortion need to be limited for certain applications.

In this chapter, which looks at the current state of compression techniques, we will focus mainly on the methods which have led to the accepted standards. Section 2.2 presents an outline of a generic compression technique and summarizes some information theory, quantization and coding tools which are required to understand the existing standards. The standards for compressing 2D still images and video are presented in detail in sections 2.3 and 2.4 respectively. We also give useful references regarding the techniques applied to 1D signals (audio, spoken word) in section 2.5, and those applied to 3D objects in section 2.6. The chapter ends with a conclusion and some thoughts on the evolution of the techniques, as well as the evolving nature of their usages. For a more detailed analysis, we refer you to [BAR 02], which looks at the compression and coding of images and video, and also the section "multimedia in computer systems" in the encyclopaedia [AKO 06].

2.2. Outline of a generic compression technique

A generic compression method can easily be represented in the form of a functional scheme composed of three blocks: reducing redundancy, quantization, and coding (see Figure 2.1). These three blocks are not always distinct or independent from each other. For example, in the case of a fractal method, the fractal model incorporates all the elements: the reduction of redundancy by detection and modeling the autosimilarity in the image, the implicit quantization using the compact fractal model and the coding of the model's parameters.

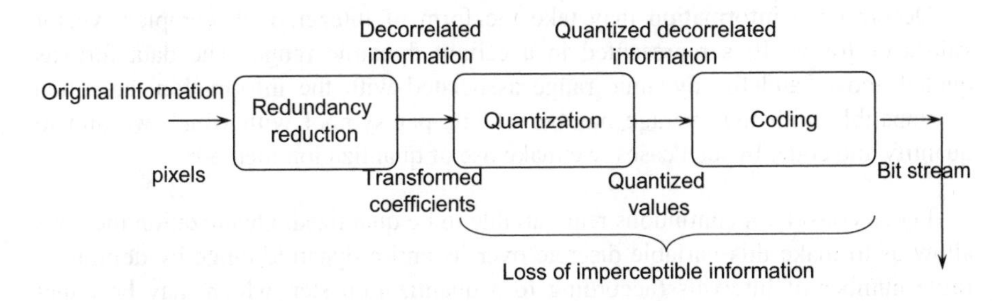

Figure 2.1. *Generic compression method scheme*

2.2.1. *Reducing redundancy*

Compression aims to quantify and code the source information using a number of bits close to the entropy of this source (the entropy is the average quantity of information contained in one of the source's messages) by exploiting the redundancy of the data representing a natural phenomenon as in medical imagery. We have Shannon [SHA 49] to thank for the mathematical definitions of these information, entropy and redundancy concepts. These definitions are looked at in detail in section 2.2.3.

Redundancy in this sense must be understood as the similarity of messages or symbols when they are analyzed one after another, or next to each other. This redundancy may be spatial (in neighboring pixels or between blocks or areas of pixels); spectral (between the different bands created by a multispectral system or the Red, Green and Blue (RGB) components); or temporal (between successive images in a video). Compression methods use these different types of redundancy and reduce the average number of bits required to code a source symbol (a pixel of the image). This step is undoubtedly that which appeals most to researchers in this field, as it involves analyzing the content of the data, detecting redundancy through the use of innovative tools adapted to the content, and then proposing a compact and decorrelated representation of the information. Although pixel-based methods do exist and can be effective, the key methods use orthogonal transform (Discrete Cosine Transform (DCT) and Discrete Wavelet Transform (DWT) most commonly) in order to change the representation space and aim at an optimal representation in terms of decorrelation/compactness with transformed coefficients. We should also note that a color transformation of RGB to YCbCr belongs to the step of the reduction of redundancy (spectral redundancy in this case).

2.2.2. *Quantizing the decorrelated information*

Decorrelated information may take the form of integer, real, complex, vector values or forms. It is represented in a certain dynamic range. The data formats quoted above, and the dynamic range associated with the information, are often incompatible with the average number of bits per symbol with which we aim to quantify and code. In such cases, we make use of quantization methods.

Let us consider a continuous real variable to be quantized. Quantization methods allow us to make this variable discrete over its entire dynamic range by defining a finite number of intervals (according to a quantization step which may be either uniform or not), and by assigning a value to each of these intervals (for example the middle value of each interval). We should note the importance of the choices of the quantization step and the value assigned to each interval. The strategy behind the quantization method will determine the optimal values of these two parameters, generally based on the statistics of the source to be quantized.

The performance of the quantization is measured in terms of minimization of global distortion (total error after quantization) for a given bitrate to allocate to this source (for example, an average of 3 bits/pixel to quantize a digital mammography numerized at 12 bits/pixel).

We can define two main classes of quantization: scalar quantization and vector quantization. The first is applied to scalars such as the intensity of a pixel or the value of a coefficient, whereas the second is applied to blocks of neighboring pixels or coefficients. It is at this stage of the quantization that the loss of information (and thus a lower quality of the restored image) is introduced into the compression process. This loss is of course irreversible. In this chapter, we have opted not to detail quantization methods. These methods can be studied in the references [MAX 60], [LIN 80] and [GRA 84]. However, when we detail the norms, the quantization stage will also be looked at in further detail.

2.2.3. *Coding the quantized values*

The values which emerge from the quantization process are generally represented by a binary code of fixed length (*N bits/symbol*). The coding stage allows us to reduce the average number of bits allocated to each quantized value.

The established techniques are based on the information theory presented below. They involve no loss of information.

It was Shannon [SHA 49] who in 1949 developed information theory, with a global vision of communication systems, and opened the door for coding techniques. The developments he made have led to an optimization in the representation of messages generated by the information source given the entropy of this source (for example, the values of pixel gray levels are the messages, the image is the information source, the entropy is calculated from the normalized histogram of the image's gray levels). This theory proposes the notion of quantity of information I associated with a symbol generated by a source S, of the entropy H_S of S and the redundancy R_S.

Let us take a source S, emitting N different symbols noted as s_i with a probability p_i. The quantity of information $I(s_i)$ associated with an emitted symbol is measured by:

$$I(s_i) = -\log_2(p_i) \qquad [2.1]$$

It is measured in bits. The lower the probability of a symbol appearing, the greater quantity of information this symbol will transmit. In order to characterize source S globally, we can calculate its entropy which estimates the average quantity of information emitted by S in bits/message or bits/symbol:

$$H_S = -\sum_{i=1}^{N} p_i \log_2(p_i) = \sum_{i=1}^{N} p_i I(s_i) \qquad [2.2]$$

The entropy is positive and restricted by $H_{max}=\log_2(N)$ which corresponds to the entropy of a source with a uniform probability density (i.e. all the symbols have the same probability of appearing). In calculating the entropy, for example on a mammography, if we get 3.5 bits/pixel, this means that theoretically we should be able to code this image's pixels with binary codes with an average length of 3.5 bits/pixel (rate) and this code will be lossless. In practice, entropy coders will go towards this minimal rate by the upper limit, without being able to reach it.

We should note that thanks to the entropy calculation and the associated information rate, we can decide upon the appropriate transmission channel. These channels must offer a capacity (bitrate) at least equal to the information rate as measured by the entropy.

Let a source S be coded on b bits with $2^b=N$ possible symbols and let there be $H_S=h$ bits/symbol, and the entropy of S. Theoretically, this means that excessive (b-h) bits/symbol have been used to code the symbols of this source: this difference identifies the redundancy R_S of S. The source coding methods will try to remove this redundancy, in order to obtain the most compact representation of the information generated by a source. Well-established coding methods called "entropic coders" will be used in order to reduce the average number of bits allocated to each quantized value. These coders are based on the *a priori* knowledge of the probability density of the source, either without taking account of the spatial context of a pixel or coefficient (coding without memory) such as the Huffman or arithmetic coder, or by integrating knowledge of the surrounding area (coding with memory or contextual coding) such as the LZW coder. This stage is lossless. The different formats used for computer file storage use these coders, particularly the LZW coder for PDF or PNG formats.

Readers can study these techniques in further detail: Huffman coding [HUF 52], arithmetic coding [RIS 76], Lemple, Zip, Welch (LZW) coding [WEL 84] and in the global works [BAT 97] and [BAR 02].

2.2.4. *Compression ratio, quality evaluation*

The compression ratio is defined as the total number of bits necessary to represent the original information divided by the total number of bits of the binary file which will be stored. In practice, we tend to use the bitrate to measure the compacting capabilities of a method. The bitrate is expressed in bits per element. The latter is a "pixel" if we are compressing a still or animated image, a "sample" if we are treating a signal or a vertex if we are dealing with a 3D chain.

In order to attain a quantitative measurement of the quality of the decoded image, the Mean Square Error (MSE) between the original image and the decoded image is given by:

$$MSE = \frac{1}{NM}\sum_{i=0}^{N-1}\sum_{j=0}^{M-1}\left[X(i,j) - \hat{X}(i,j)\right]^2 \qquad [2.3]$$

where NM is the size of the original image, with pixels varying between 0 and 255.

According to the literature, the quality tends to be expressed in dB with a peak signal-to-noise ratio PSNR:

$$PSNR = 10\log_{10}\frac{255^2}{EQM} \qquad [2.4]$$

We should note that this evaluation is purely mathematical and is not necessarily associated with the visual quality as perceived by an observer. Many works have been carried out on the estimation of visual quality, such as the ROC methods or the methods based on "consensus". They are presented in Chapter 5.

2.3. Compression of still images

For many years now, the compression of still images has been a very active area of research. The work carried out has led to the establishment of the JPEG and JPEG 2000 standards. The latter sets a very high standard in terms of rate/distortion. This has encouraged the research community to focus on this area, and improve their approach with new functionalities (watermarking, indexing and retrieval), and a greater adaptability for the relevant domains (such as medical or satellite imaging).

Usual compression methods are those which allow a controlled and invisible distortion level. They attain high compression ratios with an acceptable visual quality (a compression rate of 8-10 for the JPEG standard, and 20-30 for the JPEG 2000 standard).

The numerous different approaches can be classified into two categories: spatial methods and methods using transforms. Spatial methods directly manipulate the pixels. The best example is predictive coding or *Differential Pulse Code Modulation* (DPCM) [JAI 81]. One pixel is predicted from its neighborhood by a linear combination of the values of several pixels which are already known. Rather than the pixel, it is the difference of this prediction which is quantized and coded. The gain is due to the fact that the entropy of this difference is smaller than that of the pixel itself. This method was retained as one of the three methods in the JPEG study, along with the *Block Truncation Coding* approach (another spatial approach coding a block of pixels), and the DCT technique. The latter falls into the second category, which changes the representation space before quantization and coding. The different standards 2D, 2D+t, 3D are based on different orthogonal transform. This does not mean that methods such as DPCM have been abandoned. We find them in the standards, every time we need to reduce redundancy in a neighborhood (for example to code the average value of blocks in JPEG or to predict the spatial position of the vertices in 3D meshes in MPEG4).

In this chapter, we begin by presenting a description of JPEG. This is followed by a summary of the evolution of the wavelet transform based approaches is given, up to the JPEG 2000 standardization.

2.3.1. *JPEG standard*

The JPEG became an international standard in 1993, following a long period of research and development involving a very large scientific community. The standard gives a complete specification of the coder and decoder. Figure 2.2 shows a block diagram.

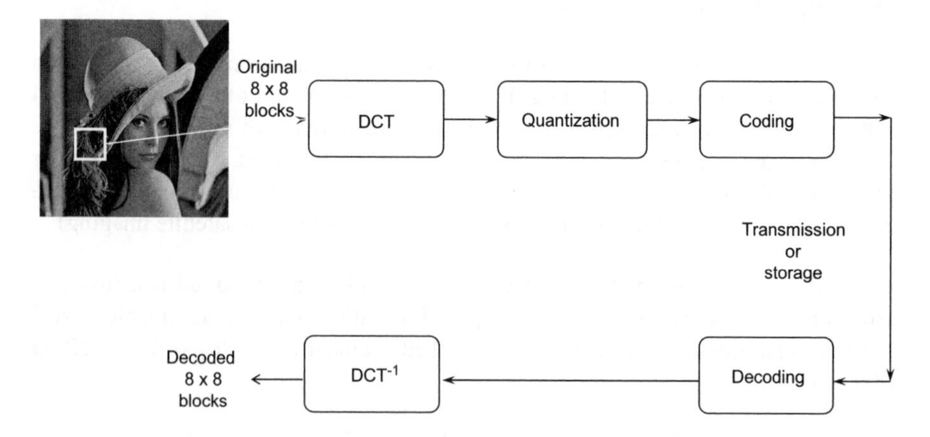

Figure 2.2. *JPEG: each 8 x 8 block is treated separately in three stages: DCT, quantization and Huffman entropic coding*

The image is subdivided into blocks of a reduced size of 8x8 in order to take into account the local properties. We should note that the large number of calculations necessary to carry out processings on the whole image is reduced thanks to this local approach. Each block is transformed by the DCT, independently of the other blocks which can be considered as a local spectral analysis using no overlapping windows. This leads to an 8x8 block of frequency coefficients. These coefficients are then qualified by a scalar quantization and are coded by Huffman entropic coding. The ".jpg" file is finally achieved by linking the bitstream associated with each of the original blocks. In order to reconstruct the decoded image, a reverse scheme is applied.

2.3.1.1. *Why use DCT?*

Like the vast majority of current standards, JPEG uses DCT. The Cosine Transform was proposed by J. Fourier (1822) and is defined as the representation of a continuous function by a linear combination of cosinus functions (the equivalent exists with sinus functions and is called sinus transformation). It is also defined as the Fourier transformation of the even part of the analyzed continuous function.

This reversible transformation has been especially known in the signal and image community since the proposition of its discrete version in 1D: DCT in 1974 [AHM 74]. Its 2D definition is given as:

$$X_{DCT}(k,l) = \frac{C(k)C(l)}{4} \sum_{i=0}^{7} \sum_{j=0}^{7} X(i,j) \cos\left(\frac{\pi k(2i+1)}{16}\right) \cos\left(\frac{\pi l(2j+1)}{16}\right) \qquad [2.5]$$

and the inverse transform is defined as:

$$X(i,j) = \frac{1}{4} \sum_{i=0}^{7} \sum_{j=0}^{7} C(k)C(l) X_{DCT}(k,l) \cos\left(\frac{\pi k(2i+1)}{16}\right) \cos\left(\frac{\pi l(2j+1)}{16}\right) \qquad [2.6]$$

where $X(i,j)$ and $X_{DCT}(k,l)$ represent the DCT pixels and coefficients. C is a normalization function. The DCT transforms a matrix of pixels into a matrix of frequency coefficients, leading to another way of representing an image's characteristics. DCT has two key advantages [RAO 90]: the decorrelation of the information by generating coefficients which are almost independent of each other and the concentration of this information in a greatly reduced number of coefficients. It reduces redundancy while guaranteeing a compact representation (see section 2.2.1). DCT's efficiency in terms of the compacting of energy has been compared to that of the Karhunen-Loève transformation (KLT) [ROS 82], known as the optimal transform, and used for the principal components analysis (PCA). In practice, for images presenting a strong inter-pixel correlation, the characteristics of decorrelation and compacity of DCT are very similar to the characteristics of KLT. Figure 2.3 illustrates these properties.

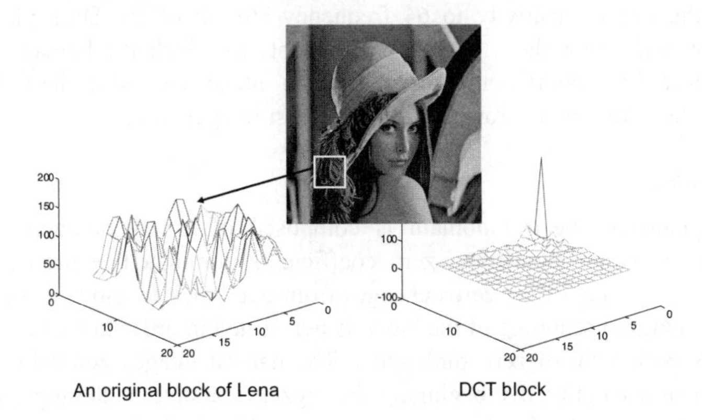

Figure 2.3. *Example of an image block with heterogeneous content and its DCT which concentrates the information in few coefficients (the peak is the coefficient associated with the mean value of the block – DC coefficient)*

2.3.1.2. *Quantization*

Each 8x8 DCT block corresponds to the local spectral content of the image. As an image is a bidimensional non-stationary signal, these local spectral signatures are very different from one another. A block with homogenous gray levels will have a DCT domain with very few significant coefficients, whereas a pixel block with high variations will lead to a DCT domain with more spectral coefficients (Figure 2.4).

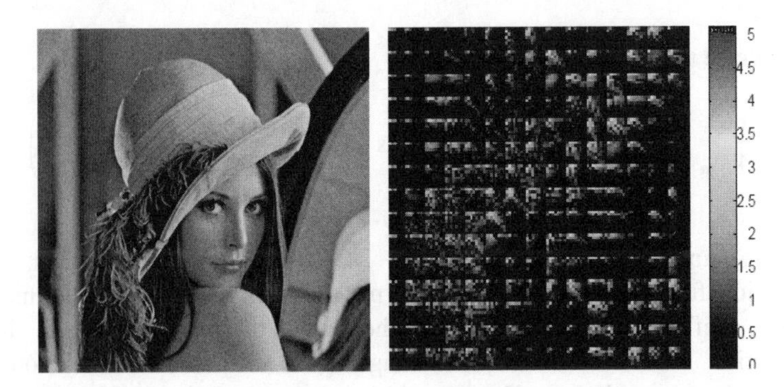

Figure 2.4. *Local spectral DCT domains of the image Lena (with absolute values of the coefficients shown in false color): we can see the real difference between the local spectral signatures*

This has led to a quantization strategy which is adapted to the content of each DCT domain in order to retain all the significant coefficients, regardless of their position or number. A normalization matrix allows this selection to be carried out. Psychovisual studies have made it possible to define the contents of this matrix with regard to the eye's sensitivity to 64 frequency stimuli of the DCT plane. Thus, quantization will favor those spectral coefficients to which the human eye is the most sensitive. Each coefficient is divided by an integer located in the same spatial position in the matrix and is rounded to the nearest integer value.

2.3.1.3. *Coding*

Each quantized spectral domain is composed of a few non-zero quantized coefficients, and the majority of zero coefficients eliminated in the quantization stage. The positioning of the zeros changes from one block to another. As shown in Figure 2.5, a zigzag scanning of the block is performed in order to create a vector of coefficients with a lot of zero runlengths. The natural images generally have low frequency characteristics. By beginning the zigzag scanning at the top left (by the low frequency zone), the vector generated will at first contain significant coefficients, and then more and more runlengths of zeros as we move towards the high frequency coefficients. Figure 2.5 gives us an example.

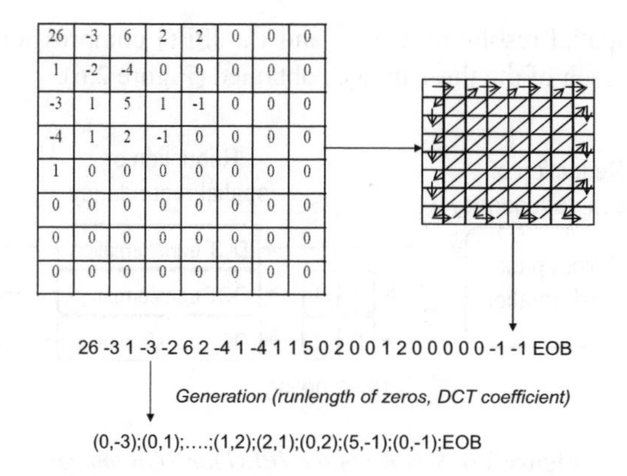

Figure 2.5. *Zigzag scanning of a quantized DCT domain, the resulting coefficient vector, and the generation of pairs (zero runlength, DCT coefficient). EOB stands for "end of block"*

Couples of (zero runlength, DCT coefficient value) are then generated and coded by a set of Huffman coders defined in the JPEG standard. The mean values of the blocks (DC coefficient) are coded separately by a DPCM method. Finally, the ".jpg" file is constructed with the union of the bitstreams associated with the coded blocks.

2.3.1.4. *Compression of still color images with JPEG*

Color images are represented by three RGB components generated by captors visualizing the same scene through filters whose bandwidths are associated respectively with the colors red, green and blue. These components are correlated for natural scenes. The first stage consists of transforming the RGB space into another representation space in which the three resulting components are decorrelated amongst themselves. In the JPEG standard, this concerns the YC_bC_r color space which is obtained through the transformation:

$$\begin{cases} Y = 0.3R + 0.6V + 0.1B \\ C_b = \dfrac{B - Y}{2} + 0.5 \\ C_r = \dfrac{R - Y}{1.6} + 0.5 \end{cases} \qquad [2.7]$$

where Y represents the achromatic channel (the luminance), while C_b and C_r are the two chromatic channels (the chrominance). This reduces the spectral redundancy. It is followed by a sub-sampling by 2 of the components C_b and C_r. Psychovisual studies show that the human eye is three times more sensitive to variation in luminance compared to variation in chrominance. It is therefore unnecessary to retain the same spatial resolution for C_b and C_r. JPEG compression is then applied independently to each of the three images obtained (Figure 2.6).

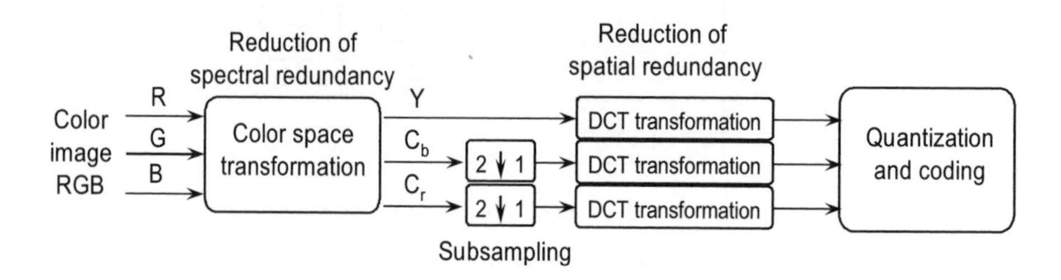

Figure 2.6. *Scheme of the JPEG for RGB images*

2.3.1.5. *JPEG standard: conclusion*

The JPEG standard is currently widely used for coding the majority of digital images available on the Internet. It is also the format used in digital cameras. It is well adapted to natural scenes and to compression ratios which do not exceed 8-10 (a rate of 0.8-1 bits per pixel). For greater ratios, artefacts (blocking effects in homogenous zones) will appear. The new JPEG 2000 standard provides a solution to this problem by combining a high visual quality with high compression ratios. This standard is detailed in the following section. For a more detailed study, see [JAI 81] for DCT, or the website http://www.jpeg.org for JPEG.

Figure 2.7. *Undesirable blocking effects appear at low resolution (0.2 bits/pixel in this example, or compression ratio=40)*

2.3.2. *JPEG 2000 standard*

2.3.2.1. *Wavelet transform*

The JPEG 2000 standard for the compression of still images is based on the *Discrete Wavelet Transform* (DWT). This transform decomposes the image using functions called wavelets. The basic idea is to have a more localized (and therefore more precise) analysis of the information (signal, image or 3D objects), which is not possible using cosine functions whose temporal or spatial supports are identical to the data (the same time duration for signals, and the same length of line or column for images). Wavelet functions enable this detailed analysis, since they have a finite support (e.g. a sinusoid modulated with a Gaussian function). Moreover, the length of this support can be modified to obtain a multiresolution analysis. Let us imagine the wavelet function dilating (a long support in time associated with a narrow frequency band or a local analysis in time corresponding to a global high frequency content). With a fixed support length, the information will be analyzed at a fixed resolution and so on to cover all the possible resolutions. We will not present in detail the relatively recent wavelet theory – rich and exciting as it is – but instead see the following works: [MEY 92], [DAU 92], [MAL 98], and [BRE 02].

2.3.2.2. *Decomposition of images with the wavelet transform*

By applying DWT, the image is decomposed into a subimage pyramid revealing the collection of details at different resolution levels. The details of an image are defined as the information difference between two consecutive resolution levels. Thus, given a resolution sequence r_j, the details at resolution r_j correspond to the information difference between the approximations of the image at resolution r_j and r_{j-1}. The subimages of "approximations" and "details" of the initial image at different resolutions are obtained via consecutive filtering and subsampling operations. Each subimage corresponds to the contents of a specific frequency band. This is why they are also called subbands [WOO 86]. Figure 2.8 presents the decomposition scheme (analysis) of the Barbara image into four subbands with a separable DWT (applied first to the lines, and then to the columns).

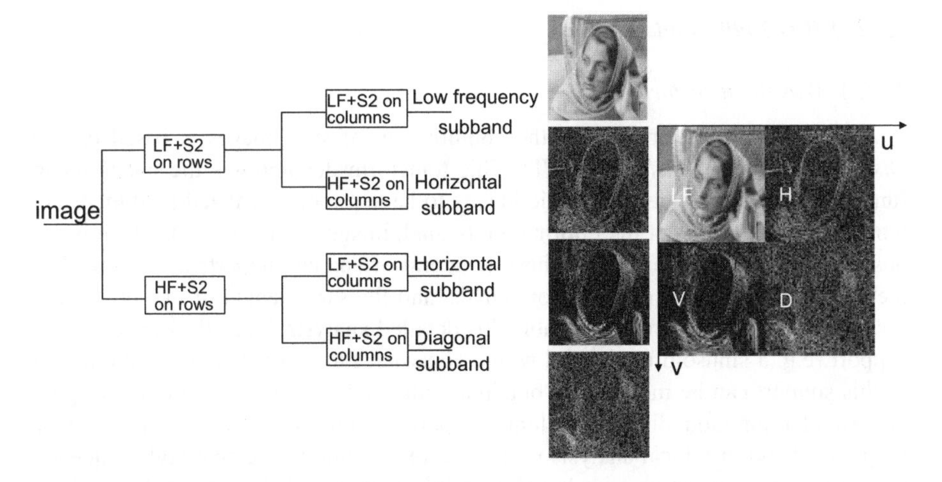

Figure 2.8. *DWT decomposition: four subimages of lower resolution, representing the contents of four complementary subbands of frequency. The LF and HF are the low frequency and high frequency analysis filters applied to the rows and columns. S2 represents the subsampling by 2*

The subband LF corresponds to the approximated image at the lower resolution r_{j-1}. The three other subbands represent the details lost in this approximation at the same resolution r_{j-1}: horizontal, vertical and diagonal variations. This scheme can be applied to the approximated image if we wish to analyze the lower resolution r_{j-2} and so on. The decomposition is of course reversible, applying the reconstruction filters to the subbands (these filters are deducible from the analysis filters). This reconstruction stage leads to a perfect reconstruction of the original image. It is indeed important to have a reversible transform for image compression.

Let us now analyze these schemes from a "digital filtering" point of view. The choice of the analysis and reconstruction filters is very important. The filters used can be characterized by several properties, in particular the exact reconstruction. Without proceeding to the quantization and the coding of the subbands, the transform used should not induce an information loss. In literature on the subject, several filters have been proposed such as *quadrature mirror filters* [EST 77], *conjugate quadrature filters* [SMI 86] and *biorthogonal filters* [ANT 92]. The latter represent some of the most frequently-used filters in wavelet theory, particularly in the field of compression.

2.3.2.3. *Quantization and coding of subbands*

A set of subbands is obtained after the decomposition. Each detail subband is composed of wavelet coefficients at different resolutions. Figure 2.9 presents an angiography and the contents of the subbands obtained at two different resolution levels.

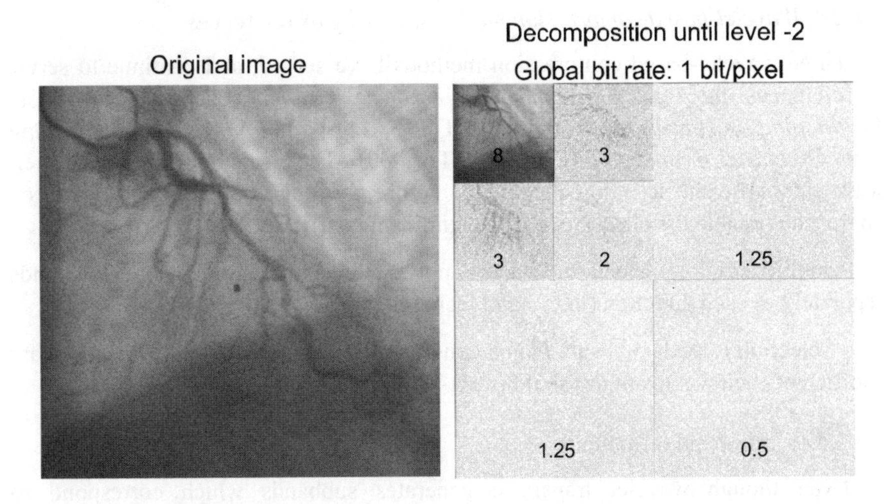

Figure 2.9. *Example of bitrate allocation in different subbands with respect to the information content of the subbands*

The problem to be resolved is as follows: given the total number of bits available to quantify and code the subbands, how many bits should be allocated to each of the subbands in order to minimize the overall distortion? Figure 2.9 gives one example of binary allocation per subband. In this example, the subband LF (image approximated at resolution *j-2*, if the original image has a resolution of *j*) is quantized and coded with 8 bits/coefficient and the diagonal subband *j-1* with only 0.5 bits/coefficient, as it contains less information.

The quantization and coding of the subbands will therefore be preceded by a binary allocation stage. This stage will lead us to the number of bits allocated to each of the subbands, which minimizes the overall distortion. The global minimization method proposed in [RAM 93] and based on a Lagrangian minimization function, is a good candidate to resolve this problem. Related algorithms adapted to the type of data (2D, 2D+t, 3D) will be used in the majority of wavelet-based compression techniques.

Once the number of bits/coefficient has been allocated to a subband, it is necessary to determine a quantization strategy, be it scalar [MAX 60] or vectorial [LIN 80], [GRA 84] (see section 2.2.2). The quantization technique can be chosen independently for each subband. It is generally followed by an entropic coding (see section 2.2.3). This is discussed in the following section.

2.3.2.4. *Wavelet-based compression methods, serving as references*

Three wavelet-based compression methods have served, and continue to serve, as references: the algorithms *Embedded Zerotree Wavelets* (EZW) [SHA 93]; *Set Partitioning on Hierarchical Trees* (SPIHT) [SAI 96] and *Embedded Block Coding with Optimized Truncation* (EBCOT) [TAU 00]. These methods differ in their wavelet coefficient selection procedure, before the quantization stage. Two key approaches enable the elimination of insignificant coefficients:

– multiresolution selection, which applies thresholds through the subbands separately in each direction (inter-band technique) [SHA 93], [SAI 96];

– selection in each subband, which consists of setting to zero all the insignificant coefficients below a given threshold (intra-band technique) [TAU 00].

2.3.2.4.1. Inter-band techniques

Even though wavelet transform generates subbands which correspond to orthogonal projections in disjoint vector subspaces, structural similarities remain in the subbands. These details are associated with consecutive resolutions and directions. This can be seen clearly in Figure 2.9. Inter-band techniques make use of this inter-subband similarity of wavelet coefficients by defining a tree of coefficients. This important development was proposed by Shapiro in 1993 [SHA 93]: the EZW algorithm (zerotree approach) groups insignificant coefficients using a tree of zeros in the horizontal, vertical and diagonal directions. This structure enables the identification of those zones of the image which contain no significant information. The EZW algorithm can be summarized as follows:

– the zerotree is constructed from significance maps showing the position of the significant coefficients for a given quantization step;

– the successive approximation of the significant coefficients allows a progressive coding according to a given rate-distortion stopping criterion;

– a dynamic arithmetic coder codes the chain of symbols.

The binary chain obtained has a very interesting property. It is an embedded code. This means that the decoding of the image at a given compression rate T can be initiated, and continue with consecutive decodings, with regularly-growing rates up to the value T (progressive transmission).

The EZW algorithm has since been improved upon by the SPIHT algorithm [SAI 96]. The latter is today considered as the reference method in the compression of still images. However, visually, certain details or textures are not always as well-preserved as in intra-band techniques, examined below.

2.3.2.4.2. Intra-band techniques

These techniques treat the subbands independently from one another. The inter-band correlation is therefore not taken into account. The selection of the coefficients is carried out by the EBCOT algorithm (selection by context) [TAU 00]. This approach is used in the JPEG 2000 standard presented below.

2.3.2.5. *JPEG 2000 standard*

The aim of this standard is to enable the compression of different types of digital documents (natural, synthetic, binary, multi-channel images), with or without loss. It offers a unified solution for the numerous needs generated by new applications (the Internet, digital photography, space imaging, medical imaging, the digitization and conservation of old documents, mobile-related services, PDAs, etc.). The standard is based on the DWT. Pyramidal representation with different resolutions leads to the progressive transmission of information. This allows the introduction of new functionalities, previously impossible with DCT:

– progressive decoding: this functionality targets transmission which may be progressive in quality or in resolution. A low resolution image of acceptable quality can be sent first to the end-user. Then, the details can be progressively transmitted in order to improve the quality;

– ability to define *regions of interest* (ROI): the user can allocate greater bitrate to a ROI (e.g. a tumour in a medical image, or a face in a video) in order to obtain a higher image quality in this ROI, while the rest of the image will be coded at a lower visual quality (Figure 2.10);

– random access to the data: this allows the decoding of only a specific zone of an image, without requiring a complete decoding.

Figure 2.10. *Coding of ROI with JPEG 2000: the image Gold is coded at a global compression ratio of 128 (ROI at the top left coded at a rate of 16); the image Lena is coded at a global compression ratio of 80 (the central ROI is coded at a rate of 16)*

Figure 2.11 presents the simplified schema of JPEG 2000. We can see the three steps of the general compression scheme: transformation, quantization and coding. The latter is applied to the whole image or on certain portions (tiles):

– each tile is decomposed into subbands;

– the wavelet coefficients of each subband are quantized in a scalar manner via a quantization step which is dependent upon the subband's dynamic (this step can be adjusted by a feedback in order to match the bitrate constraints);

– in each subband, the quantized coefficients are grouped in codeblocks with a size of either 64x64 or 32x32. Each codeblock formed is then processed independently;

– the coding of a codeblock: the codeblock is divided into bit planes ranging from the most significant (MSB) to the least (LSB). Each bit plane is coded in three passes: significance, refinement and cleanup. Each of these steps collects contextual information on the different planes. This information, associated with each binary plane, is coded by a contextual arithmetic coder;

– a second coding stage generates layers of consecutive bitstreams to build the image with an increasing rate. Each layer is composed of :

- a set of passes per bit plane,

- a set of bit planes,

- the contribution of all the codeblocks of all the subbands associated with a tile.

This fine analysis of the information allows for great flexibility in terms of compression ratio. A bitrate allocation algorithm optimizes the global rate-distortion.

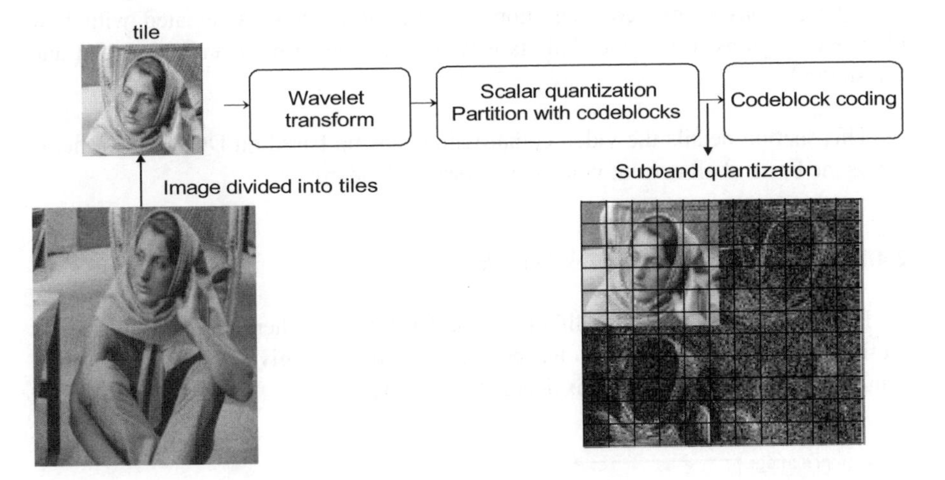

Figure 2.11. *JPEG 2000 block schema*

We should note that the bit stream associated with a codeblock can be decoded at different levels of progressive quality. Further information can be found at http://www.jpeg.org/JPEG2000.html, where all the documents linked to this standard can be accessed.

2.4. The compression of image sequences

Image sequences are made up of consecutive planes acquired at high speed, for example 25 images/second. In addition to any intra-plane spatial redundancy, video compression methods seek to detect inter-plane temporal redundancy. We can well imagine that consecutive images may look alike. Furthermore, they generally include many regions which are almost identical. There is no need to transmit these regions for every image. It is enough to code these common regions only once (in the first image of a group of images, for example). However, there is still one problem which remains: although they are identical, these regions move spatially from one image to another. The temporal redundancy detection step allows us to estimate such motions. We refer to "motion compensation". Two types of region can be distinguished:

– the regions whose motion is determined and compensated: here, we only transmit the characteristics of this motion, and any possible slight modifications to the region's content;

– the regions whose compensation is impossible: those associated with new objects or regions in a scene. This is new information to be quantized, coded and transmitted.

This section details the video compression scheme based on DCT. This scheme forms the basis of all current video compression standards.

2.4.1. *DCT-based video compression scheme*

Figure 2.12 presents a simplified generic DCT-based scheme. First proposed for the H.261 standard, this schema has constantly evolved to give us today the current standard H.264. This evolution is detailed in section 2.4.2.

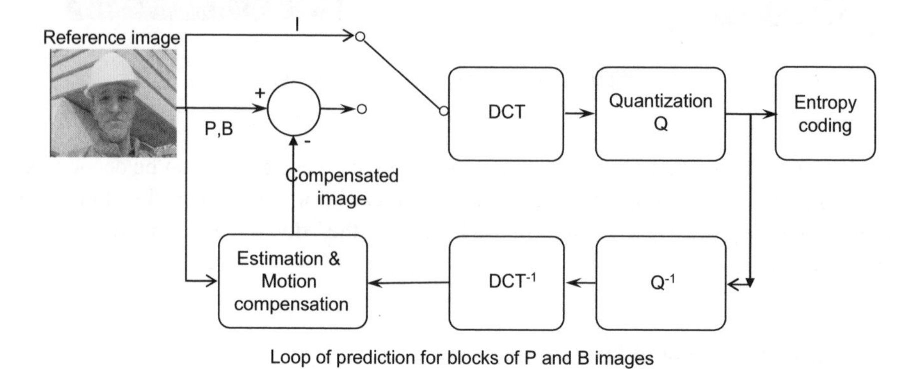

Figure 2.12. *Simplified MPEG1 scheme*

Let us explain in detail the MPEG1 approach. The sequence is divided into groups of planes (GOP). An example is given in Figure 2.13 with 12 images in such a group. Each GOP is divided into three types of images: I, P and B. Image I serves as the reference image for the group. The P images are predicted from I- or P- (mono-directional prediction) type images. The B images are predicted from I- and P- (bi-directional prediction) type images. Figure 2.13 details these two types of prediction of blocks or macroblocks.

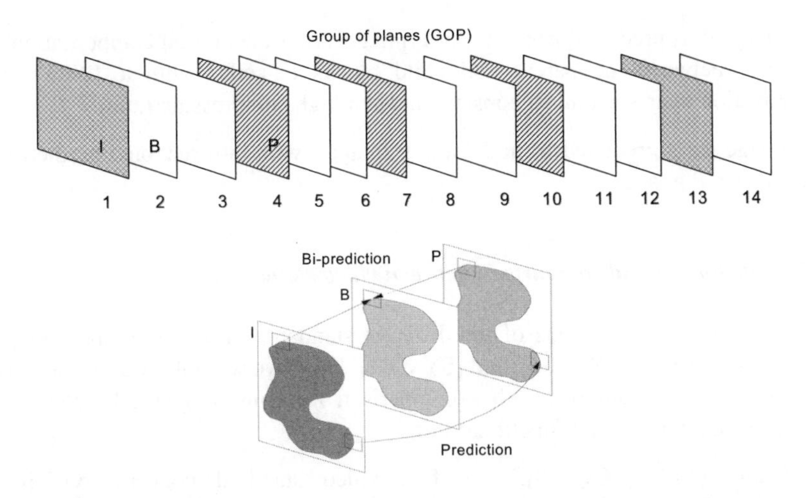

Figure 2.13. *Example of a group of 12 pictures: IBBPBBPBBPBB (at the top); prediction of a block P from a block I and bi-prediction of a block B from a block I and a block P (at the bottom)*

In order to speed up the motion prediction, we suppose it to be linear and identical for all the pixels within one block. Given a "child" block corresponding to the image $i+1$ to code, the motion is compensated by finding a similar "parent" block in the preceding image i already transmitted. The similarity is often measured by the quadratic distance. The search occurs in the neighborhood of the same spatial position of the child block. The compensation results in a motion vector. The difference between the parent and child blocks can thus be calculated. This difference corresponds to the variation in luminance and chrominance as the block of pixels is moved (due to the motion of the object including this block, or the movement of the camera). In addition to the vector, this difference is quantized and coded. This operation is carried out for each block within the image $i+1$ to code and so on for all the images in the sequence. In Figure 2.12, we can identify a feedback loop which reconstructs the preceding picture i. The original picture $i+1$ is predicted with the reconstructed plane i. The aim is to create a perfect symmetry between the coding and decoding schemes: the same prediction step in order to only use the reconstructed images for prediction.

The compression methods for each type of image are detailed below:

– image I: this reference image serves as a synchronization point. It is coded in JPEG with a normalization matrix adapted to video sequences;

– image P: is predicted from an I or P image. After the motion compensation, the difference between the parent and child blocks is coded with a JPEG whose normalization matrix is almost constant and at a high compression ratio;

– image B: is predicted from 2 I and P images, with bi-directional prediction. It is coded as a P image.

2.4.2. *A history of and comparison between video standards*

For detailed presentations of the different standards of video compression, see the studies [BAR 02], [RIC 03], and [SYM 04]. Here, we will take a brief look at the evolution of the techniques with a comparison in terms of goals, bitrates, video formats, visual quality and functionalities:

– H.261 (1993) [CCI 90]: this first video standard concerns visiophony applications for the RNIS network at multiple rates of 64 kbit/s. The image formats are the QCIF (144x176 pixels) and the CIF (352x288 pixels);

– H.263 (1995) [ITU 96]: this standard is based on H.261 and targets very low bitrates (visiophony and visioconference on RTC and RNIS). The image formats are multiples and sub-multiples of the CIF. The coder H.263 enables us to compensate motion at a precision of ½ a pixel;

– MPEG-1 (1992) [ISO 93]: a standard for compressing video and associated audio channels. It allows the storage of videos at a rate of 1.2 Mbps at a quality similar to VHS cassettes, but on CD supports. The technique is based on H.261 and improves on the motion compensation (½ pixel), and introduces the two types of prediction with the images I, P and B. The standard includes functionalities such as random access to the video, and the ability to search forwards or backwards, and to display the video in rewind (like VHS);

– MPEG-2 (1994) [ISO 95]: this standard was adopted by the Digital Video Broadcasting (DVB) consortium for digital TV services via Hertzian channels both terrestrial (DVB-T) and satellite (DVB-S). It is also used as a coding format for Digital Video Discs (DVD). It takes up the principals of the MPEG-1, while adding functionalities for televisual applications: treating interlaced formats, optimizing MPEG-1 tools (dynamic of movement vectors, etc.), controlling the stream to allow the client to adjust the video stream to their own configuration of rate and quality. This last addition is called scalability, which may be:

- temporal, acting upon the number of images,

- spatial, acting upon the size of the images,

- qualitative, acting upon the quantization step,

- regarding the hierarchy of DCT coefficients: transmission of the BF, then the HF;

– H.264 (2002) [JVT 02]: this new standard called Advanced Video Coding (AVC) is a result of the combined study of the groups ITU-T VCEG (at the origin of H.261 and H.263) and ISO MPEG (at the origin of MPEG1 and MPEG2). Still based on a DCT schema, AVC makes the key additions of:

- a directional prediction stage for the coding of intra macroblocks,

- a vertical and horizontal filtering of the compensated images in the coding scheme in order to improve the visual quality by erasing the blocking effects,

- a process of motion compensation different from those of the previous standards, with a wide variety of shapes and sizes of blocks (16x16, 16x8,…, 8x4,…, 4x4, etc.) and with a precision of up to ¼ pixel,

- a Network Adaptation Layer (NAL) which adds headers adapted to the transmission supports. AVC is the first standard to integrate this network constraint.

With all these improvements, AVC makes it possible to achieve the same visual quality as the MPEG2, for 2-3 times less bitrate. Imagine DVD quality at only 2 Mbits/s over ADSL. On the other hand, there is a price to pay for these improvements in terms of the calculations. The complexity of the calculations for AVC is four times greater than that of MPEG2. As a result, this standard is currently intended for use in offline applications such as Video on Demand (VOD) or indeed storage of films on DVD. AVC is also likely to be used for the DVD's successor: HD-DVD.

For the MPEG4 standard, see [MPE 98]. This standard introduces the notion of audio and video objects so that the user can interact with the content by manipulating these objects (cutting the "sound" object, changing the "audio language" object, moving a 3D object, changing the viewpoint, etc.). This standard makes the hypothesis of an acquisition independent of the objects, or the hypothesis of a segmentation method capable of distinguishing all the objects going to make up a scene. Segmentation remains an unanswered question in audio-visual processing. As a result, the vast majority of MPEG4 functionalities cannot currently be exploited.

Standards	Aims	Rate	Video format
H.261 H.263	- visiophony and videoconference on RTC and RNIS - low video quality	P x 64 kbits/s	- CIF (352x288) - QCIF (144x176)
MPEG1	- compression of video and audio channels - near-VHS quality	1.5 Mbits/s (1.2 Mbits/s for video)	Formats up to 4096x4096 at 30 images/s
MPEG2	- compression for digital TV, HDTV, and DVD - high visual quality	- digital TV (2 to 4 Mbits/s) - HDTV, DVD (from 1.5 to 10 Mbits/s)	Formats up to 16384x16384 at 30 images/s
H.264	- compression for DVD, HD-DVD and VOD - high visual quality	2-3 times lower than MPEG2 for the same visual quality 4 times more complex than MPEG2	Formats up to 16384x16384 at 30 images/s

Table 2.1. *Comparison of video standards*

2.4.3. *Recent developments in video compression*

DCT-based techniques have proven their effectiveness. They nevertheless remain limited in terms of scalability and do not offer flexibility of frequency, resolution or quality layers (see section 2.3.2.4 on JPEG 2000). This is possible with the recent approaches which use wavelet transform. The generic scheme of Figure 2.12 remains valid when DCT is replaced by DWT. However, as for H.264, flexibility results in a complexity of calculation which currently limits the use of DWT into the video standards.

2.5. Compressing 1D signals

Speech signals used on phones or on mobile networks, as well as all the musical signals stored in MP3 players, and sent over the Internet are digital in form. The rate constraints associated with the transmission or storage supports and the sound quality constraints due to services or uses, vary widely. In telephony, for example, a reasonable voice signal is achieved at a sampling frequency of 8 kHz and 13 bits per sample (a bitrate equivalent to 100 kbits/s). For music, the sound quality required needs wide bandwidths, as well as greater precision during the quantization of the signal's samples: for the CD format, the sampling frequency is fixed at 44.1 kHz and the resolution is at 16 bits per sample (a bitrate equivalent to 1.4 Mbits/s in stereo).

The transmission or storage supports cannot generally allow such large bitrates, hence the need to code the audio information. For further details on the different standards and recommendations for audio signals, see [LEG 00].

2.6. The compression of 3D objects

Three-dimensional objects are becoming more and more present in the fields of industry, medicine, cultural heritage, video games, etc. The initial data is very diverse in its nature: clouds of dots, surface data and volume data. Numerous models of 3D representation have been proposed to represent this variability. From the point of view of compression, these models need to be compact, easy to manipulate, and have to allow for a high quality representation for the greatest variety of 3D shapes.

The widespread use of polygon meshes for the visualization of 3D objects has led many researchers to investigate the compression of the structure of the associated data: the connectivity of a mesh vertex and its geometry (spatial positions of these vertices). Other researchers have focused on new surface models (subdivision surfaces, geometric wavelets, etc.) and volume models (superquadrics, supershapes, etc.) which are generally very compact, and thus very efficient for compression. These scientific activities combine complementary methods of two communities: image processing and geometric modeling. We will not detail these promising techniques in this chapter. Instead, you may wish to refer to [LAV 05] and [AKO 06] for a complete review of the compression of 3D objects.

Concerning the standardization, the MPEG4 standard in its second version includes various 3D mesh compression methods. The extension MPEG4 AFX (*Animation eXtension Framework*) includes a wavelet-based approach. It also proposes the use of implicit surfaces, NURBS and subdivision [MPE 03].

2.7. Conclusion and future developments

This chapter has presented the compression techniques which have become current standards. All the multimedia supports have been covered, with priority given to image and video. It is important to note that the existence of standards is an essential step towards making technological transfer more dynamic and democratizing multimedia services and usages. Today, who doesn't use a digital camera with ".jpg" files? Who does not watch DVD films coded in MPEG4? We make use of digital technology in our daily lives and, as a result, we use all the associated compression standards.

Nevertheless, multimedia research continues with as much energy and enthusiasm as ever. If research and development teams had ceased to conceive of new image compression methods in the 1980s or 1990s, perhaps because of the existence of a standard such as JPEG, then today we would not have the JPEG 2000.

The future continues to look promising: the key trend is clearly the development of new functionalities to be added to the existing compression methods. This trend is driven by innovative ideas in multimedia services and usages. It should open up many research and development directions. Principal examples include:

– taking into consideration network and transmission needs (source-channel coding) as well as receiver needs (man-machine interfaces, innovation in visualization, etc.);

– taking into consideration the growing size of databases (real-time access, real-time processing, etc.); the need for innovative approaches combining indexing and compression;

– the security of multimedia content with hybrid methods: compression-watermarking-encryption;

– collaborative work between the "compression" communities and those focusing on "knowledge modeling", in order to better understand the semantic content of images and thus to open up new perspectives for image compression.

2.8. Bibliography

[AHM 74] AHMED N., NATARAJAN T., RAO K.R., "Discrete cosine transform", *IEEE Signal Processing Magazine*, vol. 14, no. 2, p. 24–41, 1974.

[AKO 06] AKOKA J., COMYN-WATTHIAU (coordinators of the collective work), *Encyclopédie de l'informatique et des systèmes d'information, section multimédia dans les systèmes d'information*, 130 p., ISBN: 2-7117-4846-4, Editions Vuibert, 2006.

[ANT 92] ANTONINI M., BARLAUD M., MATHIEU P., DAUBECHIES I., "Image coding using wavelet Transform", *IEEE Trans. on Image Processing*, vol. IP-1, no. 2, p. 205–220, 1992.

[BAR 02] BARLAUD M., LABIT C. (eds.), *Traité IC2 : Compression et codage de l'image et de la vidéo : principes et techniques*, Hermes, 2002.

[BAT 97] BATTAIL G, *Théorie de l'information : applications aux techniques de communication*, Masson, Paris, 398 p., 1997.

[BRE 02] BREMAUD P., "Mathematical principles of signal processing, Fourier and wavelet analysis", *Springer*, ISBN: 0-387-95338-8, 269 p., 2002.

[CCI 90] *"CCITT Recommendation H.261: Video codec for audio visual services at p * 64 kbits/s"*, COM XV–R 37–E, 1990.

[DAU 92] DAUBECHIES I., "Ten lectures on wavelets", *CBSM-NSF Regional conf. Series in Applied Mathematics, SIAM*, Philadephia, PA, 1992.

[EST 77] ESTEBAN D., GALAND C., "Application of Quadrature Mirror Filters to Split Band Voice Coding Schemes", *IEEE Proc. ICASSP-77*, p. 191–195, 1977.

[GRA 84] GRAY R.M., "Vector quantization", *IEEE ASSP Magazine*, vol. 1, p. 4–29, 1984.

[HUF 52] HUFFMAN D.A., "A method for the construction of minimum redundancy codes", *Proceedings of the IRE*, September 1952.

[ISO 93] ISO/IEC, *"Information technology - coding of moving pictures and associated audio for digital storage media at up to about 1.5 mbits/s: video"*, 11172–2, 1993

[ISO 95] ISO/IEC, *"Information technology – generic coding of moving pictures and associated audio information: video"*, 13818–2, 1995

[ITU 96] ITU-T, *"Video coding for low bitrate communication"*, recommendation H.263, 1996.

[JAI 81] JAIN A. K., "Image data compression: a review", *Proceedings of IEEE*, vol. 69, no. 3, p. 349–389, 1981.

[JVT 02] *"Joint Committee Draft, JVT-C167"*, Joint Video Team of ISO/IEC MPEG and ITU-T VCEG, May 2002

[LAV 05] LAVOUE G., "Compression de surfaces, basée sur la subdivision inverse, pour la transmission bas débit et la visualisation progressive", Doctoral Thesis, 221 p., December 2005.

[LEG 00] LE GUYADER A., PHILIPPE P., RAULT J., "Synthèse des normes de codage de la parole et du son (UIT-T, ETSI et ISO/MPEG)", Annales des Télécommunications, no. 9–10, 2000.

[LIN 80] LINDE Y.L., BUZO A., GRAY R.M., "An algorithm for vector quantizer design", *IEEE Trans. on Communications*, vol. COM-28, p. 84–95, 1980.

[MAL 98] MALLAT S., *A Wavelet Tour of Signal Processing*, Academic Press, New York, 1998.

[MAX 60] MAX J., "Quantizing for minimum distortion", *IEEE Trans. on Information Theory*, vol. IT-6, p. 7–12, 1960.

[MEY 92] MEYER Y., *Ondelettes, algorithmes et applications*, Armand Colin, Paris, 1992.

[MPE 98] MPEG-4 Video Group, *"Coding of audio-visual objects: video"*, ISO/IEC JTC1/SC29/WG11 N2202, March 1998.

[MPE 03] MPEG-4 AFX, "Coding of audio-visual objects. Part 16: Animation Framework eXtension", International standard ISO/IEC/JTC1/SC29/WG11 14496-16:2003, Information technology, International Organization for Standardization, Switzerland, 2003.

[RAM 93] RAMCHANDRAN K., VETTERLI M., "Best Wavelet Packet Bases in a Rate-Distortion Sense", *IEEE Trans. on Image Processing*, vol. IP-2, no. 2, p. 160–175, 1993.

[RAO 90] RAO K.R., YIP P., *Discrete Cosine Transform – Algortihms, Advantages, Applications*, Academic Press, San Diego, CA, USA, 1990.

[RIC 03] RICHARDSON I.E.G., *H.264 and MPEG-4 Video Compression*, John Wiley & Sons, 2003.

[RIS 76] RISSANEN J., "Generalised Kraft inequality and arithmetic coding", *IBM J. Res. Dev*, no. 20, 1976.

[ROS 82] ROSENFELD A, KAK A., *Digital Picture Processing*, vol. II, Academic Press, NY, 1982.

[SAI 96] SAID A., PEARLMAN W., "A new fast and efficient image coder based on Set Partitioning on Hiearchical Trees", *IEEE Trans. on Circuits and Systems for Video Technology*, vol. 6, no. 3, p. 243–250, 1996.

[SHA 49] SHANNON C.E., *A Mathematical Theory of Communication*, University of Illinois Press, 1949.

[SHA 93] SHAPIRO J.M., "Embedded image coding using zerotrees of wavelet coefficients", *IEEE Trans. on Signal Processing*, vol. 41, no. 12, p. 3445–3462, 1993.

[SMI 86] SMITH M.J.T., BARNWELL T.P., "Exact Reconstruction Techniques for Three Structured Subband Coders", *IEEE Trans. on Acoustics, Speech and Signal Processing*, vol. ASSP-34, no. 3, p. 434–441, 1986.

[SYM 04] SYMES P.D., *Digital Video Compression*, McGraw-Hill, 2004.

[TAU 00] TAUBMAN D., "High performance scalable image compression with EBCOT", *IEEE Transactions on Image Processing*, vol. 9, no. 7, p. 1158–70, 2000.

[WEL 84] WELCH T., "A Technique for High-Performance Data Compression", *Computer*, June 1984.

[WOO 86] WOODS J.W., O'NEIL S.D., "Subband Coding of Images", *IEEE Trans. on Acoustics, Speech, and Signal Processing*, vol. ASSP-34, no. 5, p. 1278–1288, 1986.

[ZIV 77] ZIV J, LEMPLE A., "A Universal Algorithm for Sequential Data Compression", *IEEE Transactions on Information Theory*, May 1977.

Chapter 3

Specificities of Physiological Signals and Medical Images

3.1. Introduction

This chapter will broadly cover the specific features of the signals and images most commonly used in clinical procedures. Of course, we may wonder if such analysis is essential when designing compression schemes dedicated to physiological signals and medical images. In other words, are these data so different from common signals and images (e.g. audio signals or images acquired by a digital camera)?

Physiological signals should not be considered as audio signals for the simple reason that they are provided by uncommon sources, having particular features. They differ from audio signals by their frequency content, by the correlation between samples and above all, by their dynamic. In fact, the quality of audio signals can be measured through regular auditory systems; this measurement procedure however cannot be applied to physiological signals. Consequently, an efficient coding device, in terms of compression rates and diagnostical quality, which will take into account the specific features of those signals, must therefore be employed.

Chapter written by Christine CAVARO-MÉNARD, Amine NAÏT-ALI, Jean-Yves TANGUY, Elsa ANGELINI, Christel LE BOZEC and Jean-Jacques LE JEUNE.

Likewise, the specificities of medical images are found in the way that the meaning and hence the values of the pixels are represented. Most compression systems are tested on natural monochromatic or color images (e.g. the most commonly used image being the "Lena" image). The pixel intensity of those images corresponds to the reflection coefficient of natural light. In fact, images acquired for clinical procedures reflect very complex physical and physiological phenomena, of many different types, hence the wide variety of images.

This chapter will begin by introducing the characteristics of physiological signals (section 3.2) followed by the characteristics of medical images (section 3.3) and conclude by highlighting the necessity of an adequate compression system adapted to these features. Section 3.2 will first of all describe those physiological signals most commonly used in clinical procedures (electroencephalogram (EEG), evoked potential (EP), electromyogram (EMG) and electrocardiogram (ECG)). It will then outline the characteristics of the most recent acquisition systems as well as the properties of the different signals acquired thus. Section 3.3 will detail some laws of physics, the various applications used in clinical procedures as well as today's technological improvements in the principal fields of medical imaging such as radiology (X-ray imaging), magnetic resonance imaging (MRI), ultrasound (US), nuclear medicine (NM) and anatomic pathology imaging. In this same section we will then outline the most relevant features (especially for information encoding) of the images.

3.2. Characteristics of physiological signals

Although many physiological signals are used in clinical routines, in this section we have limited our analysis to the most commonly used signals, specifically those that require most storage and transmission capacity.

3.2.1. *Main physiological signals*

3.2.1.1. *Electroencephalogram (EEG)*

The EEG is a physiological signal related to the brain's electrical activity. This signal is remotely recorded using electrodes placed on the scalp. The EEG helps detect potential brain dysfunctions, such as those causing sleep disorders. It may also be used to detect epilepsies known as "paroxysmal attacks" identified by peaks of electrical discharges in the brain. Using EEG during the monitoring process is a common practice in clinical routines. In this particular process, the information is usually recorded over an extended period of time (24 hours).

3.2.1.2. *Evoked potential (EP)*

When stimulating a sensory system, the corresponding recorded response is called "evoked potential" (EP). Nerve fibers generate synchronized low-amplitude action potentials (i.e. spikes) and the sum of these action potentials provides an EP that should be extracted from the EEG, considered here as noise. Generally, EPs are used to diagnose certain anomalies linked to the visual or the auditory system or even the brain stem.

There are three major categories of evoked potentials:

– somatosensory evoked potentials (SEP), obtained through the stimulation of muscles;

– visual evoked potentials (VEP) for which a source of light is used as a stimulus;

– auditory evoked potentials (AEP) generated by stimulating the auditory system with acoustic stimuli (Figure 3.1).

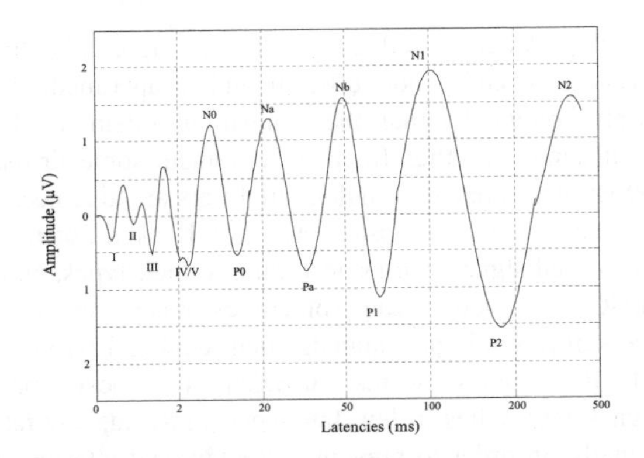

Figure 3.1. *Auditory evoked potential generated by a stimulus. Wave I: action potential; Wave II, III, IV, V: brain stem potential (BAEP); Waves N0, O0, Na, Pa, and Nb: thalamic/cortical potentials; Waves P1, N1, P2 and N2: late potentials (cortical origin)*

3.2.1.3. *Electromyogram (EMG)*

EMG is a recording of potential variations related to voluntary or involuntary muscle activity. The artefact's amplitude (5 μV) resulting from muscular contraction is higher than that of the EEG and the time period varies between 10 and 20 ms. Experiments on volunteers have shown that for muscular contractions of 40 s, muscle "spikes" (impulsional signals) appear by intermission approximately every 20 ms. This phenomenon is often modeled by a "Poisson" process.

3.2.1.4. *Electrocardiogram (ECG)*

The ECG is an electrical signal generated by the heart's muscular activity. It is usually recorded by a series of surface electrodes placed on the thorax. An electrocardiographic examination is entirely painless. It provides effective monitoring of the heart by detecting irregular heart rhythms. It is also used to prevent myocardial infarction.

Monitoring using EGG signals is a common process in clinical routines. Recordings made over a long period of time (e.g. approximately 24 hours or more) are often necessary. In some cases, recordings can take place at the patient's residence by using a "Holter" monitor. This device efficiently records all cardiac activity using "flash" memory. In other applications, remote monitoring is performed by transmitting the ECG signal (e.g. one or various leads) using a given network, such as the Internet or an intranet.

3.2.2. *Physiological signal acquisition*

Unlike acquisition systems used 20 or 30 years ago, collecting physiological signals is nowadays achieved by more efficient and cheaper methods. Before dealing with a digitized physiological signal, the acquisition system should include analog filters (e.g. anti-aliasing) or other filters to eliminate some drifts, such as those caused the 50/60 Hz interferences. Analog amplifiers are also necessary, especially for low energy signals, such as Brainstem Auditory Evoked Potentials (BEAP). The dynamic of the recorded signals varies according to their types. Nevertheless, when dealing with EGG, this range can sometimes reach several millivolts. The digitization process of physiological signals often requires 12-14 bit quantization. In fact, this level of accuracy seems sufficient for most applications. Since physiological signal energy lies in low frequencies instead, low rate samplings are then required. Finally, in order to prevent any additional interference, it is usually recommended to use cable shielding.

3.2.3. *Properties of physiological signals*

3.2.3.1. *Properties of EEG signals*

A considerable amount of energy in EEG signals is located in low frequencies (between 0 and 30 Hz). This energy is largely dominated by the following rhythms:

– the δ rhythm consists of frequencies below 4 Hz; it characterizes cerebral anomalies or can be considered as a normal rhythm for younger patients;

– the θ rhythm (5 Hz) often appears amongst children or young adults;

– the α rhythm: frequencies around 10 Hz are usually generated when the patient closes his eyes;

– the β rhythm: frequencies around 20 Hz may appear during a period of concentration or during a phase of high mental activity;

– the γ rhythm: its frequency is usually above 30 Hz, it may appear during intense mental activity including perception.

Above 100 Hz, we can note that the EEG energy spectrum varies roughly according to the function 1/f, where f stands for the frequency. As already mentioned in section 3.2.1.1, peaks and waves may appear on random epochs while recording the EEG signal. In fact, this is a common characteristic of epileptic cases. Moreover, it is important to note that other physiological signals may interfere with the EEG signal during the acquisition process. This is the case with ECG or EMG signals. The amplitude of the EEG signals varies from a few microvolts up to about 100 μV. The 10/20 system is often used for acquisition purposes. This system standardizes the position of the 21 required electrodes. In some applications, the acquisition process however does not use this standard. Instead, a higher or lower number of electrodes are used. Different EEG signals are represented in the diagram below (Figure 3.2).

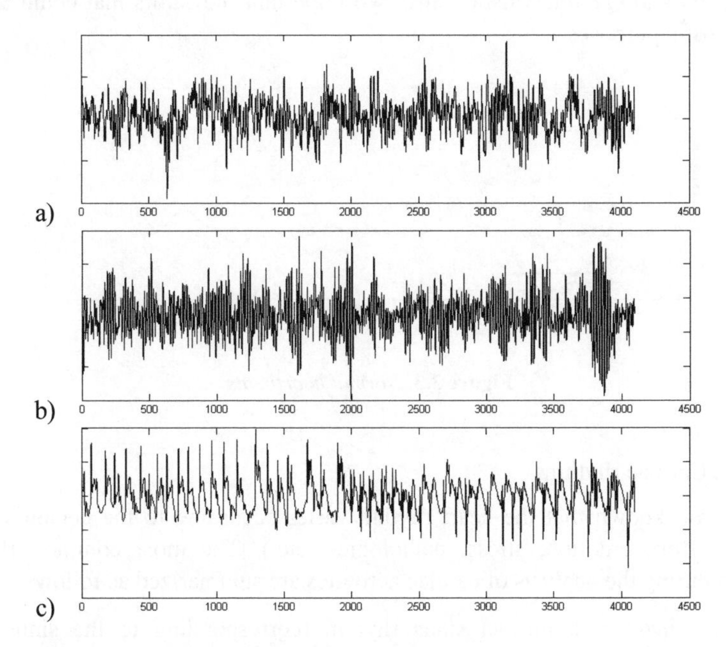

Figure 3.2. *Recorded EEG signals: a) in a healthy patient (eyes open); b) in a healthy patient (eyes closed); c) in an epileptic patient*

3.2.3.2. *Properties of ECG signals*

A typical ECG beat has 5 different waves (P, Q, R, S and T), as shown in Figure 3.3. Each wave can be defined as follows:

– the P wave corresponds to the right and left heart auricles' sequential depolarization: its amplitude is usually lower than 300 μV and its duration is of less than 120 ms, its frequency varies between the following range: 10 to 15 Hz;

– the QRS complex is produced after the depolarization process in the right and left ventricles; usually, its duration is of 70 to 110 ms and its amplitude is around 3 mV. It should also be noted that the QRS complex is often used by algorithms of automatic heart beat detection;

– the T wave corresponds to ventricular polarization: it is once again characterized by a low frequency;

– the ST segment corresponds to the time period during which the ventricles remain in a depolarized state;

– the RR interval is often used as an indicator for some arrhythmias; it may also be integrated into some ECG classification algorithms;

– the PQ and QT intervals are also two important indicators that could be used for diagnostic purposes.

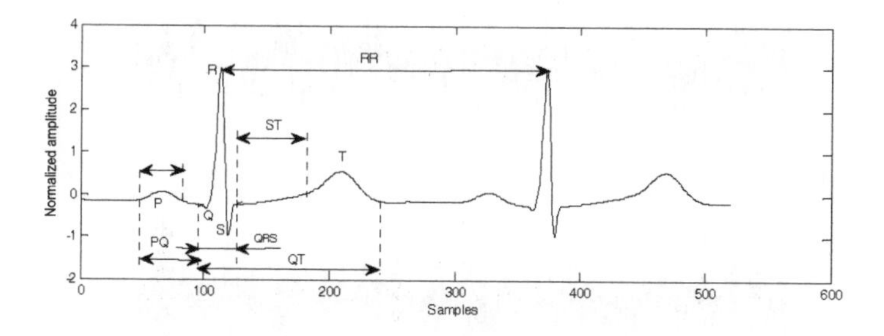

Figure 3.3. *Normal heartbeats*

3.2.3.2.1. Cardiac rhythms

It is well known that the heart rhythm varies according to the person's health (fatigue, effort, emotion, stress, pathologies, etc.). The most common rhythms observed during the analysis of cardiac activities are summarized as follows:

– *sinus rhythms:* a normal sinus rhythm (corresponding to the sinus node) usually has a frequency of 50-100 beats/minute. A low beat rhythm is known as a bradycardia; however if the rate is higher than a certain limit, it would be considered as a tachycardia;

– *extrasystole:* in some cases, the sinus rhythm mixes with an extrasystole, also called Premature Ventricular Contraction (PVC). The PVC is a premature contraction of the heart that is usually followed by a longer pause. In some cases, it can also occur between normal pulsations without affecting their regular occurrence. The origin of an extrasystole may be auricular, nodal or ventricular. Its morphology changes according to the pathology (Figure 3.4);

– *supraventricular arrhythmia:* its origin is either the atria or the atrioventricular node. The P wave is often distorted or even non-existent;

– *ventricular arrhythmia*: this heart rhythm complication arises from the ventricles including the ventricular tachycardia (i.e. characterized by extrasystoles that appear rapidly and repetitively), the ventricular "flutter" (i.e. a very rapid and regular cardiac rhythm, where the QRS complex and the T wave cannot be perceived) and ventricular fibrillation (i.e. a chaotic rhythm often followed by heart failure).

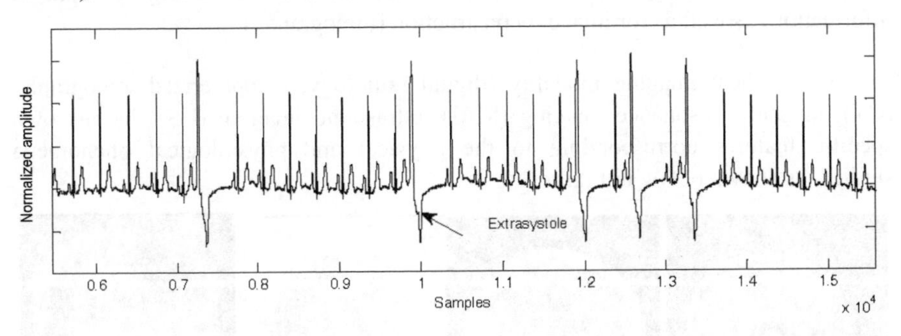

Figure 3.4. *Cardiac rhythm containing extrasystoles*

3.2.3.2.2. ECG noise

Other physical or physiological signals can often interfere while recording the ECG signal, or all other signals. For example:

– the 50/60 Hz signal: this classical interference can be removed easily by using some advanced methods like adaptive filters. For a well-known example of this, the reader can refer to the scheme proposed by Widrow *et al.*;

– electrode artifacts: these may appear when electrodes are moved on the skin's surface. Their frequencies range between 1 and 10 Hz, which might cause an overlapping with the QRS complex;

– the EMG: this signal (previously cited) can be considered a real nuisance during the ECG acquisition process. It can sometimes be eliminated by averaging methods or other approaches explained in [MOR 93] and [PAU 00];

– respiratory activity may sometimes cause changes in the morphology of QRS complexes. Some proposed techniques, found in the literature, enable the efficient estimation of the frequency of respiratory activity [LEN 03];

– etc.

More information related to the properties of physiological signals as well as processing techniques may be found in recent works [SOR 05].

3.3. Specificities of medical images

Medical imaging was initiated and developed due to the diversity of physical phenomena being used (X-rays, γ-rays, ultrasound waves, magnetic nuclear resonance). Medical imaging was further developed with the increased use of computers in the acquisition process (real-time treatment of a large amount of information) as well as for image reconstruction (tomography).

Each medical imaging modality (digital radiology, computerized tomography (CT), magnetic resonance imaging (MRI), ultrasound imaging (US)) has its own specific features corresponding to the physical and physiological phenomena studied, as shown below in Figure 3.5.

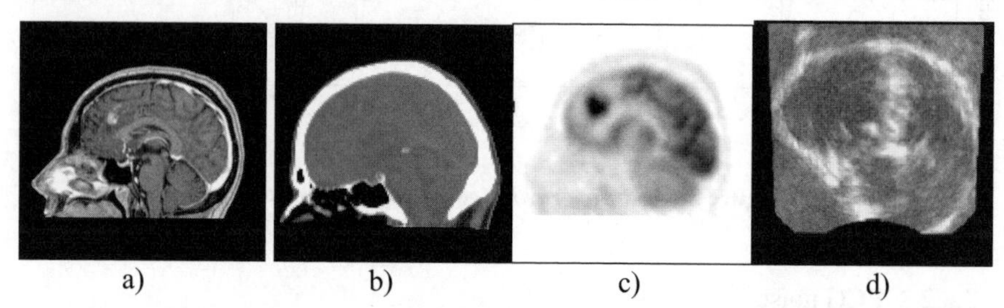

a) b) c) d)

Figure 3.5. *Sagittal slices of the brain by different imaging modalities: a) magnetic resonance imaging (MRI), b) computed tomography (CT), c) positron emission tomography (PET), d) ultrasound (US)*

3.3.1. *The different features of medical imaging formation processes*

The pixel or voxel values depend on the chemical and physical characteristics of the tissues studied. These characteristics often correspond to a physiological phenomenon. This intensity tallies with:

– an attenuation coefficient of X-rays for radiology and CT;

– a local concentration of radioactive markers in nuclear medicine giving for example access to information on the body glucose consumption or the metabolic activity within the body;

– a local concentration of contrast agent in radiology or MRIs;

– a density of protons or the speed of paramagnetic proton relaxation for MRI;

– an anisotropic movement of water molecules for diffusion MRI;

– a concentration change in oxyhemoglobin for functional MRI;

– a reflective and scattering coefficient for ultrasound;

– a local speed vector depicting blood flow during Doppler ultrasound, etc.

Imaging mechanisms based on spontaneous contrasts often provide anatomic information, while imaging mechanisms known as "functional" often use markers reflecting fluid motion or metabolic exchanges. These mechanisms efficiently depict important details on how well the body functions.

3.3.1.1. *Radiology*

Radiology is still widely used in clinical routines. However, digital radiology is gradually replacing analog radiology producing radiographic films.

3.3.1.1.1. Guiding principles

Radiology uses X-ray attenuation properties and maps out the cumulated absorption process of an X-ray during its path along the tissues (transmission imaging) [GIR 93]. X-ray absorption is very high if the body structure studied contains calcium. It is less significant if it is made up mainly of soft tissues and lower still if the body structure contains fat. It is zero in water (Hounsfield scale). In digital radiology, an image is obtained by using flat matrix panels or radioluminescent storage phosphor plates.

When the phosphor plates are exposed, the X-ray's intensity dispersion is recorded as electrons captured within semi-stable layers. Once the exposure is carried out, the panel is placed in an appropriate reader. In this reader, a laser beam scans over the panel and sends out for each pixel a luminous signal proportional to the X-ray's intensity level received on this pixel. This light transmission is then detected by a photomultiplier tube for which the signal is digitized.

Flat matrix detectors are made of one large surface active matrix. This matrix contains receptors that are able to convert X-rays into an electric charge. The electric charge at the level of each pixel is read by a low noise electronic system (thin transistor film) and then converted into numerical data. Converting X-rays into electric charges may be performed either directly or indirectly. For a direct conversion, a photoconductor must be used (amorphous selenium). Once the X-rays interact with the photoconductor, electrons are released. The quantity of electrons released is proportional to the intensity of the X-rays. For an indirect conversion, a phosphorescent layer is used to absorb the X-rays. The resulting luminous photons are then detected by a large photodetector usually made of photodiodes. Detectors such as the one described above can obtain a single projection plane of an anatomic region. The entire volume of the X-rayed body part is projected onto a plan.

To make slice images with X-rays, CT is used. CT scanners (X-ray tubes and detectors) revolve around the patient so as to obtain a variety of projections from many different angles. In practice, the X-ray beams are aimed so as to radiate over a small part of the body. The X-ray projection comes out in the form of lines. In tomography the set of projections is then used to reconstruct a map indicating all the regions where attenuation coefficients have been calculated [GRA 02].

CT delivers anatomic information but is unable to take metabolic rate into account. An iodinated substance may be injected to increase the contrast between the different tissues. The distinction will appear depending on the level of vascularization of each tissue, or depending on how fast the substance reaches the interstitial space. With multislice scanners we can obtain perfusion imaging, a slight incursion into functional imaging. However, we still need to control the amount of ionizing radiation used.

3.3.1.1.2. Images acquired in clinical routine

Digital projection radiology is widely used in clinical routines for the study of bones, as well as to examine breasts (mammography) or lungs. It is also used in cardiology for coronary angiography with a contrast agent injected in the coronaries through a catheterization process (Figure 3.6).

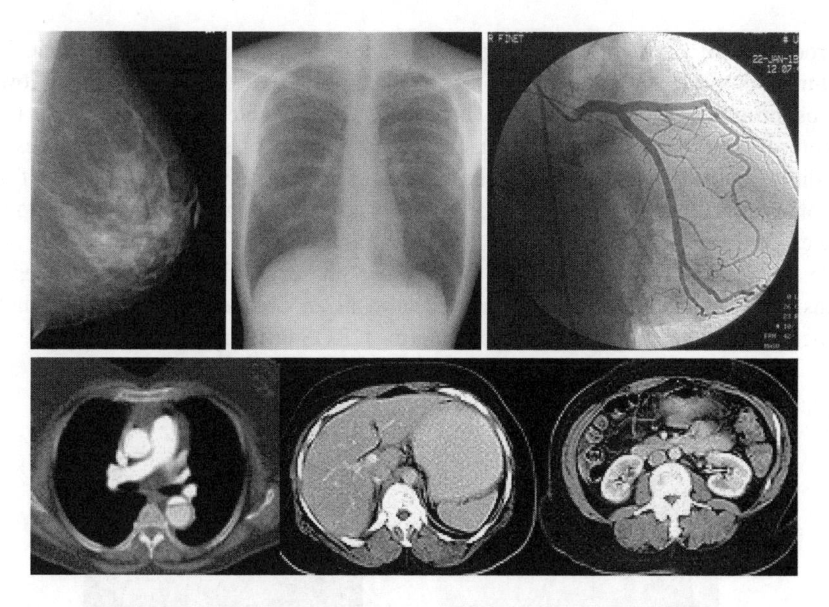

Figure 3.6. *Radiological images. 1st row: projection images; from left to right: mammography, pulmonary radiography, coronary angiography; 2nd row: thorax and abdomen CT scan (after injection of contrast agent)*

A cardiac angiography examination is made up of a dozen sequences so as to observe the ventricles and the coronary arteries at various angles. The set of sequences gathers a total of 2,000 to 3,000 images (512*512*8 bits) corresponding to about 500 to 800 MB in terms of disc storage space. Projection digital radiology is widely used on a daily basis in all imaging centers, which thus requires a large storage capacity for the considerable quantities of information transmitted and saved. For example, using this type of imaging, Angers hospital in France (1,600 beds) accumulates 4.5 TB a year, i.e. 35% of the production of all its imaging systems. Nevertheless, the increase in volume of projection images is most considerable. The CT scanner is very useful for visualizing complicated fractures, or for examining organs in the neck, the thorax or the abdomen; a contrast agent is often injected in the patient's veins so as to better distinguish the different structures in the body (Figure 3.6).

3.3.1.1.3. Recent developments

The recent upgrades of the multi-detector (or multi-slice) CT scanner considerably increased the speed at which images are acquired, and improved the quality of those images. Such developments now allow quasi-isometric 3D images on any layout plan space. The more detecting channels there are in the system, the better both temporal and spatial resolution will be. These upgrades led to the introduction of a new vascular imaging that is less intrusive and highly efficient. The volumes possible now allow us to explore the aorta and the arteries of the lower limbs through a single acquisition – hence, we are now able to visualize over a total length of more than one meter, constituted by a series of sub-millimeter slices. In cardiology, the multi-slice technique allows us to reconstruct a 3D image of the coronary tree almost as detailed as an arteriography imaging [NIK 04], thanks to the increased acquisition speed and the ECG synchronization. For example, we are now able to obtain a reconstruction of 10 or more phases, each made of 80 slices, to transfer them and observe them in real-time in an oblique reformatted or three-dimensional reconstruction (volume rendering) [ROS 06] (Figure 3.7).

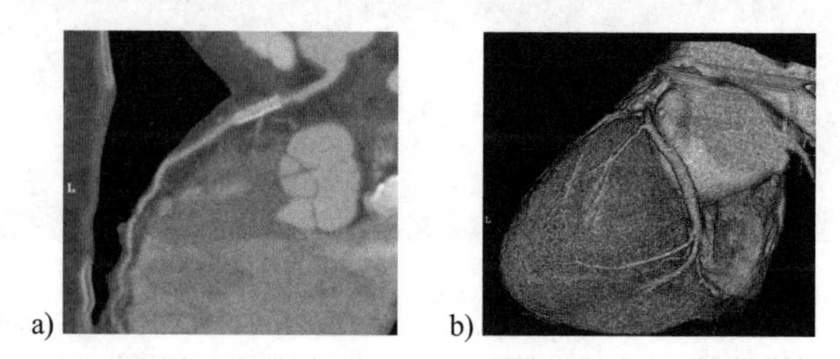

Figure 3.7. *Helical scanner: a) angioscanner of the coronary arteries (curvilinear reformatting in thick cuttings, maximum intensity point projection); b) reconstruction of a heart in volume rendering (reconstruction at 75% of the cardiac cycle)*

3.3.1.2. *Magnetic resonance imaging (MRI)*

Over 30 years, MRI has become widely used, especially to observe organ anatomy; their lesions as well as their performance. This particular type of imaging explores the cardiac or cerebral anatomy for example, in great detail. It also examines cerebral, cardiac or pulmonary activity.

3.3.1.2.1. Guiding principles

For a very precise frequency called a resonance frequency, an electromagnetic wave is sent through the tissues and topples the magnetization of some atom cores (such as that of hydrogen, a chemical element found in great quantities in all human tissues): this is known as the magnetic resonance phenomenon. The resonance frequency is proportional to the magnetic field applied (1.5 Teslas for the most commonly used appliances). During its return to a state of stable equilibrium, the magnetization of the proton is animated by a rotating movement around the magnetic field (precession). Its frequency equates the resonance frequency. If the body is placed inside a static magnetic field, by the precession process, the magnetization induces a current oscillating at the resonance frequency in a detecting coil rolled around the patient. The current's intensity is proportional to the density of the hydrogen atoms (or density of the protons) in the volume studied. By applying a varying magnetic field (gradient) in the space, a specific resonance frequency is captured for each of the volume elements. Measuring frequencies is a particularly meticulous process in physics, which is why the magnetization's distribution is measured by units smaller than millimeters. In practice, the magnetization's intensity once in a state of stable equilibrium does not allow us to obtain a sufficient contrast between the different anatomic structures. Images are therefore often measured over the relaxation time constants of longitudinal (T1 relaxation) or transversal (T2 relaxation) components, which characterize each biological tissue. The cartography of these parameters lets us obtain a highly detailed anatomic imaging (Figure 3.8).

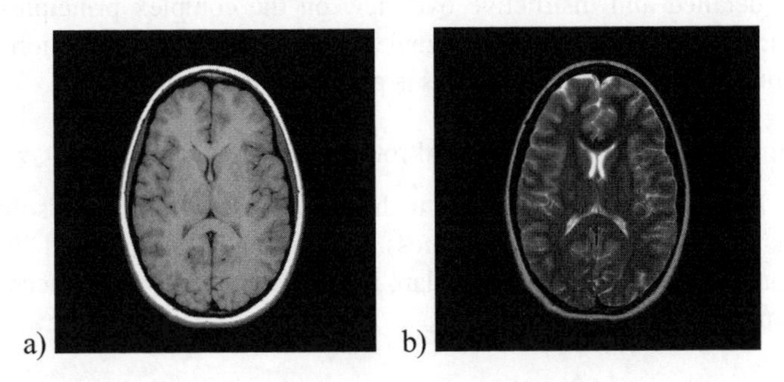

Figure 3.8. *Sagittal slice of the brain for different MRI acquisition parameters: a) T1-weighted image; b) T2-weighted image*

Phase magnetization within a volume can also provide functional information. In fact, this phase is proportional either to the speed (when applying a bipolar gradient), or the acceleration (when applying a tripolar gradient) of the voxel movement in the gradient direction. This phase is therefore measured using the Fourier transform which provides the magnetization intensity as well as the phase for each of the frequencies making up the current measures in the surrounding coil, as well as for each voxel. This is the reason why the cartography of phases becomes a map of speeds and accelerations. This is due to the fact that the reference phase is that of the fixed tissues. The microscopic movements of water can also be observed; leading us to the subject of MRI diffusion (Figure 3.9). Perfusion imaging uses a contrast agent and allows us to examine their dynamics when passing through the capillary network (Figure 3.9). Finally, functional MRI (fMRI) is aimed at observing cerebral surfaces activated when carrying out motor, sensitive, sensorial or cognitive activities. fMRI is based on the idea that activating the cerebral zone leads to a local increase in cerebral blood flow as well as an increase in oxygen consumption.

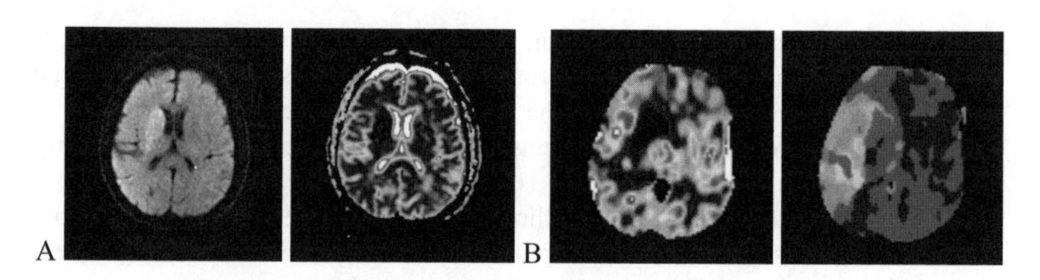

Figure 3.9. *Brain map in MRI. Patient A: echo sequence of a spin-echo planar weighted in diffusion and the diffusion's apparent coefficient map; Patient B: parametric map of cerebral blood flow and perfusion delay*

A very detailed and instructive overview on the complex principles of spatial measures and coding of the NMR signal and the image reconstruction process as well as of numerous imaging sequences is presented in [KAS 03].

3.3.1.2.2. Imaging carried out in clinical routines

MRI is an essential appliance for medical diagnosis due to its resolution being less than a millimeter, its innocuousness (in certain situations) and its ability to produce images in whichever layout plan, as well as its many sequences adapted to both anatomic and functional imaging.

MRI is now commonly used in neurology to obtain detailed images of the brain (detection and identification of cerebral injuries or lesions). Moreover, MRI diffusion dynamically probes the cerebral tissues in a microscopic scale so as to detect changes in nerve cell size. MRI diffusion also shows how nerve fibers are organized and connected to one another, which enables us to study the neuron interconnections of the brain [LEB 03].

MRI is also a reference in cardiology due to its spatial resolution. In this case, MRI is used to study the myocardial perfusion from the evolution in contrast following the injection of a small dose of an agent (gadolinium). The phase map also allows us to determine the dynamic characteristics of the cardiac mechanism (measuring cardiac flow for example). Tagged MRI methods allow us to study cardiac muscle deformation during contraction (Figure 3.10).

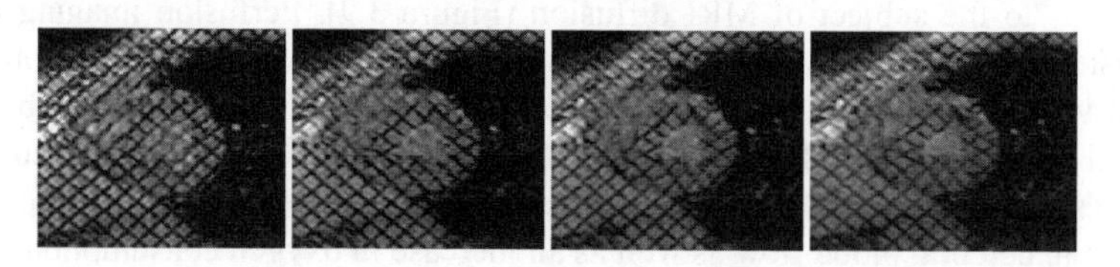

Figure 3.10. *Tagged cardiac MRI, small-axis slices*

3.3.1.2.3. Recent developments[1]

Mutli-component antennas (8, 16 or 32 components) are now available for recording systems. They reduce recording time while maintaining a high image resolution, by increasing the spatial and/or temporal resolution or the recording volume for a steady recording time span. The amount of image acquired in clinical routines has therefore increased.

Moreover, since the arrival of "high magnetic field" magnets (from 3 to 11 Teslas), the MRI potential, and especially that of MRI diffusion, will increase in terms of spatial and temporal image resolution (10 to 100 times greater than that of current images). Such systems allow us to reduce the thickness of each slice, and hence to achieve 3D viewings in clinical routines. In cardiology, it is now possible to have MRI angiographies of coronary arteries allowing us, without radiation or contrast agent, to search for coronary lesions as well as the consequences of such lesions on the perfusion and on the contractile pump (Figure 3.11).

At the same time, new MRI sequences have been developed, mainly in functional MRI (for the brain as well as the abdomen in the study of liver fats). More common sequences have also been improved such as diffusion tensor imaging for which it is now possible to observe over 150 directions (instead of 9 with the usual sequence) (Figure 3.11).

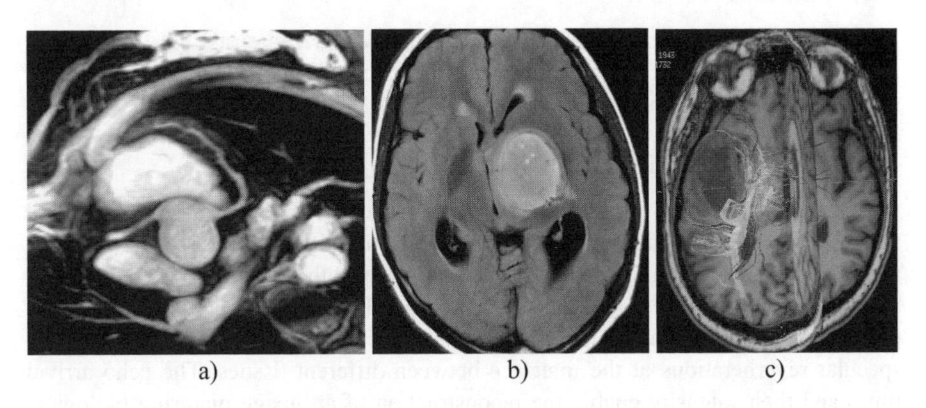

Figure 3.11. *Recent MRI developments: a) MRI angiogram of the coronary arteries; b) axial slice of the brain (presence of a tumour in the left hemisphere); c) visualization by diffusion imaging of white matter tracts (presence of lesions in the right hemisphere); image sources: General Electric Systems 3T*

1 We would like to thank Mr. Guillaume Calmon, General Electric Healthcare for all the information and images.

Finally, it is possible to acquire images with a low magnetic field using an ultra sensitive detector (SQUID – Superconducting Quantum Interference Device) working at very low temperatures. SQUIDs are used for a new type of imaging that has endured spectacular improvements in the fields of Cerebral Imaging and Magneto Encephalography (MEG), enabling us to measure low magnetic fields emitted by ionic currents running through the neurons [VOL 04] [JAN 02] (Figure 3.12).

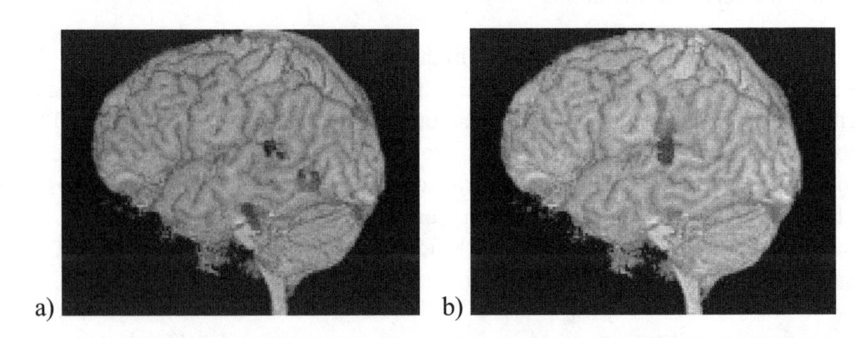

Figure 3.12. *a) MEG auditory dipole fused with 3D MRI b) MEG language dipole fused with 3D MRI [JAN 02]*

3.3.1.3. *Ultrasound*

Fetus ultrasound scans are always stirring experiences as the first visual encounter between the future parents and the coming baby. However, ultrasound scans are also used in many other cases.

3.3.1.3.1. Guiding principles

Ultrasound imaging probes the acoustic properties of body tissues. A sound wave (ultrasound) propagates inside the body along with local pressure variations, and is partly absorbed and partly reflected along its path, depending on the tissues' biological compressibility and density [BUS 01]. This sound wave is generated by piezoelectric captors which also measure the reflected waves (echoes) generated by specular reverberations at the interface between different tissues. The echo arrival times and their intensity enable the reconstruction of an image picturing biological issues up to a depth of a few centimeters and with a resolution down to a few millimeters. The intensity of echoes at tissue interfaces depends on the difference in acoustic impedance (a function of the biological properties of the tissue). The range of ultrasound frequencies used in medicine extends between 1 and 15 MHz for axial resolution ranges between 100 and 1,500 micrometers and for an ultrasound depth of several centimeters (e.g. 5 MHz used for a scan depth of 15 cm). Lower frequencies (bearing less attenuation) are mainly used for the study of innermost organs in the body. Higher frequencies (delivering a better spatial resolution) are used for the study of superficial organs. Absorption and dispersion phenomena (non-specular reflection) alleviate and deviate ultrasound waves. Moreover, when running through soft tissues, heterogenities produce diffusion echoes that create an ultrasound textured image and introduce some “speckle” noise.

The transducer is the active part of the ultrasound probe and is made of piezoelectric material (ceramic or polymer) that changes shape and generates an acoustic wavelength [CHR 88] when placed under an electrical potential. This same device receives the wave’s echoes and transforms the ultrasound waves into an electrical signal. An array of piezoelectric elements with simultaneous transmission or recording of several ultrasound waves is used to create a wavefront, which can be oriented and focused at various depths.

Ultrasound scans also enable us to measure the speed of blood flow in cardiac cavities and large vessels by measuring changes in frequencies by the Doppler Effect between the ultrasound wave transmitted and the reflected echoes [JEN 96]. During an ultrasound exam, red blood cells are sufficiently large (spheres of about 7 μm in diameter) compared to the dimension of the wavelength (150-770 μm for frequencies of 2-10 MHz) and are thus seen as echo reflectors producing spherical waves. A particular speed of flow in a vessel will create, using the Doppler Effect, echoes with modified frequencies. This effect is characteristically defined by the power spectrum density, directly proportional to the speed of blood flow. The range of frequency changes measured physiologically extends between 1-26 kHz for a range of ultrasound frequencies of 2-10 MHz and a range of blood flow speeds of 50-200 cm/s. The Doppler signal is often represented in three different forms:

– a spectrogram signal reproducing the speed of blood flow, in a given direction; through time and according to the energy of the signal received (Doppler power);

– an anatomical color image visually overlaying the information on speed values (Doppler imaging);

– an audio signal when the Doppler range of frequency shift is within the audible spectrum (Doppler audio).

3.3.1.3.2. Imaging produced in clinical routines

In a clinical routine, ultrasounds are used to analyze the morphology of organs as well as to detect potential anomalies. Each type of ultrasound examination is carried out with a specific ultrasound probe whose shape, size and working frequency is adapted to the organ examined. The echoes received in different orientations of auscultation, defining a sector (auscultation zone), are displayed on a fan-shaped diagram, whose echo lines need to be reconstructed in order to go from a series of registered measures with spherical coordinates to a regular pixel display grid with Cartesian coordinates. The reconstructed ultrasound images can be viewed on a video screen in A (amplitude according to a fixed direction), B (gray intensity according to a set of directions creating an image) or M (combination of modes B and A to portray organ movements) mode.

Each sector must function at a rate of 50 images per second so as to provide images in real-time. The spatial axial resolution, directly related to the frequency at which the ultrasound probe functions, sits between 100 and 1,500 µm. The lateral resolution depends on the echo's depth as well as on the focal systems used, and varies between 1 and 5 mm. However, the image's thickness varies between 2 and 12 mm, according to the size of all active elements in the ultrasound probe.

In cardiology, ultrasound imaging (echocardiography) is widely used. It is in fact the only type of cardiac imaging that operates in real time, while being transportable and not too expensive to purchase or operate. An echocardiography examination includes a series of acquisitions for the orientation of different probes and different acoustic windows. For each position of the ultrasound probe, a series of images is obtained to cover several cardiac cycles.

Widely used in obstetrics, ultrasound imaging allows us to diagnose potential morphological anomalies (Figure 3.13). By measuring morphologies on different parts of the skeleton and on certain internal organs such as the heart, we are able to determine the fetus' gestational age and to follow its development.

Abdominal ultrasound scans can be used to examine various soft organs located in the abdomen (liver, spleen, kidneys and ganglions) as well as organs containing liquid (gall bladder, principal biliary tracks, urinary bladder). Such scans can also be used to examine organs that are harder to observe due to the gases that they contain (stomach, small intestine, colon and appendix). When it comes to pathologies, ultrasound scans can be used to specify the size and echo structure determining benign or malignant tumours, cysts, abscesses and liver cirrhosis. An ultrasound scan of the pelvis helps us evaluate the general state of the organs it contains (urinary bladder, uterus, prostate and genital organs), to measure their dimensions and to detect the presence of potential anomalies (tumours, cysts, fibroma, haematomas).

Finally, Doppler ultrasound is used to study arterial hypertension and vascular anomalies (thrombosis, aneurysm, venous angioma and haematoma). Exploration by Doppler ultrasound enables the analysis of arterial walls by telling us whether they are homogenous and regular and how to characterize artherosclerosis plaques. It is also used to carry out and monitor vascular interventions in the case of a stenosis (Figure 3.13). A Doppler ultrasound can be combined with morphometric analysis evaluating the minimal diameter of the vessel with blood flow speed at the location of the stenosis, during a full cardiac cycle.

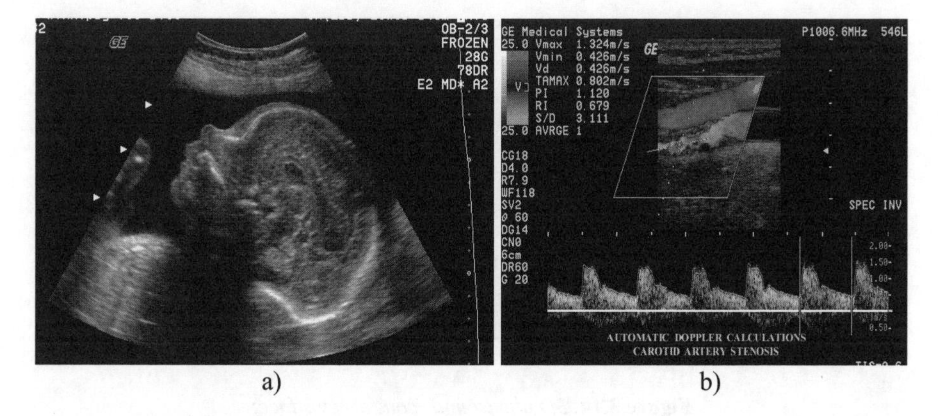

Figure 3.13. *Ultrasound images: a) fœtus; b) Doppler of the carotid artery with a colored image and a power spectrum (General Electric Healthcare)*

Some organs cannot be visualized by ultrasound, the reason being that ultrasound waves cannot travel through air or bones. It is therefore impossible to examine the lungs or the brain using this type of imaging. Ultrasound brain imaging can only be performed on infants through their fontanels.

3.3.1.3.3. Recent developments

There also exists 3D ultrasound imaging techniques originating from a series of 2D acquisitions, either using specific volumetric medical probes, or with common ultrasound probes equipped with spatial positioning sensors. With these probes we are able to examine 3D volumes of organs, with a limited temporal resolution (1 to 5 volumes/second), lower than for traditional 2D examination. New real time 3D ultrasound medical probes have also recently been developed. They are made of a matrix of active components. By using these, we can acquire a volume of information with the probe in a fixed position (Figure 3.14). The rate of acquisition for these medical probes, for a traditional cardiac examination, varies from 10 to 20 volumes/second, for an information output of 45 MB by examination and a resolution of 1 mm^3 per voxel.

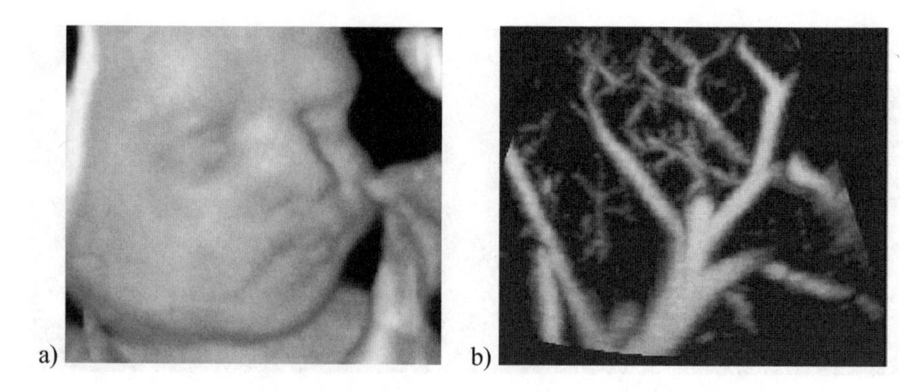

Figure 3.14. *3D ultrasound scan: a) fetus face; b) hepatic vascularization (Philips Medical Systems)*

Other recent advances to ultrasound acquisition methods allow additional information of clinical significance to be stored. Hence, elastrography imaging (tissue Doppler, measures of displacements or relative compressions) and high-frequency imaging are all major improvements in terms of physiological information, even if it means that the total volume of information has largely increased. High-frequency ultrasound scans unveil structural anomalies in cells that could be due to a growth of the membrane or to the nuclei dimensions. With the generation of shear waves we are now able to analyze tumour structures within a certain depth of tissue material [BER 04]. It is also possible to record non-stop ultrasound information in the form of radiofrequency wavelengths, so as to deal with the signal before the images are formed. A recent innovation of great importance is that of ultrasound harmonic imaging. When echoes are created in the human body the ultrasound wavelength includes the main interfering wavelength as well as a harmonic frequency component. Harmonic imaging works by filtering the echo so as to record and enhance this harmonic component. Using this technique we are able to obtain images benefiting from reduced reverberation, visual distortion or border effect, as well as a better image quality and a better penetration.

3.3.1.4. *Nuclear medicine*

Nuclear medicine gives us information on the function and the metabolism of the body, which is why it is known as a functional imaging.

3.3.1.4.1. Guiding principles

The purpose of nuclear medicine is to detect γ rays emitted by radioactive tracers and administered either through the veins or orally (emission imaging). Such tracers produce photons that are detected by gamma-cameras, we are therefore talking about Single Photon Emission Computed Tomography (SPECT). Other tracers produce positrons which annihilate in 511 KeV photons that are detected in PET (Positron Emission Tomography) imaging. For this type of imaging, a radio-pharmaceutical tracer made of functional molecules is used. The tracer is adapted to the organ and/or the pathology to be studied

The gamma camera detector is made up of a collimator (a lead plate pierced by several holes that only allows those gamma rays through that are perpendicular to the collimator), a sparkling crystal that converts γ photons into UV or visible photons as well as a photomultiplicator pipe that then converts them into electric signals.

In PET imaging, the events that are to be detected are not the γ photons produced by the radionuclide but those resulting from the encounter between the positrons and electrons. The collision between a positron and an electron leads to the transformation of both particles into two gamma photons that move apart in diametrically opposed directions, along a feedback line. A ring of detectors (PET camera) detects each photon pair arriving simultaneously on opposite detectors. The system treats the multiple coinciding couples so as to reconstitute the different pairs that are then assembled into an image, after a mathematical analysis (tomographic reconstruction [GRA 02]). Since the energy level of the photons is considerably high, the sparkling crystals must be dense and have a powerful output as well as a rapid relaxation so that they can detect new elements.

A lot less information is required in order to make up an image in nuclear medicine (also known as scintigraphy) than in radiology or MRI imaging. This leads to an important statistical blur. Moreover, the spatial resolution of the cameras is rather mediocre (sitting between 7 and 10 mm). Scintigraphic images are therefore not viable for a morphological study of the organ, as shown by the images in Figure 3.15. They are useful however for what is functional or pertains to the metabolism. [CHE 03] gives a detailed overview of the basic principles of radioactivity and detection methods, measurements or ways of analyzing information in nuclear medicine.

3.3.1.4.2. Imaging produced in clinical routines

Scintigraphy applications vary greatly according to the radiopharmaceutical given to the patient. Figure 3.15 displays some of these applications.

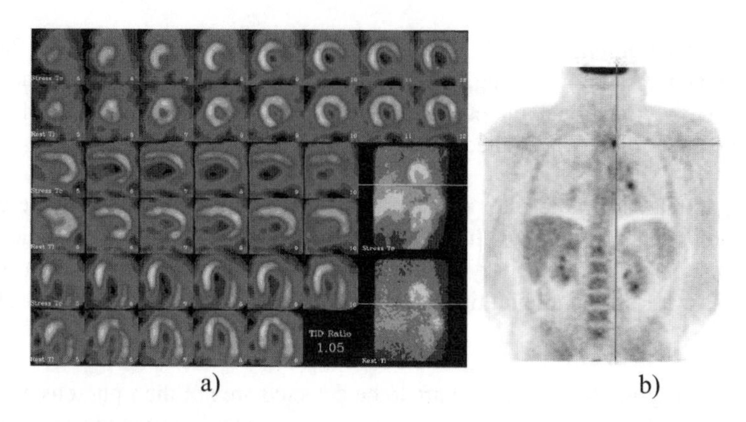

Figure 3.15. *Clinical applications of nuclear medicine: a) cardiac scintigraphy, long and short axis slices (myocardial ischemia); b) PET whole body (malignant concentration in the mediastinum) – General Electric Healthcare*

In cardiology, scintigraphy is the main imaging technique used to examine myocardial perfusion (Figure 3.15). The radioactive tracer used in this case is technetium 99 harnessed by cells according to tissue perfusion; the more efficient the perfusion is, the more tracers are fixed. Cardiac PET helps us obtain images of the metabolism at the cellular level. For example, to study the sugar consumption in a cell and analyze the tissue's viability (using glucose stained by fluor 18).

In cancer research, the PET with fluorodeoxyglucose (FDG) (cancerous cells consume more glucose than other cells) is used to detect small tumours (Figure 3.15). It is today the most sensitive examination to detect lung cancers, colon cancers, breast cancer and lymphoid cancer. Moreover, the FDG-PET exam is an essential tool for the premature evaluation and follow-up of chemotherapy treatments and radiotherapy.

In fact, sintigraphic imaging is used together with anatomic imaging. Both types complete one another. Scintigraphic imaging allows us to detect dysfunction before a morphological anomaly appears, since all pathologies start with such abnormal events, first in the molecules, and then in the cells.

3.3.1.4.3. Recent developments

PET and SPECT detectors are now more often used along with X scanners. The reason is that this allows us to obtain both anatomic and functional information in one single examination. The individual under examination must stay still, and both images overlay one another. The physician can then determine the location of the tracer fixation on PET images by using the anatomic image provided by the CT (Figure 3.16).

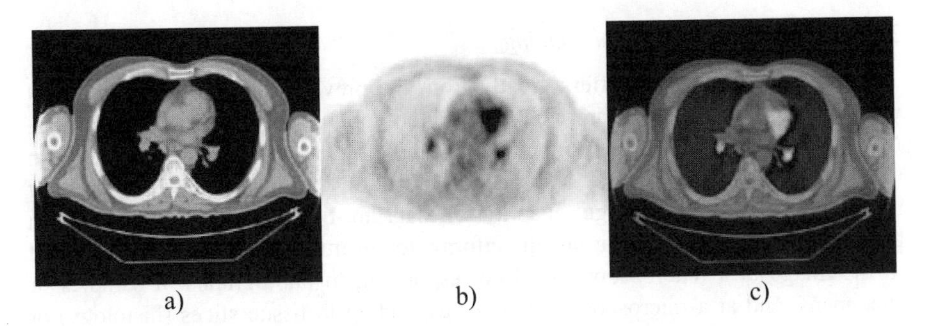

Figure 3.16. *Axial slice at heart level: a) CT scan (anatomic information); b) PET (functional information); c) fusion of both images; General Electric Healthcare*

Numerous research projects still aim at improving each key component of emission tomography. Thus, researchers are seeking more specific tracers of physiological phenomena able to detect pathologies. Moreover, technologies that involve detectors keep improving, with greater spatial resolution and higher sensitivity. Finally, important advances have been brought to 3D tomographic reconstruction, including a series of *a priori* information available when using bimodal PET/CT or SPECT/CT systems (Figure 3.17). All these medical developments in nuclear medicine create a functional imaging of good quality, now being widely used.

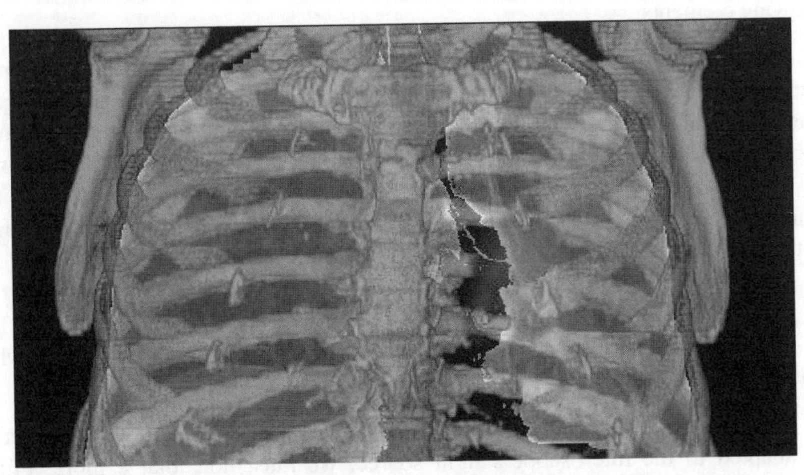

Figure 3.17. *Fused 3D PET images/CT scan (pulmonary tumour) – GE Healthcare*

3.3.1.5. *Anatomopathological imaging*

This type of imaging differs greatly from the previous one we have just dealt with, because the images are colored and are obtained using a CCD camera. Nevertheless, we have chosen to include it in this chapter partly due to its interest in terms of diagnostic process, and also because of the recent developments of techniques providing a large quantity of data to be inserted in a PACS. In anatomopathological applications, the information useful for the diagnosis is located in the image at a macroscopic level corresponding to photographs of samples or specimens, and at a microscopic level corresponding to tissue slices (histology) or cell layers (cytology) examined with a microscope.

3.3.1.5.1. Guiding principles[2]

The acquisition method for macroscopic images is similar to the replication process of a document using a white panel lit up by a white light and equipped with a photo camera.

There are two different digitization methods for microscopic images: stations using both a motorized microscope and a camera, and a slide scanner. This system works mainly through a lighting system (spectrum, open field, lighting, contrast) and its multiple lenses each having a different numerical aperture, zooming effect, as well as a choice between a flat field and apochromatism. The camera translates the observer's vision through the microscope, by favoring resolution aspects, restoring undertones to the maximum and restoring shades. It is characterized by the following elements:

– the resolution compromise of details/noises: detail resolution requires small-sized pixels in the sensor. On the other hand, these generate an unfavorable signal-to-noise ratio interfering with the image undertones, especially in shady areas. With today's CDD technology, this compromise is set at about 7 microns. We must note that this compromise is coherent with high zooming power lenses, and that resolution is insufficient when using a low zooming power lens, and lenses of a high digital aperture;

– the number of pixels in lines or columns determining the field of view;

– the pixel intensity's dynamics: capacity to reduce light and dark zones while avoiding saturation;

– the sensitivity curve of color components (red, green, blue) which is more or less adapted to the microscope's light source; we must note that a more refined approach has not been explored extensively yet; such a multispectral approach, would allow us to drop coloring under certain circumstances [YAG 05].

---

2 We would like to thank Jacques Klossa for the technical information.

3.3.1.5.2. Images produced in clinical routine

Anatomopathological examination is used to observe, identify and describe structural changes in tissues in order to diagnose some pathology. Anatomopathological imaging thus has an incredibly important role in cancer research, dermatology, gynaecology and forensic medicine.

Nowadays, most anatomopathological labs using digital imaging are equipped with a single station producing fixed images used in staff discussions or when requesting opinions and advice. Only some sites have tried out a strategy of systematic screening (Figure 3.18) and sharing of image records in the laboratory or the health institution managed by the lab's administration services.

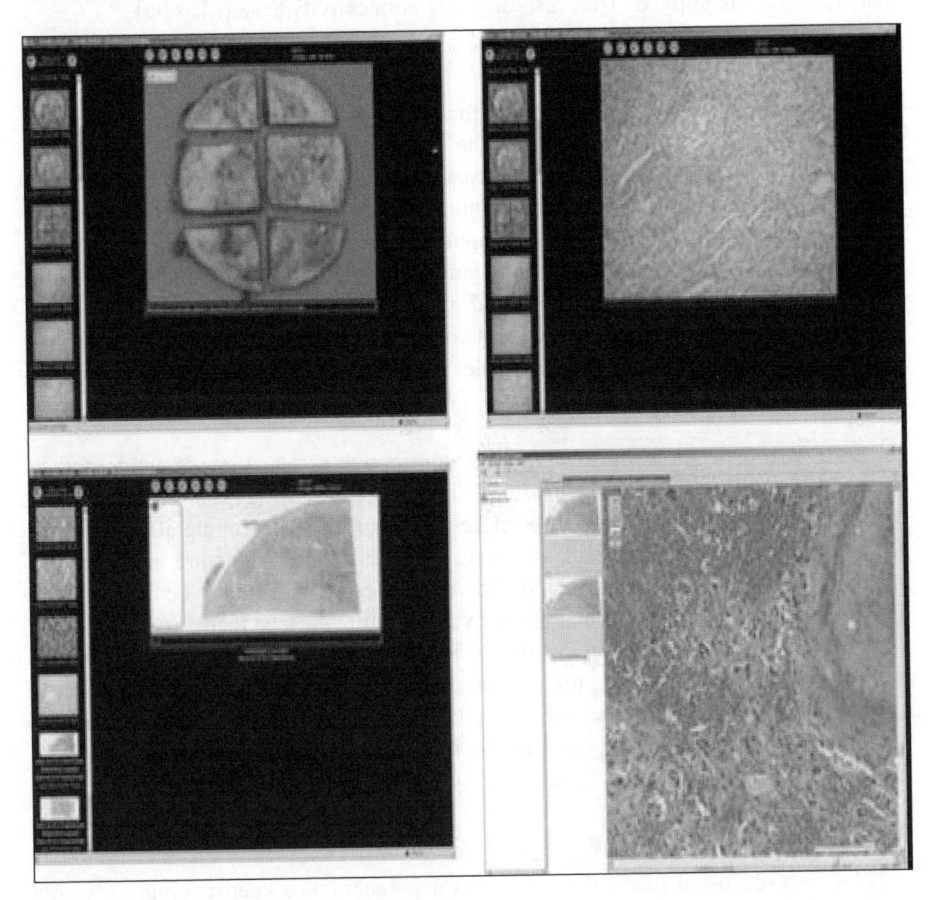

Figure 3.18. *Anatomopathological image record: macroscopy, microscopy and virtual slide*

Nevertheless, some studies to define sampling and acquisition procedures are necessary [KLO 04]. Once the images are digitized, additional technical, clinical and biological information must be included to create an image record. The definition procedure of such information is regulated by international rules set by a DICOM work group dedicated to pathological anatomy (DICOM WG26). Moreover, we must first adopt exchange standards (such as the DICOM format presented in Chapter 4) dedicated to PACS, for image recording [GAR 04] [LEB 04]. In France, the Association for the Development of Informatics in Cytology and Pathological Anatomy promotes the use of international standards within the framework of the International IHE (Integrating the Healthcare Enterprise) initiative. IHE participants, users and sellers of information systems to health departments, coordinate the way that standards are applied. They also organize connectivity tests [LEB 06].

3.3.1.5.3. Recent developments

In digital microscopic imaging, the procedure of selecting key images corresponding to areas of interest on the slide has long been deemed necessary. However, there are limits to this approach. This selection process considerably lengthens the imaging production time and introduces (by selecting certain elements over others) the risk that certain important morphological structures are not digitized. This situation often leads to an additional further digitization of other images on the slide. The limits of "fixed microscopic images" have restricted their use in diagnosis or expertise. Recent technologies of "virtual slide" production have helped to partly overcome the obstacles of fixed images such as the lengthy production time and the risks of selection [WEI 05].

Slide scanners allow us to digitize a continuous series of wide fields and to reconstruct a "virtual slide" (or a whole slide image (WSI)) (Figure 1.2 of Chapter 1) that can be examined on its entire surface and has various zooming effects, thanks to its appropriate interface. Slide scanners use the same architecture as motorized microscopes and a camera combined. However, they are different when it comes to their system of linear sensors that brush over the entire surface that is to be digitized after having taken its marks on the area of interest through a macroscopic view. The computer unit receives rough information that it is then intended to treat according to the users' particular needs: lighting, color or noise correction by averaging the images, inter-image processing to improve image resolution or the depth of the field, creating a "mosaic" to increase the field of view.

3.3.1.6. *Conclusion*

Not only do usual medical *in vivo* imaging techniques keep evolving towards more rapid imaging techniques of better spatial and/or temporal resolution, but additional methods are also being developed, such as:

– optical imaging based on the study of light propagation in tissues; amongst all these new methods we can name a few:

- optical coherence tomography (OCT) uses non-diffused photons or ballistics. This type of imaging allows us to visualize structures present in a layer of a few millimeters, with an image of a high resolution that often reaches micrometers when carried out in real time and without contact [DUB 04],

- retinal imaging using adaptative optics that works with distorting mirrors and improves the resolution of images so as to observe the retina at a cellular level (retinal photoreceptors as well as the blood cells flow in the retinal capillaries become visible *in vivo*) [DOB 05],

- imaging using diffuse optics works by diffusing photons, and allows us to observe in greater depth; most imaging systems using diffuse optics function solely by measuring variations in absorption coefficients linked to the chemical composition of the tissue; this technique offers a very high temporal resolution up to a hundredth of a second [GIB 05],

- acoustico-optics imaging helps us acquire optical images at greater depth (a few centimeters) with a resolution of a few millimeters thanks to a marking by the luminous ultrasound waves coming from a given region in the screened tissue [WAN 04];

– imaging using T-rays (terahertz waves) sensitive to the tissue malignancy and to the mineral content of tissues; it allows us to detect skin cancers and tooth cancers at their earliest stage [WAL 04].

These new imaging methods are not only harmless, non-invasive and not too expensive, but they are also sensitive to other types of contrasts and constitute additional visual *in vivo* imaging techniques.

Along with these new imaging techniques come new tracers that target the genetic information in cells, or intelligent tracers that guide the action of a therapeutic substance in a patient.

All of this leads to a multiplicity of specific techniques. It is now very common to combine various imaging techniques so as to improve the diagnosis. Medical imaging systems keep evolving, which increases the amount of information that can be acquired in a clinical routine.

3.3.2. *Properties of medical images*

The properties of medical images vary greatly from one image to another depending on the method of acquisition used, the organ studied and the acquisition protocol applied. Here, we will only deal with the specificities linked to compression. We will therefore be considering the size of images, the amount of coding bits, the spatial and temporal resolutions and the noise present in those images. The image files are commonly available in DICOM format on acquisition systems, detailed in Chapter 4.

3.3.2.1. *The size of images*

Table 3.1, as a rough guide, presents the average sizes of image and image files (made up of numerous images during voluminous and synchronized acquisitions) acquired during clinical routine using different methods and for different organs.

Modality	Organ	Image size	Number of bits per pixel	Number of slices	File size
Radiography	Thorax	2,060*2,060	16	-	8 MB
CT	Brain	512*512	16	≈ 300	150 MB
	Abdomen	512*512	16	≈ 500	250 MB
	Heart	512*512	16	126*16 frames	1 GB
MRI	Brain	512*512	16	≈ 20*6 sets	10 to 60 MB
	Abdomen	512*512	16	≈ 30	15 MB
	Abdomen 3D	512*512	16	104	50 MB
	Heart	256*256	16	20*20 frames	50 MB
MPET	Whole body	256*1024	16	-	0.5 MB
PET	Whole body	128*128	16	350	10 MB
	Heart	128*128	16	47*16 frames	24 MB
	Brain	256*256	16	47	6 MB
US	Standard	512*512	8	50 images/sec	12.5 MB/sec
	Doppler	512*512	(RGB) 3*8	50 images/sec	37.5 MB/sec
Anatomo pathology	Virtual slide	Around 15,000*20,000	(RGB) 3*8	-	858 MB

Table 3.1. *Table summarizing the typical properties of images and image files acquired in clinical routine through various methods and on different organs*

Medical images have a larger range of gray levels than in natural images. The amplitude of each pixel or voxel is coded on 16 bits while often in radiology only 12 bits are used (4,096 gray levels). This is an important point when talking about compression. In fact, some widely known standards (such as most JPEG versions) cannot compress monochrome images coded on 16 bits. The question is why do we use so many different gray levels, knowing that the eye is unable to differentiate as many tones and that most visualization screens are 8-bit LCD screens? In fact, for any diagnosis with silver film, radiologists are used to dealing with the entire range of 4,096 levels of gray. When the radiologist uses 8-bit screens, he carries out a windowing of the image gray levels so as to increase the contrasts on the organ or the pathology and display it prominently, as illustrated in Figure 3.19.

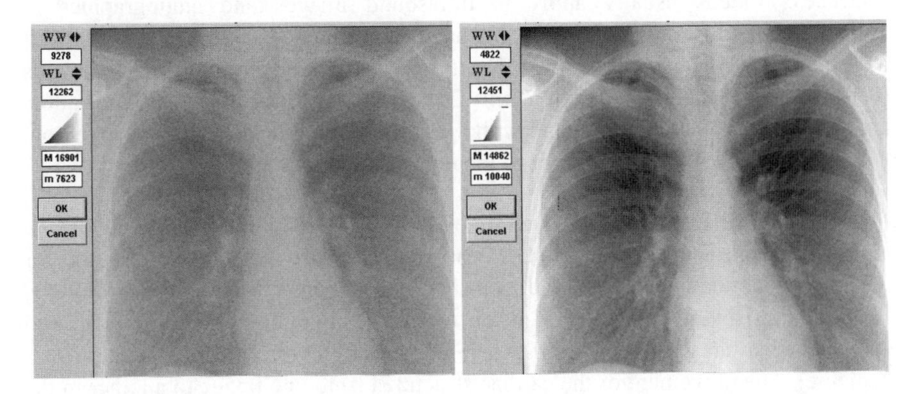

Figure 3.19. *Using gray level windowing so as to increase the contrasts on the pulmonary structure. Values are indicated in the left hand corner of the image*

A recent study has compared the performance in terms of interpretation time and in terms of the radiologists' diagnosis reliability when projected on 8-bit screens and 11-bit screens for various types of digital radiology [SID 04]. Authors have concluded that 11-bit monitors improve the reliability of the diagnosis, especially diagnoses related to soft tissue on X-rays. Interpreting an image on 11-bit screens is also a lot faster.

3.3.2.2. *Spatial and temporal resolution*

The spatial resolution of an image is largely linked to the acquisition method used: down to centimeters in nuclear medicine, and below millimeters for MRIs, or even less (100 micrometers) for ultrasound or X-rays. During volume acquisition procedures, the voxels are rarely isotropes. Often, the volume is divided into a set of not very thick slices that cannot always be linked to one another. In fact, the relation between the signal and the noise is proportional to the slice thickness and the acquisition time is proportional to the number of slices. The thickness of the slice and the space between each slice should be taken into account especially in 3D compression [SIE 04]. In ultrasound examinations, information is obtained under a spherical geometry (angular depth and positions) with axial and lateral resolutions that depend on the depth of each acquisition. It is therefore necessary to carry out an interpolation stage on a Cartesian grid of pixels to form images. In order to respect average spatial resolutions (measured in millimeters) obtained during acquisition procedures, the pixel resolution is chosen randomly. A “speckle” noise inevitably interferes with that resolution level. This prominent “speckle” noise produces a block effect in homogenous structures.

For temporal sequences we can distinguish between videos and synchronized sequences. Videos usually apply to ultrasound images and angiographies in radiology. For ultrasounds, a series of images are obtained in real time at the fast rate of about 50 images per second. 3D ultrasounds with electronic control allow us to obtain a volume of information in only 2 seconds. Real time 3D ultrasounds allow us to obtain 15 volumes per second with a resolution of almost 1mm^3 per voxel, but on a limited field of view (a quarter of the myocardium for example).

Synchronized sequences however are often used to study the heart through MRI, nuclear medicine and with multi-slice scanners. They are also sometimes used for dynamic examination of cerebrospinal fluid during MRI. In cardiology, synchronizing the acquisition system on electrocardiograms helps to split up and analyze the cardiac cycles and its many frames (usually 20 in MRI and 16 in nuclear medicine). The movement of the cardiac structures from one image to another in the sequence largely depends on the number of frames captured as well as the number of frames studied (the myocardium changes shape very rapidly during contraction) as illustrated in Figure 3.20.

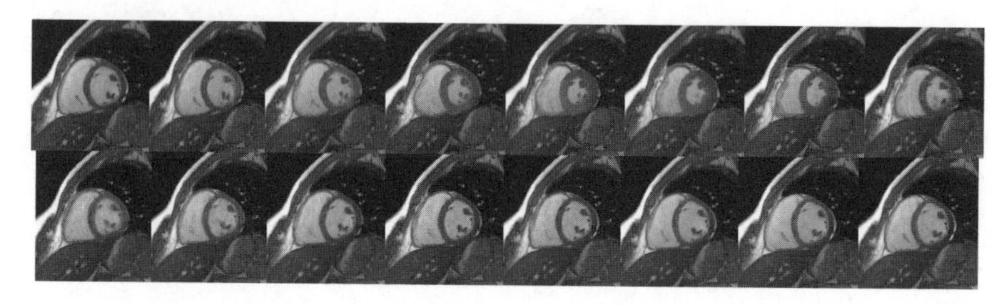

Figure 3.20. *Temporal cardiac MRI sequence on a cardiac cycle (16 phases)*

3.3.2.3. *Noise in medical images*

The amount of noise in medical images depends on the acquisition method used, as well as on the parameters and acquisition rules applied. This noise is not only due to the random nature of the physical phenomena studied, but also, to a lesser extent, to the set of acquisition and reconstruction systems used. When looking at the physical phenomenon, the noise may be qualified as Poisson in nuclear medicine, Rayleigh in MRI and Rayleigh or Gamma in ultrasound.

Physicians are used to this noise and would doubt the acquisition quality if that noise was turned down, for example by increasing the wavelet compression. The images would then be considered as "washed" images [CAV 01]. Nevertheless, we must not mix up image noise with image texture or artefacts that appear with the pathology in the studied organ and lead to a physiological change such as increased roughness. Usually, the relationship between signal and noise depends on the spatial and temporal resolutions. This relation can be improved during the acquisition process by averaging spatial and temporal resolutions out of numerous acquisition procedures (compounding) so as to diminish the risks of random interferences.

3.4. Conclusion

The characteristics of physiological signals (frequency content evolving according to the signal being studied but also to the pathology, the noise, the repetition of a signal-type modified by alterations (ECG)) must be exploited so as to develop optimal compression systems, the best in terms of compression rate and in terms of the diagnotic quality of the reconstructed signal (these systems are explained in Chapter 6). It is important to note however that protecting the physiological signal against that contaminating noise is essential, so that components containing no information would not be coded. This protection can be applied differently for each acquisition method, either technically by reinforcing the system or by treating the signal itself (digital filtering).

Taking the characteristics of medical images into account also enables us to define an optimal use of all compression systems, as we will explain in Chapters 7, 8 and 9. Moreover, medical image systems keep improving, leading to an increase in the data quantity obtained in clinical routine. This progress in the volume of information acquired may be qualified as exponential (as indicated in Figure 1.1). This can also be said of the technical developments brought to archiving and storage systems. Figure 3.21, taken from Erickson's article [ERI 02], clearly shows that over the last 25 years, both elements have evolved in parallel. Erickson [ERI 02] concludes that it is not sufficient to rely solely on technical progresses for the reduction of storage and transmission costs.

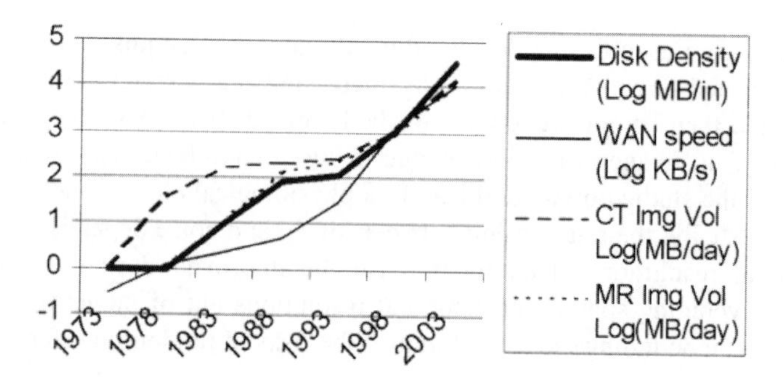

Figure 3.21. *Comparison of increases in disk density (thick solid line), speed of wide area networks (thin solid line), the number of bytes of image data produced by CT (dashed line) and MRI (dotted line) in a working day*

Compression is thus a useful additional tool and a low-cost alternative to increase the capacity of archiving and storage systems and transmission networks [BAN 00]. Nevertheless, we must evaluate all deteriorations caused by the compression system before applying it in clinical routine (evaluation methods are described in Chapter 5). Moreover, this system will have to include a standard exchange format such as the DICOM format (Chapter 4).

3.5. Bibliography

[BAN 00] BANKMAN I.N., *Handbook of Medical Imaging-Processing and Analysis*, Academic Press, New York, 2000.

[BER 04] BERCOFF J., Tanter M., FINK M., "Supersonic shear imaging: a new technique for soft tissue elasticity mapping", *IEEE Trans. on Ultrasonics, Ferroelectrics and Frequency Control*, vol. 51, no. 4, p. 396-409, April 2004.

[BUS 01] BUSHBERG J.T., SEIBERT J.A., LEIDHOLDT JR. E.M., BOONE J.M., *The Essential Physics of Medical Imaging*, 2nd Edition, Lippincott Williams & Wilkins, 2001.

[CAV 01] CAVARO-MÉNARD C., GOUPIL F., DENIZOT B., TANGUY J.Y., LE JEUNE J.J., CARON-POITREAU C., "Wavelet compression of numerical chest radiographs: a quantitative and qualitative evaluation of degradations", *Int. Conf. on Visualization, Imaging and Image Processing*, Marbella, Spain, p. 406-410, September 2001.

[CHE 03] CHERRY S.R., SORENSON J.A., PHELPS M.E., *Physics in Nuclear Medicine*, 3rd Edition, W.B. Saunders/Elsevier Science, Philadelphia, 2003.

[CHR 88] CHRISTENSEN D.A., *Ultrasonic Bioinstrumentation*, John Wiley & Sons, 1988.

[DOB 05] DUBOIS A., GRIEVE K., MONERON G., LECAQUE R., VABRE L., BOCCARA C., "Ultrahight-resolution full-field optical coherence tomography", *Applied Optics*, vol. 43, no. 14, p. 2874-2883, May 2004.

[DUB 04] DOBLE N., "High-resolution, in vivo retinal imaging using adaptive optics and its future role in ophthalmology", *Expert Review of Medical Devices*, vol. 2, no. 2, p. 205-216, March 2005.

[ERI 02] ERICKSON B.J., "Irreversible compression of medical images", *Journal of Digital Imaging*, vol. 15, no. 1, p. 5-14, March 2002.

[GAR 04] GARCIA-ROJO M., GARCIA J., ROYO C., CARBAJO M., "Working with virtual slides: DICOM-3 as a standard in Pathology", *7th European Congress on Telepathology and 1st International Congress on Virtual Microscopy*, Poland, July 2004.

[GIB 05] GIBSON A.P., HEBDEN J.C., ARRIDGE S.R., "Recent advances in diffuse optical imaging", *Physics in Medicine and Biology*, vol. 50, no. 4, p. R1-R43, February 2005.

[GIR 93] GIRON J., JOFFRE F., *Bases physiques et évolution de l'imagerie radiologique*, Abrégés d'Imagerie Radiologique, 1993.

[GRA 02] GRANGEAT P., *La tomographie médicale : imagerie morphologique et imagerie fonctionnelle*, Traité IC2, Série traitement du signal et de l'image, Hermes, Paris, 2002.

[JAN 02] JANNIN P., MORANDI X., FLEIG O.J., LE RUMEUR E., TOULOUSE P., GIBAUD B., SCARABIN J.M., "Integration of sulcal and functional information for multimodal neuronavigation", *Journal of Neurosurgery*, vol. 96, no. 4, p. 713-723, 2002.

[JEN 96] JENSEN J.A., *Estimation of Blood Velocities Using Ultrasound, A Signal Processing Approach*, Cambridge University Press, 1996.

[KAS 03] KASTLER B., VETTER D., PATTAY Z., Germain P., *Comprendre l'IRM*, 5e Edition, Collection Imagerie médicale, Diagnostic, Masson, Paris, 2003.

[KLO 04] KLOSSA J., LE BOZEC C., MARTIN E., CORDIER J.C., LUSINA D., MARTELLI M., TROUSSARD X., "TELESLIDE: a multipurpose collaborative platform dedicated to morphological studies", *7th European Congress on Telepathology and 1st International Congress on Virtual Microscopy*, Poland, July 2004.

[LEB 03] LE BIHAN D., "Looking into the functional architecture of the brain with diffusion MRI", *Nature Reviews Neuroscience*, vol. 4, no. 6, p. 469-480, June 2003.

[LEB 04] LE BOZEC C., THIEU M., ZAPLETAL E., JAULENT M.C., HEMET J., MARTIN E., "DICOM Interchange format for pathology", *7th European Congress on Telepathology and 1st International Congress on Virtual Microscopy*, Poland, July 2004.

[LEB 06] LE BOZEC C., HÉNIN D., FABIANI B.; BOURQUARD K., OUAGNE D., DEGOULET P., JAULENT M.C., "Integrating anatomical pathology to the healthcare enterprise", *Medical Informatics in Europe*, Maastricht, August 2006.

[LEN 03] LEANDERSON S., LAGUNA P., SÖRNMO L., "Estimation of respiration frequency using spatial information from the VCG", *Medical Engineering and Physics*. vol. 25, no. 6, p. 501-507, 2003.

[MOR 93] MORTARA D., "Source consistency filtering", *Journal of Electrocardiology*, vol. 25, p. 200-206, 1993.

[NIK 04] NIKOLAOU K., FLOHR T., KNEZ A., RIST C., WINTERSPERGER B., JOHNSON T., REISER M.F., BECKER C.R., "Advances in cardiac CT imaging: 64-slice scanner", *International Journal of Cardiovascular Imaging*, vol. 20, no. 6, p. 535-540, December 2004.

[PAU 00] PAUL J. S., REDDY R., KUMAR V., "A transform domain SVD filter for suppression of muscle noise artefacts in exercise ECG", *IEEE Transactions on Biomedical Engineering,* vol. 47, no. 5, p. 654-663, 2000.

[ROS 06] Rosset A., Spadola L., Pysher L., Ratib O., "Informatics in radiology (infoRAD): navigating the fifth dimension: innovative interface for multidimensional multimodality image navigation", *Radiographics*, vol. 26, no. 1, p. 299-308, February 2006.

[SID 04] SIDDIQUI K.M., SIEGEL E.L., REINER B.I., MUSK A.E., HOOPER F.J., MOFFIT R., "11bit versus 8bit monochrome LCD for interpretation of 12bit DICOM images", *RSNA'04*, Chicago, USA, SSJ21-01, p. 496, November 2004.

[SIE 04] SIEGEL E.L., SIDDIQUI K.M., REINER B.I., MUSK A.E., JOHNSON J., CRAVE O., "Thinner but less compressible: effect of slice thickness on image compressibility", *RSNA'04*, Chicago, USA, SSM21-02, p. 591, November 2004.

[SOR 05] SÖRNMO L., LAGUNA P., *Bioelectrical Signal Processing in Cardiac and Neurological Applications*, Elsevier Academic Press, 2005.

[VOL 04] VOLEGOV P., MATLACHOV A.N., ESPY M.A., GEORGE J.S., KRAUS JR. R.H., "Simultaneous magneto encephalography and SQUID detected nuclear MR in microtesla magnetic fields", *Magnetic Resonance Medicine*, vol. 52, no. 3, p. 467-470, September 2004.

[WAL 04] WALLACE V.P., TADAY P.F., FITZGERALD A.J., WOODWARD R.M., CLUFF J., PYE R.J., ARNONE D.D., "Terahertz pulsed imaging and spectroscopy for biomedical and pharmaceutical applications", *Faraday Discussions*, vol. 126, p. 255-263, 2004.

[WAN 04] WANG L.V., "Ultrasound-mediated biophotonic imaging: a review of acousto-optical tomography and photo-acoustic tomography", *Disease Markers*, vol. 19, no. 2-3, p. 123-138, 2004.

[WEI 05] WEINSTEIN R.S., "Innovations in medical imaging and virtual microscopy", *Human Pathology*, vol. 36, p. 317-319, 2005.

[YAG 05] YAGI Y. http://telepathology.upmc.edu/lecture/tvconf2/ms/July27_files/frame.htm "Multi Spectral Imaging in Pathology: Digital Stain", University of Pittsburgh, Center for Pathology Informatics, website visited in 2005.

Chapter 4

Standards in Medical Image Compression

4.1. Introduction

A standard (or norm) is a specification which has been adopted by those working within a particular field, in order to describe a process in an organized manner. In the case of data compression, the use of standards is particularly important as the compression process aims at the creation of an intermediary form of the information, which is more compact, and therefore easier to send over communication networks, to store and to receive. In other words the compressed form is not an end in itself; it is only an intermediary between a creation process and usage. It is therefore essential that this intermediary form or, if required, the means of access to this form conforms to specifications which ensure a smooth interaction between creation and usage.

A second motivating factor is the necessary life-span of the information. The existence of standards recognized by international bodies is a guarantee of the endurance of compressed data, and the continuation of the programs which create and read them.

Here, it is of interest to look specifically at medical data (see Chapter 3), compared to other computerized data. Does this field require the creation of specific standards, uniquely for the compression of medical data, in particular images? We can anticipate two different views on this matter.

Chapter written by Bernard GIBAUD and Joël CHABRIAIS.

Firstly, from a theoretical perspective, it seems desirable to keep the use of specific standards to a minimum, both for reasons of efficiency – to avoid repeating work which has already been carried out in part in other research circles – and also to minimize the cost of components. The latter argument is the more pertinent: the most general information treatment mechanisms lead to uses on a very large scale, in the form of specialized processors, with optimal performance, reliability and cost, which is impossible to achieve with products specific to one particular field.

Secondly, in practice, choices are clearly not made based on purely rational criteria, but rather emerge from a wider context. A determining factor concerns in particular the ability to identify common needs within a given context. Thus, for example, it is to be expected that the standards within medical image compression should be tackled by DICOM (Digital Imaging and COmmunications in Medicine), and that the solutions most widely-used today should have been defined within this particular context, bringing together the key industrial and academic professionals in the field, rather than in a wider circle such as the ISO (International Standards Organization). In this chapter it will become apparent, however, that adopting specific solutions by no means excludes the use of the most general standards and technology.

The implementation of standards is therefore essential in this field, in order to meet the needs for interoperability and life-span which arise from the healthcare sector. This brings with it certain side-effects. It can be frustrating for researchers working in data compression to know that if the advances they make are to be widely-used, not only will they have to offer a significant advantage over currently existing standards, but they will also have to be recognized by the standardization bodies, a process which can take several years. Nevertheless, this research is essential in order to achieve progress within the standards (JPEG 2000 became a standard only after considerable fundamental research into discrete wavelet transform).

In this chapter, we begin with a look at standardization and the bodies which set these norms in the field of medical data. Here, we will introduce the DICOM standard and will consider some key ideas in order to achieve a thorough appreciation of the implementation of image compression in this standard. The following section will detail the different types of compression available in the DICOM standard, as well as the methods of accessing compressed images. Finally, we will conclude by highlighting the key points within the use of image compression standards, particularly within the context of healthcare networks.

4.2. Standards for communicating medical data

4.2.1. *Who creates the standards, and how?*

We can generally distinguish three categories of standards.

The *de jure* standards are those created by official standardization bodies such as the International Standards Organization (ISO) or in Europe, the European Committee for Standardization (Comité Européen de Normalisation or CEN). The representation of the principal players is organized by country. This means that the selected experts represent the official standardization body of their country, for example, AFNOR (Association Française de Normalisation) for France. The issues in question are therefore defined at this level by "mirror groups" of the international committees concerned. This approach has the advantage of guaranteeing free access to the creation of a standard by all interested parties, whether they are from industry, from academia, consumer associations, public bodies, etc. One of the two main criticisms of this approach is that it leads to a rather long standardization process – around 5 to 10 years – despite the fact that in fields such as IT and communications, technological advances happen at a very fast rate, meaning that a standard, once approved, can be out of pace with market needs, simply due to technological advances. Another criticism is that this approach does not give a sufficiently important role to the industrial players within the field. In fact, many norms are defined without ever being used in products. Some see this as the consequence of the unnecessary complexity of the solutions offered, imposed by academics or consultants, who are more concerned with the scientific quality of the solutions than the economic viability of products which use these norms. The European Community's procurement rules require that contracting bodies are obliged to define their specifications with reference to European standards, where available.

Industrial standards are created by associations of developers or academic bodies such as, for example, the World Wide Web Consortium (W3C), the Internet Engineering Task Force (IETF) – two bodies who play a key role in Internet standards – or the DICOM committee regarding the DICOM standard. These organizations set industrial standards, usually via very well-defined procedures, which are accepted by national standards bodies. These associations operate on a voluntary basis. In general, free access to standardization records is available online.

Unlike those produced in either of the two contexts given above, "*de facto*" standards do not arise from a formal agreement between interested parties, but rather from a process of market selection. Into this category fall Word and the Rich Text Format (RTF, Microsoft™), the Portable Document Format (PDF, Adobe™), which have become standards only as a result of endorsement from huge communities of users, which has led to their standardization.

4.2.2. *Standards in the healthcare sector*

4.2.2.1. *Technical committee 251 of CEN*

Within Europe, the technical committee 251 of CEN, "Health Informatics" (http://www.centc251.org) was founded in 1991 to develop a group of standards for the exchange of health data in conditions which guarantee interoperability, security and quality. This committee is organized into four working groups focusing on information models (WG I), terminology and knowledge representation (WG II), security and quality (WG III), and technologies of interoperability (WG IV). This committee has produced a significant number of technical reports, experimental norms (ENV), as well as some European Norms (EN). The most important is ENV 13606 "Electronic healthcare record communication", itself divided into four parts: 1) architecture, 2) domain term list 3) distribution rules and 4) messages.

We can also cite norms or pre-norms on communication security (ENV 13608 – 1 to 3), the recording of coding systems (ENV 1068), messages for the exchange of information on medical prescriptions (ENV 13607), blood transfusion related messages (ENV 13730 – 1 and 2) and a system of concepts to support the continuity of care (ENV 13940).

4.2.2.2. *Technical committee 215 of the ISO*

Technical committee 215 of the ISO (website accessible from http://www.iso.org), also called "Health Informatics", was created in 1998, with a very similar objective to Technical Committee 251 of CEN, but for the world stage. This committee is organized into eight working groups: data structure (WG 1), information exchange (WG 2), semantic content (WG 3), security (WG 4), health cards (WG 5), pharmacy and medical products (WG 6), devices (WG 7) and business requirements for electronic health records (WG 8). The key standardization documents produced since the creation of Technical Committee 215 of the ISO concern the communication between medical devices (ISO/IEEE series 11073), the interoperability of telemedicine systems and networks (ISO/TR 16056 1-2), public key infrastructure (ISO/TR 16056 1-2), web access for DICOM persistent objects (ISO 17432) and patient healthcare data (ISO 21549 1-3).

4.2.2.3. *DICOM Committee*

A recap of its history: the DICOM Committee (http://medical.nema.org) in its current organization has existed since the early 1990s. This Committee is the successor of the ACR-NEMA Committee, formed in 1983 by the American College of Radiology (ACR) and the National Electrical Manufacturers Association (NEMA), with aims of internationalization. Today it incorporates around 40 players from academia and industry, working in the field of biomedical imagery (Table 4.1).

Industrial Bodies	Academic Bodies
AGFA U.S. Healthcare	American Academy of Ophthalmology
Boston Scientific	American College of Cardiology
Camtronics Medical Systems	American College of Radiology
Carl Zeiss Meditec	American College of Veterinary Radiology
DeJarnette Research Systems	American Dental Association
Dynamic Imaging	College of American Pathologists
Eastman Kodak	Deutsche Röntgengesellschaft
ETIAM	European Society of Cardiology
FujiFilm Medical Systems USA	Healthcare Information and Management Systems Society
GE Healthcare	Medical Image Standards Association of Taiwan
Heartlab	Societa Italiana di Radiologia Medica
Hologic	Société Française de Radiologie
IBM Life Sciences	Society for Computer Applications in Radiology
Konica Minolta Medical Corporation	Canadian Institute for Health Informatics
MatrixView	Center for Devices & Radiological Health
McKesson Medical Imaging Company	Japan Industries Association of Radiological Systems (JIRA)
MEDIS	Korean PACS Standard Committee
Merge eMed	National Cancer Institute
Philips Medical Systems	National Electrical Manufacturers Association
RadPharm	
R2 Technology, Inc.	
Sectra Imtec AB	
Siemens Medical Solutions USA, Inc.	
Sony Europe	
Toshiba America Medical Systems	

Table 4.1. *List of organizations present in the DICOM Committee*

All the key players in medical imagery actively contribute to the development of the standard. Today there are 26 working groups, bringing together around 750 technical or medical experts. These groups are listed in Table 4.2.

DICOM Committee working groups	
WG-01: Cardiac and Vascular Information	WG-14: Security
WG-02: Projection Radiography and Angiography	WG-15: Digital Mammography and CAD
WG-03: Nuclear Medicine	WG-16: Magnetic Resonance
WG-04: Compression	WG-17: 3D
WG-05: Exchange Media	WG-18: Clinical Trials and Education
WG-06: Base Standard	WG-19: Dermatologic Standards
WG-07: Radiotherapy	WG-20: Integration of Imaging and Information Systems
WG-08: Structured Reporting	WG-21: Computed Tomography
WG-09: Ophthalmology	WG-22: Dentistry
WG-10: Strategic Advisory	WG-23: Application Hosting
WG-11: Display Function Standard	WG-24: Surgery
WG-12: Ultrasound	WG-25: Veterinary Medicine
WG-13: Visible Light	WG-26: Pathology

Table 4.2. *List of the DICOM Committee's working groups*

The DICOM 3.0 standard was published in 1993, following preliminary work carried out over the previous decade, and which led to the ACR-NEMA standards 1.0 and 2.0 (published in 1985 and 1988 respectively).

Field covered: the DICOM standard [DIC 06] covers many areas, including:

– the communication of related images and data (over a network or using physical media), for almost every existing technique (image modality);

– the printing of images onto physical media;

– the communication of reports on imaging procedures;

– the management of activities related to the acquisition, treatment and interpretation of images, through the management of "work lists";

– the security of data exchange, via a service called "storage commitment", and various mechanisms for digital signature of documents;

– the coherence of image rendering on hardcopy and softcopy (monitor display).

The standard is modular, so as to effectively meet general or specific needs concerning the numerous modalities within imagery and medical specialities making use of imagery. It is organized into 18 relatively independent sections (Figure 4.1).

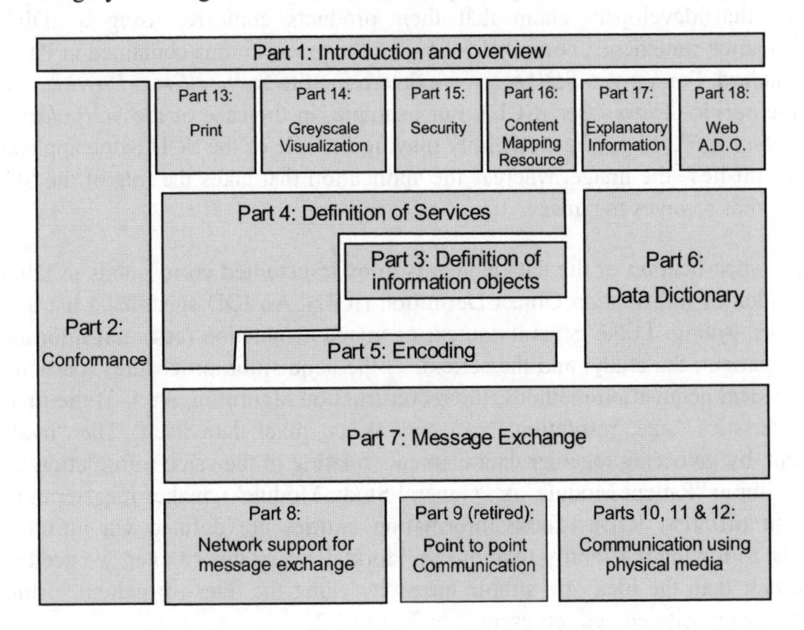

Figure 4.1. *The different sections of the DICOM standard*

Key principles of the DICOM standard: the following elements focus on the key notions which must be grasped in order to understand the implementation of compression in the DICOM standard. For a more detailed look at the standard, see [CHA 04].

DICOM is, above all, a protocol for image exchange, over a network or with the aid of physical media (CD ROM, DVD, etc.). Taking into account the diversity of image modalities, the standard is organized in such a manner as to allow both the specificities of each modality to be respected, and to create common ground between many data elements.

Thus, the standard is organized in a modular manner, particularly through the "*Service Object Pair*" (SOP) principle, linking a class of images of a certain kind, for example CT (Computed Tomography) images or X-ray tomography images, to a specific exchange service (for example the "image storage" service). This idea of "classes", inspired by the "object" paradigm, leads us to discover the notion of an "*SOP class*" – an abstraction of all the images of a certain type, CT for example –

and the notion of an "*SOP Instance*", which refers to a concrete image example, identified thanks to a unique identifier. One essential function of this notion of *SOP class* concerns conformance to the standard. It is by referring to the different *SOP classes* that developers claim that their products conform, using a "DICOM conformance statement", compiled following the prescriptions contained in Part 2 of the standard. For a given *SOP class*, we determine the *Service Class Provider* (SCP) and the *Service Class User* (SCU). For example, in the case of the *SOP class* "*CT Image Storage*", the application entity playing the role of the SCU is the application which "pushes" the image, whereas the application that takes the role of the SCP is the one that receives the image.

The specification of the data elements to be transmitted corresponds in DICOM to the idea of Information Object Definition (IOD). An IOD specifies a list of data elements, giving: 1) the general context of image acquisition (essential information on the patient, the study, and the series), 2) the acquisition procedures (particularly the physical acquisition methods, the reconstruction algorithm, etc.), 3) the image's characteristics (size, resolution, etc.) and 4) the pixel data itself. The "module" concept, by gathering together data elements relating to the same information entity (for example "Patient Module" or "General Study Module"), makes it easier to reuse them in different IODs. These information entities are defined via information models, following the "entity-relationship" formalism. In this chapter, we need think no further than the idea of a simple hierarchy along the lines of: patient – study – series – composite object, represented in Figure 4.2.

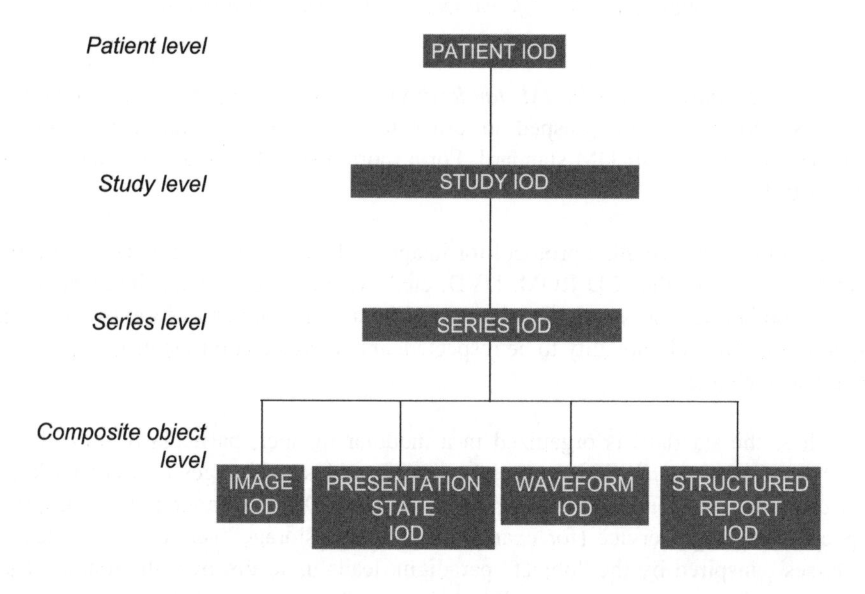

Figure 4.2. *Hierarchical model of DICOM entities*

This notion of composite objects, initially introduced to express the aggregation of information which concern different real-world entities, allows us to generalize, in management terms, all DICOM entities which have a persistent character. This concerns in particular images, but also "presentation states", which allow us to choose a specific appearance for the images (for example, a particular window or zoom rate), structured reports, able to reference images and waveforms, used to represent physiological signals such as the ECG.

As for the image data itself, DICOM decided from the outset to give a classical definition of the image: as a simple 2D representation of values, each image in a series representing an individual dataset, and carried by an individual message. Thus a series of 100 CT images results in the sending of 100 separate messages. Meanwhile, the principle of a "multi-frame" image was conceived, particularly when representing ultrasound images. Today it is becoming more widespread, with the specifications of Enhanced MR and Enhanced CT objects, designed to cope with the growing needs of MRI and CT imagery.

4.2.2.4. *Health Level Seven (HL7)*

HL7 (http://www.hl7.org) was initially an American organization for standards definition, accredited by the American National Standards Institute (ANSI), and which is now internationally recognized, with national groups in many different countries. The name of this standard refers to the seventh application layer of the OSI (Open System Interconnection) model of the ISO. HL7 was created to develop standardized messages allowing health information system applications to communicate between one another. Table 4.3 summarizes the different versions which have been published. The current version is 2.5 (March 2005). HL7 uses Arden Syntax, a syntax devoted to the representation of medical information in professional systems. Version 1.0 was developed and published in 1999 by the American Society for Testing and Materials (ASTM), before being taken up by HL7. HL7 has also integrated the Clinical Context Object Working Group (CCOW), whose aim is to synchronize desktop applications with a given context (user, patient, etc.). Alongside HL7 v3, HL7 has developed a Clinical Document Architecture (CDA), which in its second version, CDA release 2, has been an ANSI standard since April 2004. This architecture was initially in competition with that proposed by the CEN (ENV 13606), but now the groundwork has been carried out alongside the CEN experts, to merge the two architectures. Aside from its collaboration with the CEN, HL7 works closely with the ISO/TC 215, a pilot project which has been approved with the aim of having certain HL7 standards approved by the ISO.

<table>
<tr><th colspan="2">Version</th><th>Publication date</th><th>Notes</th></tr>
<tr><td colspan="2">Version 1.0</td><td>1987</td><td>Prototype standard</td></tr>
<tr><td rowspan="7">Version 2.x[1]</td><td>2.0</td><td>1988</td><td rowspan="7">Acceptable standard, arbitrary data conditioning.
No information model use.</td></tr>
<tr><td>2.1</td><td>1990</td></tr>
<tr><td>2.2</td><td>1994</td></tr>
<tr><td>2.3.1</td><td>1997</td></tr>
<tr><td>2.4</td><td>12/2000</td></tr>
<tr><td>2.5</td><td>06/2003</td></tr>
<tr><td>2.6</td><td>2008</td></tr>
<tr><td colspan="2">Version 3.0</td><td>10/2003
for the first elements</td><td>Formal methods, based on information model, the Reference Information Model or RIM.
Modular standard; publication in successive blocks since 2003.</td></tr>
</table>

Table 4.3. *The different versions of the HL7 standard*

4.2.2.5. *Synergy between the standards bodies*

Each standards body develops its standards in relation to its own objectives and rules. Thus, we note that the official organizations (ISO, CEN) have a *top-down approach* to problems, whereas industry associations generally have shorter-term approaches, with more pragmatic attitudes, guided by the market (a *bottom-up approach*) [GIB 98]. We also often observe a widening of the area covered by a body, going beyond its initial remit. Thus, the DICOM standard, which was very much focused on imagery when it first began, has come to take an interest in the structure of medical documents, such as structured reports, thereby coming across problems encountered by other organizations, in this case the CEN (ENV 13606 part 1), and particularly HL7, with the architecture of CDA documents. This leads, therefore, to collaboration between standards bodies, by various different means. For example, there was an active collaboration from 1994 to 1997 between DICOM and CEN TC 251, serving to define collaboratively DICOM extensions affecting imagery workflow (managing working lists, storage commitment, etc). In the end, the entirety of the DICOM standard was recognized by the CEN as a European norm (CEN EN 12052) [CEN 04]. Similarly, an active collaboration was begun shortly after between CEN TC 251 and HL7 for the creation of Version 3 of this standard,

---

1 The differences between the sub-versions arise from the longevity of the standard and its permanent ability to meet needs (addition of missing messages or fields, or deletion of obsolete ones).

concerning in particular the Reference Information Model (RIM) to draw all possible lessons from the work on the pre-norms ENV 13606. There is also a joint working group between DICOM and HL7 (DICOM's WG 20), to harmonize data exchange concerning imaging procedures and reports (structured or not) between these two standards. Lastly, there is also a type A liaison between the ISO TC 215 and the DICOM Committee, which renders official the acknowledgement of the fact that biomedical imagery is specifically studied in the DICOM Committee, then formally adopted by the ISO via a fast-track process. Thus, part 18 of the DICOM standard on the access to DICOM persistent objects has been adopted by the ISO (ISO 17432-2004).

4.3. Existing standards for image compression

4.3.1. *Image compression*

The general standards in the field of data compression can be divided into two categories:

– those relying on a particular spatio-temporal organization (2D image, or series thereof);

– those which do not make such a hypothesis.

The first have been developed chiefly by the ISO, to meet the needs encountered in photography (still images), cinema and television (animated images): these are the JPEG and MPEG families of standards, from the names of the working groups created by the ISO to work on these issues, the Joint Photographic Experts Group and the Moving Picture Experts Group respectively. The applications targeted at the outset were mainly linked to e-commerce (online catalogues), the press and tourism. The expansion into professional imagery occurred naturally in the fields using visible light (satellite imagery, the controlling of industrial processes, CCTV, etc.).

Common ground between general standards and specific standards: emerged perfectly organically. Two main approaches can be distinguished, as shown in Figure 4.3. The first approach uses compression techniques making no hypothesis about the nature of the information to be coded; into this category fall the compression tools distributed with UNIX/Linux such as "compress" or "gzip", or used by programs such as "WinZip™" (Figure 4.3, approach 1). The second approach focuses the compression task on the data of the signal image (in the widest sense), exploiting the redundancy present in the different spatial or temporal dimensions. The representation of this data can either be based on the general standards of data compression (as in Figure 4.3, approach 2a and 2b), or not (see Figure 4.3, approach 2c).

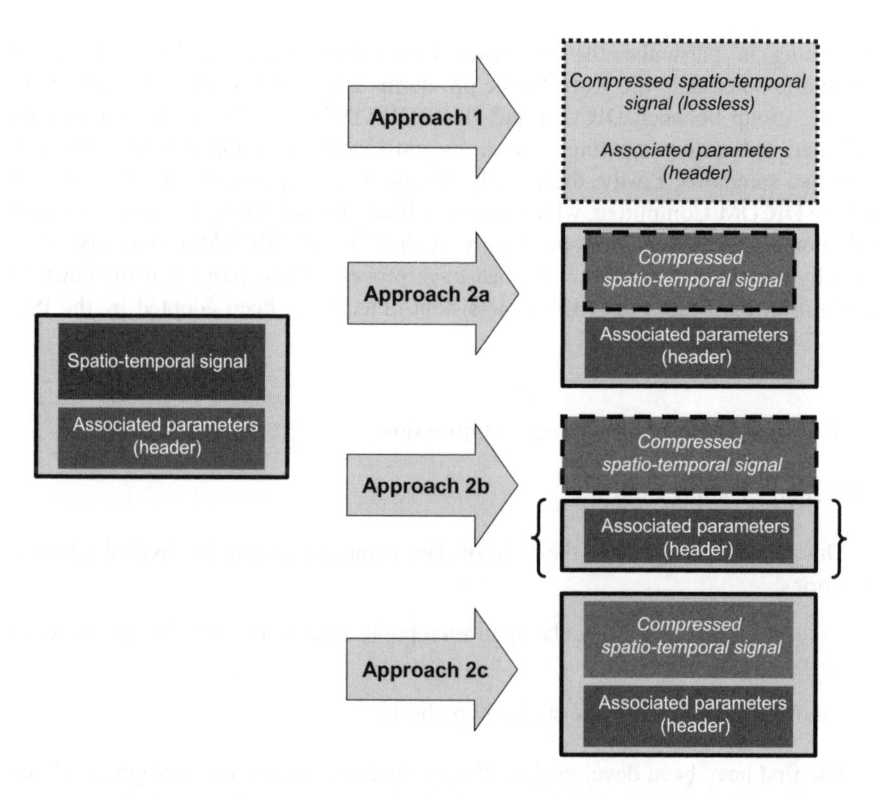

Figure 4.3. *General and specific approaches to image compression; the left-hand part of the figure represents the image data and header; the right-hand part describes each of the four possible approaches, numbered 1, 2a, 2b, and 2c respectively. The borders of the boxes represent the type of standard containing the data, i.e. general standards (e.g. JPEG, MPEG), represented by a dotted border or specific coding (e.g. DICOM), represented by a continuous line*

The advantages and disadvantages of the two approaches are summarized in Table 4.4. The first approach is used for lossless coding. It has the distinct advantage of being unspecific, i.e. it can be applied to any file, whatever its format or size. The disadvantage is that of limited effectiveness, dependent upon the data itself. Thus, noisy image data (without large uniform areas) is difficult to compress at a compression rate greater than 3:1. In contrast, the second approach can use image spatial and temporal structure in order to optimize the elimination of redundancy. The resulting compression rates – when lossy coding is acceptable – can be very significant, with factors ranging between 8:1 and 20:1, or even higher. The use of standard compression methods and formats makes the use of general decompression and visualization programs (e.g. web browsers) much easier.

	Advantages	Disadvantages
Approach 1 "general compression" (e.g.. gzip, compress)	Generality Ease of operability Very inexpensive	Poor performance
Approach 2a "general image compression" by encapsulation (e.g. JPEG, MPEG)	Reuse of existing applications for compression/decompression and image visualization Very good performance Takes medical context into account (header containing patient name, acquisition methods, etc.)	May not be well adapted to very specific needs, or sub-optimal performances
Approach 2b "general image compression" (e.g. JPEG, MPEG)	Makes wide distribution easy (outside imaging departments, throughout the hospital and to out-patients) and cost-effective (web browsers)	No header taking medical context into account
Approach 2c "specific image compression"	Can allow optimal performance to be achieved, due to a very good adaptation to the structure of the data	Development costs high due to specific nature

Table 4.4. *Advantages and disadvantages of general versus specific standards*

4.3.2. *Image compression in the DICOM standard*

As we have seen, the DICOM standard plays a dominant role in the field of medical imagery. DICOM included very early on in its standard a possible recourse to data compression methods. The main aim of this section is to show how DICOM integrates the use of image compression, and particularly how DICOM has integrated the general approaches to compression provided by the ISO. For ease of comprehension, we will begin with a few reminders about the coding of data in the DICOM standard.

4.3.2.1. *The coding of compressed images in DICOM*

General aspects: the coding of data elements is described in part 5 of DICOM, called "coding". It is carried out in binary, following a structure of the "type – length – value" type. The type field is expressed in the form of a pair of unsigned integers (represented on 16 bits) called a Tag. This is a simple unique ID (UID) for the element in question. For example, the Tag (0028,0010) Rows represents the number of lines in an image. The field length is represented in unsigned binary on two bytes

– it denotes the number of bytes occupied by the value field. The value field contains the value of the data element in question.

The data elements are defined in part 3 of DICOM, at the semantic level, in tables corresponding to the different modules. Each line of these tables details:

– the Tag of each element;

– its name;

– its mandatory or optional nature: type 1: mandatory (i.e. present and not empty), type 2: present (but possibly empty), type 3: optional;

– the free text definition of the element.

This information is complemented in Part 6 of DICOM (the data dictionary), which defines for each data element:

– its Tag;

– its name;

– its value representation (VR); DICOM defines (at the start of Part 5, "coding") 27 types of element, including: *Unsigned Long* (UL), *Unsigned Short* (US), *Person Name* (PN), etc.;

– its multiplicity value (the number of possible occurrences).

The data elements are then listed in increasing order to form a dataset, the body of the message. An example is given in Figure 4.4. There are different transfer syntaxes, taking into account different coding options, the main ones being:

– whether or not the data element type is explicitly given (Explicit Value Representation, for example);

– the order of the representation of the bytes: *big-endian* (the most significant byte is represented first, and then the others in decreasing order of significance), or *little-endian* (the less significant byte is represented first, and the others are listed in growing order of significance);

– the data compression usage, which we will look at in detail below.

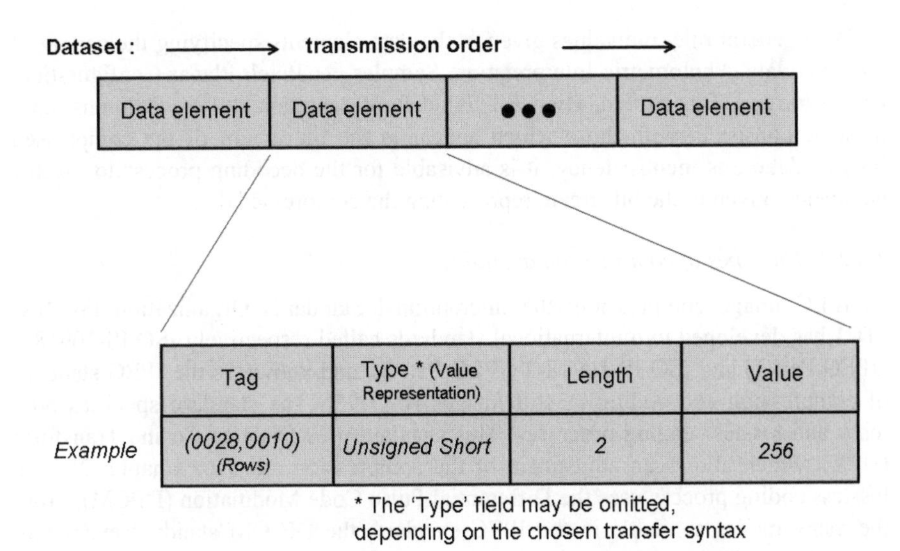

Figure 4.4. *The organization of data elements, following the Tag – length – value triplet*

The pixel data is represented in the data element (7FE0,0010) Pixel Data, which can be of the *Other Word String* (OW) or *Other Byte String* (OB) type. As a general rule, the data is packed taking into account the elements (0028,0100) *Bits Allocated* and (0028,0101) *Bits Stored.* We can in fact distinguish two different cases: the first concerns the native format (without compression), and the second is the format with encapsulation (with compression). In the first case, the pixel data is subject to compacting in which the last bit of a pixel stored is immediately followed by the first bit of the following pixel, following a constant order (the pixels follow on from left to right, and from top to bottom, meaning that the first pixel transmitted is that situated at the top left, followed by the rest of the first line, then the first pixel in the second line, and so on). Depending on whether the OW or OB type is used, the coding will or will not (respectively) be affected by the choice of a big-endian or little-endian ordering.

Encapsulation of compressed images: coding by encapsulation involves the inclusion in the data element (7FE0,0010) *Pixel Data* of the bit stream expressing the compressed image. In this case the transfer syntax used has to be of the explicit VR type, which means that the specification of the element type is present in the bit stream. Also in this case, the data element (7FE0,0010) *Pixel Data* is necessarily of the type OB, and the ordering has to be little-endian. Different compression techniques can be used, using the corresponding choice of transfer syntax.

As a general rule, the values given in the data elements specifying the coding of the pixel data (Photometric Interpretation, Samples per Pixel, Planar Configuration, Bits Allocated, Bits Stored, High Bit, Pixel Representation, Rows, Columns, etc.) must be consistent with those which appear in the bit stream of the compressed image. If there is inconsistency, it is advisable for the decoding process to use the parameters given in the bit stream representing the compressed data.

4.3.2.2. *The types of compression available*

JPEG image compression: the International Standards Organization ISO/IEC JTC1 has developed two international standards called respectively ISO/IS-10918-1 (JPEG Part 1) and ISO/IS-10918-2 (JPEG Part 2), and known as the JPEG standard of compression and coding of still images [ISO 95]. The standard specifies both lossy and lossless coding processes. This standard uses Discrete Cosine Transform (DCT), which allows an adjustment of the compression rate (see Chapter 2). The lossless coding process uses the Differential Pulse Code Modulation (DPCM). From the many modes available in the JPEG standard, the DICOM standard eventually retained four. Their principal characteristics are shown in Table 4.5.

UID of the transfer syntax	JPEG coding process	Description
1.2.840.10008.1.2.4.50	1	“baseline”, lossy
1.2.840.10008.1.2.4.51	2 (8 bits), 4 (12 bits)	“extended”, lossy
1.2.840.10008.1.2.4.57	14	lossless, non-hierarchical
1.2.840.10008.1.2.4.70	14 (Selection Value 1)	lossless, non-hierarchical, first-order prediction

Table 4.5. *Transfer syntaxes implementing the JPEG standard*

The compression modes are referenced thanks to four different transfer syntaxes. The first corresponds to the baseline mode, applied to images on 8 bits, lossy and using Huffman coding. The second corresponds to the JPEG 2 and 4 modes, known as “extended”, also lossy, which are applied respectively to images on 8 and 12 bits. The third corresponds to the JPEG 14 mode, lossless, based on the DPCM method, still with Huffman coding. Lastly, the fourth differs from the third in that it involves an order 1 prediction (i.e. horizontal).

In order to facilitate the interoperability of applications using JPEG transfer syntaxes, the standard explicitly states that:

– applications using lossless JPEG must support JPEG mode 14 (the fourth given in Table 4.5);

– applications using lossy JPEG for images on 8 bits must support JPEG mode 1 (the first given in Table 4.5);

– applications using lossy JPEG for images on 12 bits must support JPEG 4 mode (the second given in Table 4.5).

Lastly, the developer must note in his or her DICOM conformance statement whether it is only capable of receiving compressed images, or whether it can also treat them.

Run Length Encoding (RLE) compression is a very simple coding algorithm based on the removal of repetitive patterns, used in the TIFF 6.0 format, called *PackBits* [TIF 92]. Note that in this case the data element (0028,006) Planar Configuration takes the value color-by-plane, in order to maximize the size of the uniform patterns. The corresponding transfer syntax carries the UID 1.2.840.10008.1.2.5. It can be used for both single images and *multi-frame* images. In the latter case, each frame leads to a separate fragment in the bit stream.

JPEG-LS image compression: JPEG-LS, i.e. ISO/IS-14495-1 (JPEG-LS Part 1) presents another standard proposed by the International Standards Organization ISO/IEC JTC1 to represent compressed still images, whether lossy or lossless [ISO 99]. It specifies a unique compression mode, founded upon a predictive method using a statistic model, modeling the differences between pixels and their neighborhood. This method is considered to be more effective than that given in JPEG, i.e. ISO 10918-1. It should also be noted that JPEG-LS can treat images up to a depth of 16 bits. Two DICOM transfer syntaxes have been defined: the first, which has the ID 1.2.840.10008.1.2.4.80 references the usage of JPEG-LS in lossless mode. The second, with the ID 1.2.840.10008.1.2.4.81, references lossy JPEG-LS usage, with absolute error limited to a precise value, specified in the bit stream.

JPEG 2000 image compression: JPEG 2000 is the most commonly-used name for the standard ISO/IEC 15444 (JPEG 2000), still dedicated to the representation of compressed still images [ISO 04a] [ISO 04b]. It introduces new compression schemes based upon Discrete Wavelet Transform and multi-component transforms, notably applicable to color images. Part 2 of the standard (ISO/IEC 15444-2) complements Part 1 (ISO/IEC 15444-1) by extending the multi-component transforms ICT (Irreversible Color Transform) and RCT (Reversible Color Transform). These extensions represent in part DPCM-type prediction schemes, and also more complex transforms such as the Karhunen-Loève Transform. All these schemes are adapted to black and white or color image compression, up to a depth of 16 bits; signed or unsigned, lossy or lossless.

DICOM references this standard thanks to four transfer syntaxes, the first two based on ISO/IEC 15444-1, and the other two based on ISO/IEC 15444-2:

– the first, which carries the ID 1.2.840.10008.1.2.4.90 references the usage of JPEG 2000 Part 1 in lossless mode. It uses a compression scheme using Discrete Wavelet Transform or a multi-component transform in reversible mode, without quantification;

– the second, which bears the ID 1.2.840.10008.1.2.4.91 references the usage of JPEG 2000 Part 1 in lossy mode. This can use either reversible or irreversible transforms, with or without quantification.

– the third and fourth, which respectively bear the IDs 1.2.84.10008.1.2.4.92 and 1.2.840.10008.1.2.4.93, extend the possibilities of the first two transfer syntaxes, making use of the possibilities offered by JPEG 2000 Part 2. There is a generalization of the multi-component coding, which is applied in Part 1 of JPEG 2000 to color images, considering that any image sequence can be seen as a multi-component image. A flexible mechanism allows for the organization and grouping of the components into component groups, for optimum efficiency. Applied to multi-frame DICOM images, these syntaxes therefore allow for the elimination of inter-image redundancy, independently from the semantic associated with this third dimension (space variable for 3D sequences, time variable for temporal sequences, etc.). These should therefore be used more and more extensively, with the diffusion of the new *Enhanced* CT and *Enhanced* MR IODs, which make extensive use of multi-frame images.

JPIP progressive image compression: this possibility was created in response to the need to send images progressively, i.e. allowing data display with growing precision as the transfer progresses. It therefore allows the user to see the image before the transfer is complete, or even to stop the transfer, if it is no longer what the user requires. The implementation of this mechanism is based on the Interactive Protocol proposed with JPEG 2000 (JPEG *Interactive Protocol*, or JPIP).

It is used in DICOM to replace the bit stream usually present in the data element (7FE0,0010) *Pixel Data* with reference to a data provider for this bit stream, given in data element (0028,7FE0) *Pixel Data* Provider URL, for example:

Pixel Data Provider URL (0028,7FE0) = http://server.xxx/jpipserver.cgi?target=imgxyz.jp2

The JPIP server must return a dataset of *Content-type image/jp2*, *image/jpp-stream* or *image/jpt-stream*. It is also possible to specify a particular bit stream sub-set thanks to the modular nature of the coding, for example, the following URL allows for frame number 17 of a multi-frame image sequence to be restored at a resolution of 200x200:

Pixel Data Provider URL (0028,7FE0) = http://server.xxx/mframe.jp2?fsiz=200,200&stream=17

These possibilities use two particular transfer syntaxes, JPIP referenced transfer syntax, with the ID 1.2.840.10008.1.2.4.94, and JPIP referenced deflate transfer syntax, with the ID 1.2.840.10008.1.2.4.95. The difference between the two lies in the fact that the second also adds a lossless coding of the JPIP bit stream, based on the deflate algorithm (RFC 1951).

MPEG2 image compression: the MPEG2 standard (ISO/IEC 13818-2) was developed by the ISO for the compression of video or animated images, and any associated sound signal [ISO 00] (see Chapter 2). Reference to this standard uses a unique transfer syntax, with the ID 1.2.840.10008.1.2.4.100. It references MPEG2's MPEG option MP@ML (*MainProfile@MainLevel*). MPEG2 MP@ML uses a source code in 4:2:0 reducing the input data rate to 162 Mbits/sec. The *Main Profile* (MP) indicates an MPEG sequence composed of images which may be intra (I), predictive (P) or bidirectional (B) and the *Main Level* (ML) at a definition equivalent to television standards. The output rate is not set by the standard – between 1.5 and 15 Mbits/sec.

This can be applied to single- or multi-component data represented on 8 bits (which can present a problem for the compression of medical images, often represented on more than 8 bits); in the case of single-component data, the data element (0028,0004) *Photometric Interpretation* has to take the value MONOCHROME2, whereas for multi-component data, it has to take the value YBR_PARTIAL_420. In both cases, the MPEG bit chain includes both a luminance signal and two chrominance signals. The spatial and temporal resolution of the images, i.e. the data elements (0028,0010) *Rows*, (0028,0011) *Columns*, (0018,0040) *Cine Rate* and (0018,1063) *Frame Time* must be in conformity with the values specified for MP@ML, shown in Table 4.6.

Video type	Frame rate	Frame time	Max. number of rows	Max. number of columns
525-line NTSC	30	33.33 ms	480	720
625-line PAL	25	40 ms	576	720

Table 4.6. *Spatial and temporal resolution of images in the MPEG2 standard MP@ML; in practice, it is advised to follow a 4:3 ratio*

4.3.2.3. *Modes of access to compressed data*

There are four exchange modes offered in the DICOM standard: 1) network exchange, using the STORAGE and QUERY & RETRIEVE services; 2) exchange using physical supports, for example CDROM or DVD; 3) email attachments; and 4) WADO (Web access to DICOM persistent objects). They were successively introduced into the standard to meet the needs of different applications. This section gives a brief introduction to the specificities of each of the exchange modes, and gives the part of the standard which details them.

Network exchange: this exchange mode (with the STORAGE and QUERY & RETRIEVE services) was already present in the 1985 and 1988 versions of the ACR-NEMA standard, and was taken up again in the DICOM 3.0 standard of 1993. It involves "pushing" the images one by one, by means of a C-STORE type message (defined in Part 7 of DICOM). These simple transfer operations can be integrated into wider transactions including search capabilities via criteria such as patient name or study number (QUERY & RETRIEVE): this uses other services defined in Part 7 of DICOM, i.e. the services C-FIND, C-GET and C-MOVE.

The use of exchanges over a network between two application entities requires the prior negotiation of services (*SOP Class*) and the transfer syntaxes to be used. This is called association negotiation. This negotiation is initiated by the requestor of the association, which gives the list of *SOP Classes* which it supports, specifying the role of each (SCU or SCP) as well as the potential transfer syntaxes. This *SOP* triplet *Class – Role – Transfer Syntax* is called the presentation context. The second application entity replies, giving in turn the supported presentation contexts, so that this negotiation can serve to define the list of services and syntaxes which can be used in the exchange.

It is at this stage therefore that the use of transfer syntaxes involving image compression techniques can be introduced. Next we must determine the ways of identifying the images, according to their original or compressed nature, as well as the rules which govern the conversions between an original, uncompressed format and a compressed format.

A general principle (below we will look at how it can be modified) is that the image exists independently of its encoding. Thus if an image is transferred from an image source to an image server using a transfer syntax A, this same image can be restored – *via* QUERY & RETRIEVE for example – by a work station using a transfer syntax B. All it takes is for transfer syntax B to have been given preference over syntax A during the association negotiation between the work station and the image server. We should remember that the choice of transfer syntax is determined by the order of presentation contexts proposed and accepted by those involved in the exchange, and it is not a specific choice made during a C-GET or a C-MOVE.

However, we do have to add some caveats. In the case of images compressed with a lossy method, it is possible that several copies of one image could be managed by an image server application in order to differentiate between the uncompressed images and the images compressed using various different compression techniques and therefore represented in different transfer syntaxes. There is therefore a data element (0028,2110) Lossy Image Compression, which when positioned at the value 01 indicates that the image has been subjected – at some point in its life – to a lossy compression. In this case the value 1 (i.e. the first field) of the data element (0008,0008) Image Type has to contain a DERIVED value, to indicate that it is a derived image. The application which creates such a derived image has to create a new instance of the image, called "derived", giving it a new unique *SOP Instance* UID. A mechanism exists so that, following a QUERY & RETRIEVE query about an original image (not compressed), the service provider can indicate, via element (0008,3001) *Alternate Representation Sequence*, the existence of another (compressed) version of the image, in case the original image has not been saved.

Exchange via physical media: the communication of images via physical media was introduced into the DICOM standard in 1995, with the publication of supplements 1, 2 and 3 (Parts 10, 11, and 12 of the standard). Naturally, in this case no negotiation of the transfer syntaxes is possible. Thus, the copy of the image present on the physical medium explicitly states the transfer syntax used during the encoding. The choice of potential transfer syntaxes is determined by the "application profiles", particular to a clinical field and the technology of a given physical medium. These are defined in Part 11 of the standard, Media Storage Application Profiles. Thus, for example, the profiles STD-XA1K-CD and STD-XA1K-DVD give the *SOP Classes* and the transfer syntaxes to use for exchange by CD and DVD respectively for angiographic images up to resolutions of 1024x1024 on 12 bits.

Email attachments: The ability to send images in the DICOM format as email attachments was introduced into the standard in 2001 with the publication of supplement 54. This extension defines a new application profile STD-GEN-MIME allowing the use of all DICOM composite objects and all existing transfer syntaxes. A group of DICOM files is contained in a new MIME (Multipurpose Internet Mail Extensions) entity called the DICOM file set, of the Multipart/mixed or Multipart/related type. Each file is coded in the form of a MIME component called DICOM File, of the Application/dicom type. It is advisable to use the extension ".dcm". These specifications are dealt with in the RFC 3240 [RFC 02].

WADO access: Web access to DICOM persistent objects meets the increasingly present need to retrieve – via the Internet protocols http and https – DICOM persistent objects, whatever they may be (images, structured reports, etc.). It also tackles the need to reference these information objects very easily in the form of URL/URI (Uniform Resource Locator/Identifier) in all sorts of text or hypertext documents. This tool was introduced into the DICOM standard in 2003 with supplement 85, in the form of part 18 of the standard. It was also recognized by the ISO TC 215 (ISO 17432) [ISO 04c].

The DICOM persistent objects concerned can be divided into four categories: (1) images, (2) multi-frame images, (3) text objects and (4) other objects. Table 4.7 gives an exhaustive list of the MIME types that can be used in each of these four cases.

	Single images	**Multi-frame images**	**Text objects**	**Others**
MIME types that can be used	application/dicom image/jpeg image/gif image/png image/jp2	application/dicom video/mpeg image/gif	application/dicom text/plain text/html text/xml application/pdf text/rtf application/x-hl7-cda-level-one+xml	application/dicom

Table 4.7. *MIME types that can be used in response messages with DICOM WADO*

As we can see in this table, it is not only a question of allowing the use of access methods other than the traditional DICOM exchange protocols (QUERY & RETRIEVE), but also of making data retrieval easier in general syntaxes such as JPEG, GIF or MPEG, without encapsulation into the traditional DICOM syntaxes.

The format of the requests is based on the standard format of URL/URI specified in the RFC 2396. The main parameters are as follows:

– *requestType* (obligatory value: WADO);

– *studyUID, series UID, objectUID*, corresponding to the three levels of DICOM's "study – series – composite object" hierarchy;

– *contentType*, containing the list of MIME types which can be used.

Other optional parameters can also be defined, including:

– *charset*, to determine the set of characters to use, in order of preference (this is just as relevant to text objects as DICOM objects represented with the MIME type Application/dicom);

– *anonymize*, to state that the object must be anonymized;

– *annotation,* to state that annotations (concerning the patient and the technique) must be burnt into the pixel data (only relevant for images, and with a MIME type other than *Application/dicom*);

– *rows, columns, region, windowCenter, windowWidth, frameNumber*, to specify the image or section of image to return.

4.4. Conclusion

Bearing in mind the large volumes of data which result from medical imagery, the use of compression techniques – lossy or lossless – is clearly desirable. The need to share these images – within a hospital, but especially over a healthcare network – creates a strong need for standards, to guarantee the interoperability of applications and give long life spans to the data in question. Lastly, the context of national-level usage of the Personal Medical File will lead to the fundamental question of the role of medical imagery within these files. It is evident today that having standards adapted to the representation of compressed images is essential for the images to be correctly represented in these files. Faced with these needs, numerous possibilities exist, offering performance levels which can be qualified as satisfactory today, in both lossy and lossless compression [CLU 00].

Nevertheless, the ever-increasing volume of image data in existence represents a challenge for the future. This increase is in part due to new multi-slice CT scanners, but also to the development of dynamic imagery (X-ray imagery, ultrasound, dynamic MRI, endoscopy) and to the progress of digitization in anatomopathology (virtual slices). This evolution always calls for an improved performance within the fields of image compression and the communication of compressed data. Therefore, the ball is in the court of technical experts working in the field of image compression, who will undoubtedly rise to the challenge. The experience of the past 10 years has shown that the algorithmic progress achieved in lab results in international standards, both via the ISO and other standards committees such as DICOM. This requires time and effort, but it is essential if a standard is to be recognized and benefit from wide-spread usage in the industry.

4.5. Bibliography

[CEN 04] CEN EN 12052 – Health Informatics – Digital Imaging – Communication, workflow and data management, 2004.

[CHA 04] CHABRIAIS J., GIBAUD B., "DICOM: le standard pour l'"imagerie médicale", *EMC-Radiologie*, vol. 1, p. 577-603, 2004.

[CLU 00] CLUNIE D.A., "Lossless compression of grayscale medical images – Effectiveness of traditional and state of the art approaches", *Proceedings of SPIE: Medical Imaging 2000*, vol. 3980, p. 74-84, 2000.

[DIC 06] Digital Imaging and Communications in Medicine (DICOM) – National Electrical Manufacturers Association, Parts 1 to 18, 2006.

[GIB 98] GIBAUD B., GARFAGNI H., AUBRY F., TODD POKROPEK A., CHAMEROY V., BIZAIS Y., DI PAOLA R., "Standardisation in the field of medical image management: the contribution of the MIMOSA model", *IEEE Transactions on Medical Imaging*, vol. 17, no. 1, p. 62-73, 1998.

[ISO 94] ISO/IEC 10918-1:1994 – Technologies de l'information – Compression numérique et codage des images fixes de nature photographique : principes et lignes directrices, JTC 1/SC29, 185 pages, 1994.

[ISO 95] ISO/IEC 10918-2:1995 – Technologies de l'information – Compression et codage numériques des images fixes à modelé continu : tests de conformité, JTC 1/SC29, 62 pages, 1995.

[ISO 99] ISO/IEC 14495-1:1999 – Technologies de l'information – Compression sans perte et quasi sans perte d"images fixes à modelé continu : principes, JTC 1/SC29, 66 pages, 1999.

[ISO 00] ISO/IEC 13818-2:2000 – Technologies de l'information – Codage générique des images animées et du son associé : données vidéo, JTC 1/SC29, 230 pages, 2000.

[ISO 04a] ISO/IEC 15444-1:2004 – Technologies de l'information – Système de codage d"image JPEG 2000 : système de codage noyau, JTC 1/SC29, 200 pages, 2004.

[ISO 04b] ISO/IEC 15444-2:2004 – Technologies de l'information – Système de codage d"image JPEG 2000 : extensions, JTC 1/SC29, 337 pages, 2004.

[ISO 04c] ISO 17432:2004 – Informatique de santé – Messages et communication – Accès au web pour les objets persistants DICOM, TC 215, 18 pages, 2004.

[RFC 02] RFC 3240 – Digital Imaging and Communications in Medicine (DICOM) – Application/dicom MIME Sub-type Registration, Internet Engineering Task Force (IETF), 2002.

[TIF 92] Adobe Developers Association, TIFF Version 6.0, 1992.

Chapter 5

Quality Assessment of Lossy Compressed Medical Images

5.1. Introduction

Lossy compression techniques do not leave an original medical image unimpaired. It is therefore necessary to evaluate the degradations caused on the image. For natural images, coding techniques must keep to one single criterion relating to visual quality of the reconstructed image. For medical images, it is essential that the compression avoids any distortions that could modify the diagnostic interpretation of the image and the value of anatomic and/or functional parameters that are supposed to indicate the state of the organ being studied. The American College of Radiology pointed out in its practical guide on radiology that this compression must be carried out without losing any information useful to the diagnosis [AME 05].

Defining the amount of distortion accepted that could preserve the reliability of the diagnosis of the reconstructed image is a complex problem and an open debate in the medical imaging field. In fact, the eligible compression rate does not only change according to the compression method applied, but also largely depends on the characteristics of the image being studied; characteristics that are linked to the gathering techniques as well as to the nature of the organ being explored and to the pathology itself [ERI 02].

Chapter written by Christine CAVARO-MÉNARD, Patrick LE CALLET, Dominique BARBA and Jean-Yves TANGUY.

In this chapter, we will start by outlining the consequences of any degradations generated by the two compression norms, JPEG and JPEG 2000, in medical imaging. We will then describe in further detail the different subjective evaluation methods that are most often used in medical imaging, as well as the more recent objective methods.

5.2. Degradations generated by compression norms and their consequences in medical imaging

5.2.1. *The block effect*

It is widely known that compressing an image using the JPEG norm causes the appearance of blocks. This is a direct consequence of the structure of the algorithm that cuts an image into blocks before independently processing each one of them. In the case of medical images, various problems may arise:

– discontinuities in linear details: such as the fine lines of a rectangular interstitial syndrome that could appear in a pulmonary X-ray (Figure 5.1);

– bad visibility of nodular details; for the pneumoconiosis disease, the reticular syndrome of micro-nodular type is made up of millimetric nodular images that may be less visible when cut by juxtaposing two or more blocks;

– change of texture; trabecular bones form a reticular network of thin infra-millimetric meshes with an alignment which is dominant to a varying degree depending on the bone type. Their image on an X-ray or a CT scanner, must maintain the same visual aspect as it is important for the diagnosis of a fracture or of a tumorous lesion for example (Figure 5.2).

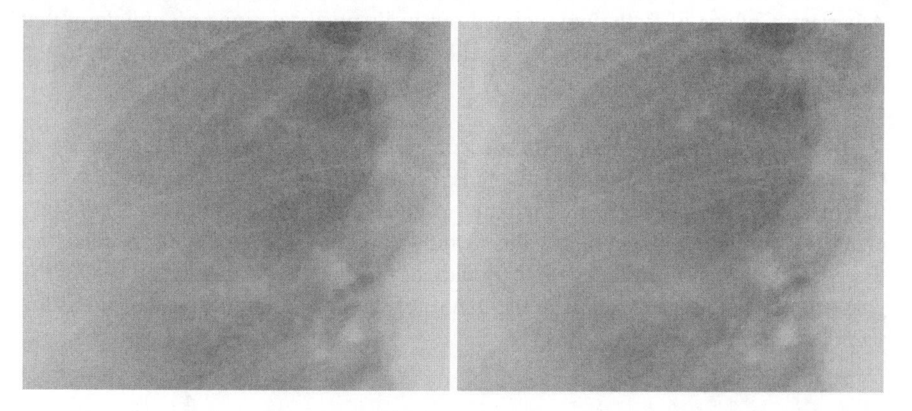

Figure 5.1. *Effects of JPEG compression on a linear detail observed on a pulmonary X-ray. On the left: original image, on the right: compressed image with CR=10:1 (small fissure: line formed by the separation between two right-lung lobes)*

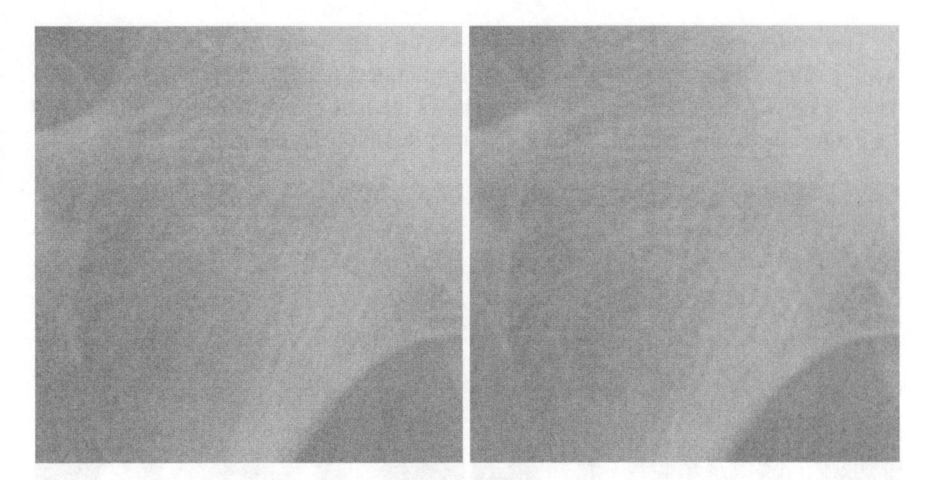

Figure 5.2. *Trabecular texture change of neck of the thighbone by 12:1 JPEG compression blocks (on the right). It would be impossible to detect the presence of a bone fracture*

5.2.2. *Fading contrast in high spatial frequencies*

Some compression methods (such as those based on a decomposition of the image signal by wavelets, the JPEG 2000 norm for example) cause a fade in contrast in the high spatial frequency zones of the image. This reduces the clarity of the image's contours or its linear details (fibrous aspects). The final product represents a smoothed surface on certain areas of the image which sometimes causes the texture of the image to change, such as that of zones representing lesions, for example the "salt and pepper" aspect of a glomic tumor, or the finely striated, radial or granular aspect of tumourous tissue. The spatial resolution may be insufficient compared to the amount required in order to render all details of the image correctly. When the size of a particular detail is smaller or equal to that of the pixel, the only thing visible is the layout, for example the local linear layout made up of a repeating series of pixels, one next to the other and of close amplitudes. If a particular treatment reduces the noise of the image, it may also reduce the visibility of the details in the image. It is therefore better to preserve the noise of an image and its spatial structure that covers high two-dimensional frequencies, rather than reduce the structured textures of the fine tissues being observed that interfere with the use of pathological objects when establishing a diagnosis. The texture of these details is not always structurally described in radiological semiology studies. It is taken into account however, perhaps unconsciously, by experienced physicians, when analyzing and detecting the existence of a lesion.

The following image is a frontal slice of the brain showing an olfactory meningioma. This image illustrates how visible linear details are, knowing that their width is close to that of the pixel. It also shows the loss of clarity in the radial plot at the base of the lesion, after a JPEG 2000 compression (Figure 5.3).

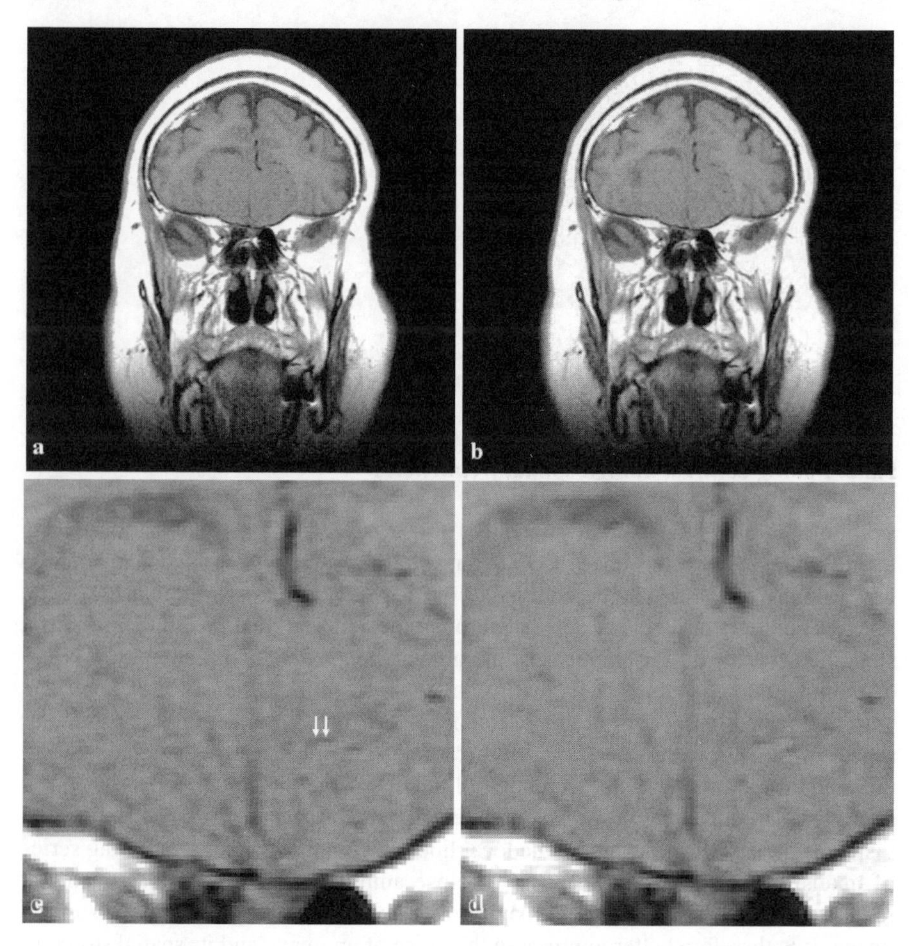

Figure 5.3. *Olfactory meningioma on a frontal slice of a T1-weighted spin echo MRI (acquired on a 1.5 Tesla system): (a) original image; (b) smoothed image after 21:1 JPEG 2000 compression; the 21:1 rate of compression preserved a good global visibility of the lesion; (c) original image, zoom x5, the arrows point at linear details linked to the fibro-vascular architecture of the lesion, the width of which is close to the pixel width; (d) JPEG 2000 21:1 compression, zoom x5: reduced visibility of details*

5.3. Subjective quality assessment

Subjective quality assessment methods are run by a group of experts who rate the quality of the diagnosis as well as the visual quality of the image. They work in the adapted conditions, similar to those applied during clinical routine, allowing them to observe all the necessary detail.

In fact, the choice of assessment method depends on the task that is to be evaluated. In other words, it depends on the information that can be extracted from the images themselves. Usually, there are two different types of task:

– detection tasks that call upon a binary answer (presence or absence of pathology);

– estimation tasks that lead to an estimation of a quality grade according to specific diagnosis criteria.

Thus, for every detection task, the appropriate assessment method is the one which allows for effective calculation to test the sensibility and specificity of the technique to be evaluated. In this case, the most common and most efficient method used is one based on the ROC (*Receiver Operating Characteristics*) analysis. For estimation tasks, the appropriate assessment approach requires the regression study or the Bland and Altman type of study, of inter- and intra-observer analysis. Two important subjective assessment approaches are then suggested to judge on the quality of compressed medical images:

– the assessment of diagnosis reliability, classified as an assessment of detection task;

– the assessment of the diagnosis criteria quality, classified as an assessment of estimation task.

5.3.1. *Protocol evaluation*

Images must be evaluated by at least three physicians (a practical compromise between what is feasible and the strength of statistical tests) selected from amongst senior radiologists specialized in the analysis of the organ being studied. The more motivated the physicians are on the subject of compression, the more efficient, rational and productive the assessment will be. The number of images used must be sufficient for a statistical evaluation (minimum 30). For the diagnosis quality assessment, the pathologies being studied will be selected according to the probability that they can represent different degrees of subtlety, and the probability that they are to be influenced by degradations caused by the compression system (smoothing, block effect, etc.). For example, in thoracic digital radiology, the most commonly studied pathology is interstitial syndrome (subtle anomalies characterized in the frequency-plane by high frequencies) [CAV 01] [SUN 02].

In order to avoid drifts during the statistical analysis of results, the conditions of observation must be standardized and remain close to the conditions applied during clinical practice. Assessment sessions are therefore often carried out in the form of a double-blind test, unlimited by time. Certain actions such as zooming, contrast changes and luminosity changes are not allowed for the simple reason that they are specific actions directed by the physician. They may also have an impact on the visibility of compression artifacts. These factors, usually specific to the physician, are not reproducible and may therefore bias the analysis of results. In order to reduce the contextual effect during an evaluation, images are presented in a random order. Moreover, to obtain stable results, it is best to initiate a practice session on images that will not be incorporated in the final study, so that all experts agree on the anatomic and/or pathological criteria applied, on the psycho-visual evaluation scale, and on the reading conditions.

5.3.2. *Analyzing the diagnosis reliability*

During a diagnosis otherwise known as a binary decision (patients being either normal or pathological), four different situations may occur, depending on whether the observer takes one or the other decision according to the established reality of the gold standard[1]. These four situations are summarized below in Table 5.1.

		Disease	
		Present	**Absent**
Physician's Answer	**Positive**	*TP*	*FP*
	Negative	*FN*	*TN*

Table 5.1. *Diagnostic test with TP the True Positive fraction (disease correctly classified as positive), FP the False Positive fraction, FN the False Negative fraction and TN the True Negative fraction*

1 Gold standard: diagnostic test indicating the formal diagnosis of each patient in the experiment.

In order to analyze the diagnosis reliability, it is not enough to simply count the percentages of correct answers, for two different reasons:

– in case of low prevalence of the disease, for example; for a disease affecting about 5% of the population, a systematic negative response would judge the test as correct in 95% of all cases;

– the percentage of correct responses gives no indication whatsoever on false positive or false negative fractions that play an important part in clinical practice.

Four diagnostic indices have been defined that characterize how correct the answers on classification are: sensitivity *Se*, specificity *Sp*, the Positive Predictive Value *PPV* and the Negative Predictive Value *NPV*.

$$Se = \frac{TP}{TP + FN} \quad [5.1]$$

$$Sp = \frac{TN}{TN + FP} \quad [5.2]$$

$$PPV = \frac{TP}{TP + FP} \quad [5.3]$$

$$NPV = \frac{TN}{TN + FN} \quad [5.4]$$

All the above values hold between 0 and 1. The *Se, Sp, PPV* and *NPV* values are directly related to the prevalence of the disease, according to the Bayes theory. They therefore remain coherent even when the disease is only slightly present.

Nevertheless, every assessment method using these four diagnostic values requires a gold standard that provides the reference diagnosis. This is not always an easy task, especially when the images being studied normally form the gold standard (such as X-ray images of the coronary arteries and numerous MRI, etc.). To solve this problem, a commonly used approach is to call upon a group of physicians who are then asked to establish, according to consensus, a diagnosis on the images being studied. It is also possible to ask these physicians to give their individual opinions on the image, and then keep only the images on which all physicians raise a similar diagnosis.

5.3.2.1. *ROC analysis*

ROC analysis works by observing the entire set of images coded by different compression rates. Radiologists perform this observation task and must then issue a diagnosis (normal or pathological) from these images. We must note that this method is also used to judge the general diagnostic performances of new protocols or new clinical parameters.

In the medical field, there is no set way to distinguish a normal subject from a pathological one. In practice, the physician works spontaneously using a density guide outlining the probabilities of having normal or pathological cases according to a law looked upon as a Gaussian function, as indicated in Figure 5.4.

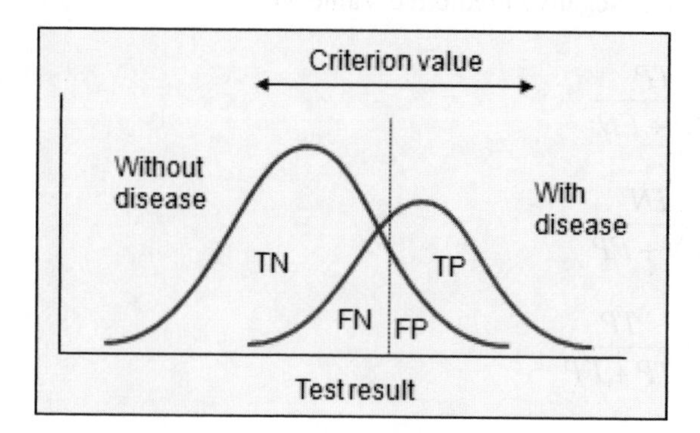

Figure 5.4. *Distribution of medical diagnosis*

The sensitivity and specificity values vary according to two different phenomena: where the decision threshold settled by the physician lies and the detailed precision of the pathology (the more subtle it is, the more overlapping there is between the Gaussian probability density curves (in Figure 5.4)). The ROC approaches solve the problem concerning the decision threshold for the simple reason that they specifically allow the evaluation of a system without considering the chosen decision threshold. In other words, the ROC approach consists of representing the sensitivity value as a function of the specificity for all threshold values possible, and then joining these two points on the curve, as shown in Figure 5.5. Each point on the curve therefore represents a compromise between sensitivity/specificity corresponding to a particular decision threshold. The ROC curve summarizes the entire range of sensitivity/specificity compromises for all the different threshold values.

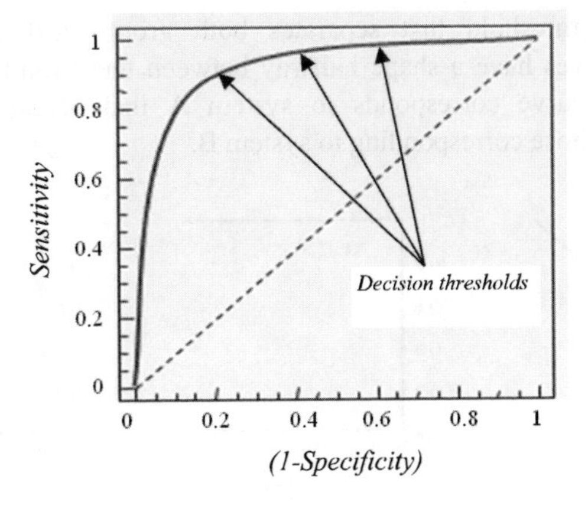

Figure 5.5. *ROC curve*

In order to artificially vary the decision threshold, the most common technique in medical imaging is to ask a question relating to how confident the physician is in noting the presence or absence of the disease in the image: the disease is most certainly present (1), probably present (2), the case is being disputed (3), the pathology is probably absent (4), certainly absent (5). There are thus 5 possible levels of answer. Amongst these responses, 4 couples (Se, 1-Sp) are created by simulating different decision thresholds that a physician could have had if the answer was strictly binary. For example, when a physician answers (3), the responses number (1), (2) and (3) can all be considered positive in the context of a binary answer, while the responses number (4) and (5) will be negative. Adding answers (1), (2) and (3) gives us the sensitivity level and adding (4) and (5) gives the specificity. A point (*Se, 1-Sp*) corresponding to the level of answer number (3) is then established. The four couples now being determined, we can add them to the couples (0,0) and (1,1) to obtain the points defining the ROC curve. Numerous solutions enable the curve to pass through these 6 different points. The most commonly used approach is a parametric approach: it is based on the method of maximum likelihood so that it corresponds to a binormal curve [MET 89][2].

It is possible to interpret an ROC curve in a qualitative manner. The shape of the curve only characterizes the detection performances. A curve that has merged within the diagonal of the two axes (Figure 5.6a) corresponds to a non-discriminating system. In other words, the answer given by the observer is in no way related to the presence or absence of an anomaly. A curve shape like the one represented in Figure 5.6.b corresponds to a perfectly discriminating system, for which there exists a

2 This approach is at the root of the free software "Rockit", available on the following website: http://www-radiology.uchicago.edu/krl/KRL_ROC/software_index.htm.

specific decision threshold that separates both groups distinctly. In practice, effective ROC curves have a shape halfway between these limits. For example in Figure 5.6c, the curve corresponds to system A that reveals better detection performance than those corresponding to system B.

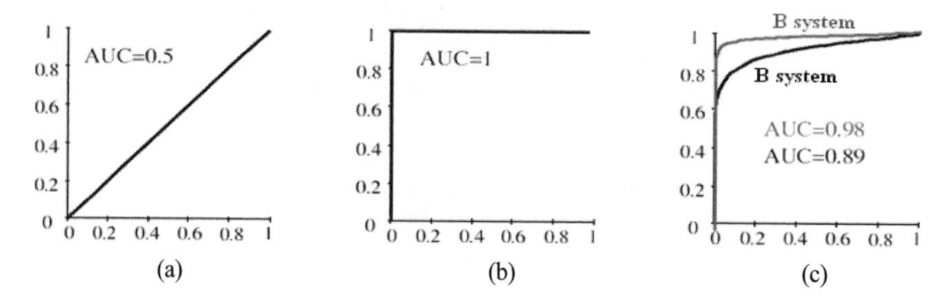

Figure 5.6. *Interpretation of ROC curves: (a) non-discriminating system, (b) perfectly discriminating system, (c) usual system (system A > system B)*

When we come to classifying the compression methods being studied by increase in performance, it is then possible to objectively compare the areas under the ROC curves otherwise known as the AUC (*Area Under Curve*) areas. The AUC represents the probability that we can correctly identify the image containing an anomaly when both an image with and another without an anomaly are simultaneously presented to the observer.

During the evaluation of a compression system, a ROC curve is first established from the set of original images, and then compared to a second curve drawn from images compressed at a given rate. If the compression causes a reduction in the diagnostic performance, the second curve appears under the first. Various ROC curves corresponding to images compressed at different rates may be compared as shown in Figure 5.7.

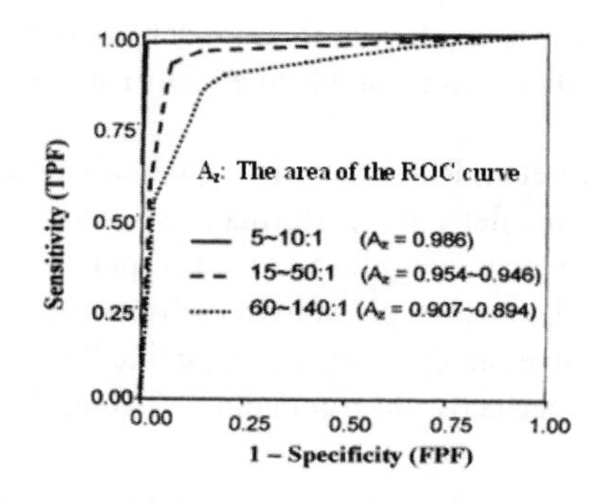

Figure 5.7. *Example of ROC curves; from [SUN 02]*

The ROC analysis is widely used to validate compression algorithms on medical images. This method has the advantage of receiving a general consensus, and is often considered as a reference amongst radiologists for quality assessment [SUN 02][ZHI 05].

Nevertheless, ROC analysis is appropriate mainly for binary diagnostic assignments (normal/pathological). The traditional ROC analysis therefore cannot apply in the case of a diagnosis involving multiple pathologies on a same image. With the FROC (*Free Response Operating Characteristic*) and AFROC (*Alternative FROC*) approaches it is possible to analyze the location of potential anomalies by indicating on the ordinate the number of anomalies correctly detected and located, and on the abscissa, either the average number of false positives in each image for the FROC curves [PEN 05] or the fraction of images that contained at least one false positive (abscissa of ROC) for the AFROC curves [CHA 90]. However, there is no specific statistical test allowing for the comparison of two FROC curves.

5.3.2.2. *Analyses that are not based on the ROC method*

If the diagnostic task is not binary (determining the amount of nodules for example), it is also possible to statistically analyze the diagnostic indices *Se* and *PPV* according to the compression rate of the images [COS 94].

In the case of cardiac angiograms, not only are the sequences considered as gold standard in the study of coronary cardiopathies, but the diagnostic task also combines the location (depending on a rigorous lexicon), the detection, as well as the classification of the anomaly. Beretta thus suggests that the evaluation be based on the estimation of intra and inter-observer concordances between all the visual interpretations of the sequences with or without compression [BER 97]. Since the gold standard is absent, that concordance is statistically estimated by a kappa test [COH 60] if the answers characterize a certain category. All these concordances must be associated with a threshold of statistical significance often marked as p or α: if $p < 0.01$, the concordance between them is not the result of pure chance. The inter-observer concordance obtained on the original images is then compared with the one obtained on the compressed version of these images. Similarly, the intra-observer concordance tested between two replicas of the original image is compared to the one tested between the two replicas of the compressed image.

5.3.3. *Analyzing the quality of diagnostic criteria*

The analysis of the diagnosis reliability is often rather “general”. This is because it refers to a set of test images. This analysis is also “unique” because it focuses on analyzing one specific pathology. In clinical practice, each diagnosis is very specific and may deal with more than one pathology. Thus, all local information (often anatomic information) in the image is analyzed according to specific criteria in order to best integrate the diagnostic process. Evaluating local attributes of an image is therefore an important part of the diagnosis that reflects the clinical practice and is often useful in the decision process (by physicians) on whether to use irreversible compression methods in clinical routine.

For this type of analysis it is crucial to define both the grading scale and the diagnostic criteria that are to be evaluated. Moreover, they must be defined very precisely (as indicated in Figure 5.8) in order to avoid inaccuracies. The criteria selected are specific to the organ being studied. The notation scale is also adapted to the characteristics of the acquisition system.

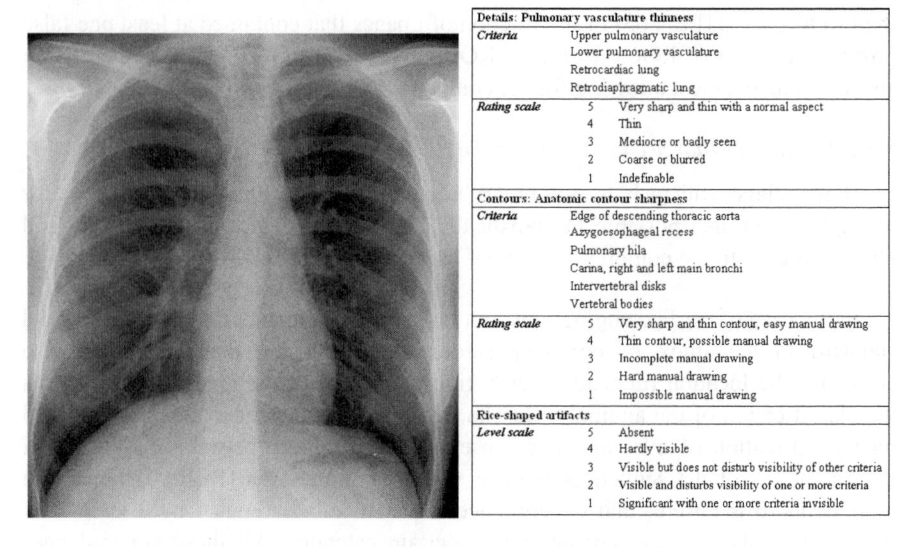

Details: Pulmonary vasculature thinness		
Criteria	Upper pulmonary vasculature Lower pulmonary vasculature Retrocardiac lung Retrodiaphragmatic lung	
Rating scale	5	Very sharp and thin with a normal aspect
	4	Thin
	3	Mediocre or badly seen
	2	Coarse or blurred
	1	Indefinable
Contours: Anatomic contour sharpness		
Criteria	Edge of descending thoracic aorta Azygoesophageal recess Pulmonary hila Carina, right and left main bronchi Intervertebral disks Vertebral bodies	
Rating scale	5	Very sharp and thin contour, easy manual drawing
	4	Thin contour, possible manual drawing
	3	Incomplete manual drawing
	2	Hard manual drawing
	1	Impossible manual drawing
Rice-shaped artifacts		
Level scale	5	Absent
	4	Hardly visible
	3	Visible but does not disturb visibility of other criteria
	2	Visible and disturbs visibility of one or more criteria
	1	Significant with one or more criteria invisible

Figure 5.8. *Anatomic criteria defined in [CAV 01] for quality assessment of normal PA chest radiographs*

Since the answers are evaluated by a quantitative score; we are able to measure the proportion of concordance. The correlation between all observations can be evaluated qualitatively by graphical analysis and by displaying the answers of the first observation on one axis and those of the second observation on the second axis, as shown in Figure 5.9.

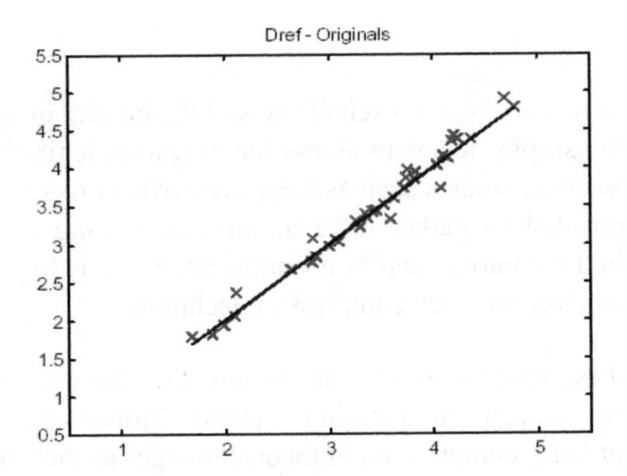

Figure 5.9. *Example of a graphical analysis of the correlation between all observations [BER 97]*

The greater the difference between the first and the second observation, the more dispersed the points will appear around the straight line representing the response couples. The concordance correlation coefficient defined by Lin [LIN 89] then allows us to quantify the degree by which these pairs of measurements are on the straight line (first diagonal). This coefficient can be estimated using the bootstrap technique, when the overall dispersion of measurements is not Gaussian and/or when the number of studied images is relatively small [CHE 99].

Regarding the quality assessment of compressed images, the first observation corresponds to the results obtained on the original images by one physician and the second observation corresponds to the results obtained on the same images after compression by the same physician. The correlation coefficient can then be compared to the coefficient obtained first by the intra-observer analysis (the original images are evaluated twice by the same observer at sufficiently distant times so as to avoid what is often called the learning effect) and then by the inter-observer analysis (the original images are evaluated independently by at least two observers).

This notation of anatomic criteria useful to any diagnosis (defined as *Diagnostic Quality Pattern* or DQP in [PRZ 04]) allows us to determine a local score (and thus a quality map) or a global score (an average), for every compression rate. This average score is often referred to in literature as the "Mean Opinion Score" or MOS [YAM 01].

5.3.4. *Conclusion*

In order to analyze the diagnosis reliability and the quality of specific diagnostic criteria, we need to ask physicians to assess the diagnosis legibility on a particular set of images. In practice, studies such as these are difficult to set up [PRZ 04]. It is indeed very complicated to gather different images making up a representative sample of the studied pathology, and/or a sample enabling us to judge the result on each type of criteria given for each compression technique.

Nevertheless, these studies are still, up to this day, the only method on which physicians agree to evaluate compression systems. Moreover, these studies help physicians to accept lossy compression of medical images as they are often surprised to see no difference between an original and a compressed image (under specific conditions of course). In the case of cardiac angiogram compression, Beretta notes that physicians cannot determine the difference between MPEG compressed sequences at a compression rate of 12:1 and the original corresponding sequences [BER 97].

5.4. Objective quality assessment

Objective quality assessment methods are carried out by calculating a value representing the local or global quality of an image using mathematical algorithms known as "objective quality criteria" and without any human intervention. This value must best reveal the visual appreciation of a human observer. For this purpose, this value may be defined as a linear combination of various distortion factors (based either only on a difference between the amplitude of pixels, or preferably on a human visual system model). Adjustment of the weights of each factor could be performed by linear regression, in order to best reproduce the MOS [MIY 98] [PRZ 04].

In literature, three main objective approaches are used in medical imaging when judging the quality of reconstructed images:

– measuring the signal-to-noise ratio (SNR) or any other measurement based on pixel values;

– calculating a set of parameters based on a model version of the Human Visual System (HVS);

– analyzing the change of clinical semantic parameters, knowing that they are calculated automatically or semi-automatically (for example, the degree of stenosis).

5.4.1. *Simple signal-based metrics*

The Mean Square Error (MSE) evaluates the importance of the distortions in a reconstructed image after an irreversible coding, by an Euclidean distance. The SNR metric is another widely used technique directly linked to the MSE. It evaluates the quality of image reconstruction through a quality measurement based on contrast. We must note that these two measurements are full reference metrics (called FR metrics), as they need the original image version (non-distorted by the coding process) corresponding to the coded image to be evaluated. The coded image must be perfectly spatially aligned with the original to make sure that the measurements are not biased.

Nevertheless, during the quality assessment of an image it is very often found that these two measurements fail [ERI 02] [PRZ 04]. The reasons for this will be further explained in this chapter. We must note however that the HVS does not directly compare the original and coded values of each individual pixel. Instead, the HVS uses a more complex way of representing the observed images. It is important to know that the observation conditions are essential when looking at an image: ambient lighting, visualization screen, observing distance, etc. In addition, for medical imaging and other types of imaging, the image field must cover both regions of interest and the peripheral zones that are considered less useful. However, the PSNR and MSE criteria are global average measures that do not take into account all local variations of interest. In medical imaging, if important anatomic details have been degraded by compression while the rest of the image is generally well reproduced, then the MSE will remain low, but the medical expert will evaluate the image as bad in terms of quality. Finally, during a diagnostic task, the physician uses on one hand his knowledge of anatomy, function and potential pathologies of each organ, and the acquisition protocol characteristics on the other hand.

5.4.2. *Metrics based on texture analysis*

Some studies have aimed at evaluating one particular type of degradation by texture analysis. The works shown next have looked at the loss of contrast in high spatial frequency on medical images (indeed, this degradation largely interferes with the diagnosis, as explained in section 5.2.2).

The Moran test consists of calculating a coefficient (a Moran coefficient) corresponding to the spatial autocorrelation observed between the pixels in an image block [CHE 06]. For a smoothed area, when the pixel intensities are close, there will be a high coefficient. The histogram of normalized Moran coefficients therefore presents a high peak while the spatial correlation (and hence the smoothing) will increase (Figure 5.10).

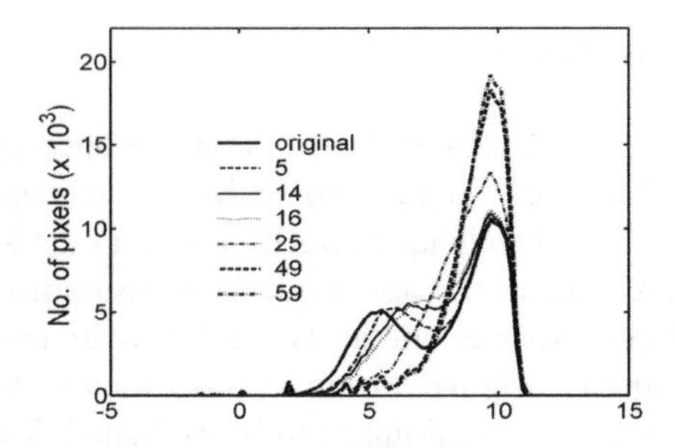

Figure 5.10. *Histogram of Moran coefficients for different compression rates according to the JPEG 2000 norm [CHE 03] (the higher the compression rate is, the more significant the smoothing will be)*

In [HIS 02], a frequency analysis reveals the smoothing being introduced by a JPEG 2000 compression chain. The radial information extracted from the Fourier spectrum of a compressed image (by adding the frequency coefficients located on a given radius circle – and thus a given frequency) outlines the significant loss of high frequencies after compression (Figure 5.11). In order to characterize the low-pass filter, a transfer function is calculated for each compression rate according to the following formula:

$$20\log\left(\frac{FFT_{rad}(\text{Reconstructed Image}(\text{CR}=\text{i}:1))}{FFT_{rad}(\text{Original Image})}\right) \quad [5.5]$$

The cut-off frequency at -0.5 dB (because of the low initial gain) of each filter quantifies the smoothing effect (Figure 5.11).

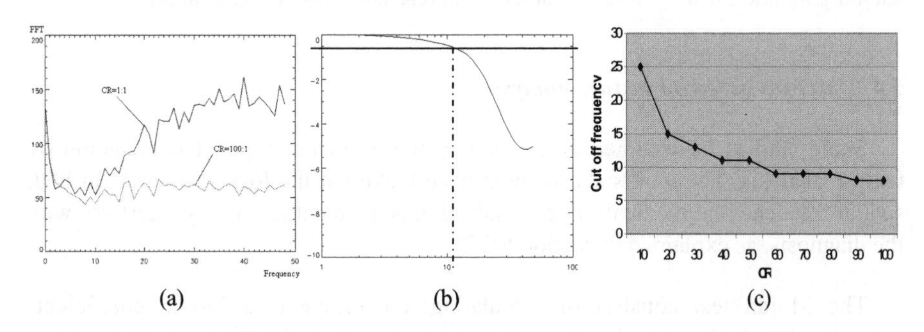

Figure 5.11. *(a) Frequency analysis of radial information for two different compression rates; (b) transfer function for a given compression rate; (c) variation of the cut-off frequency at -0.5 dB of the filters according to CR (significant gradient between 20:1 and 30:1) [HIS 02]*

5.4.3. *Metrics based on a model version of the HVS*

Metrics based on the HVS usually use an approach based on a visual discrimination model. Two images are therefore analyzed simultaneously by an algorithm providing a Just Noticeable Difference (JND) map between these two images. The JND values vary according to the observing conditions (general lighting, screen, observation distance, etc.). It is then necessary to gather all of these visual differences to establish an overall score for the image quality. Transforming the JNDs into an overall score is not an easy task, and must be performed while taking into consideration both the observation context and the final task to be accomplished. In medical imaging, whether we are dealing with detection or estimation tasks, this last step in the process remains an open field of investigation that requires high-level treatment by the observers themselves. In fact, this synthesis procedure depends on the shapes that have been identified or that are to be identified in the image as well as on the semantics of the content being observed, according to the expert. In medical literature, this spatial pooling of visible errors is often performed using a simple approach such as a Minkovski summation (Lp norm with p sometimes equal to 2, and often greater than 2). We will therefore be looking only at the best-known approaches that detect JNDs and leave the pooling stage aside.

Many physiological studies of the human visual system have led to numerous properties and a variety of assumptions. All of these properties must be taken into account when trying to understand how our visual system translates and represents all the information it comes across. Nevertheless, this system remains a complex one, and we have not understood every single one of its aspects. Thus, the models based on our visual system are still often incomplete. In this section, we review the main visual factors relating to the quality assessment of an image. We will then outline the criteria present in the literature.

5.4.3.1. *Luminance adaptation*

The human visual system is able to adapt to a great dynamic of different light intensities, thanks to the anatomic structure of the eye and its retina. This capacity to adapt to different luminosities enables humans to distinguish light variations, even if the luminosity varies considerably. There are three different mechanisms by which the system adapts to luminosity: size variation of the pupil by the iris (response time of a few seconds), variations in the concentration of chemical products in photoreceptors (response time is very slow: a few minutes) and adaptation of neurons involving all the cells of the different layers of the retina that adapt to changes in light intensity by increasing or reducing their exit signal (fast adaptation: less than a second).

5.4.3.2. *Contrast sensivity*

It is well known that the human visual system is more sensitive to local light variations compared to the average luminosity than it is to full variations. The Weber-Fechner theory is best adapted to explain this dependence. It confirms that a luminous stimulus' sensibility to differences in light intensity is proportional to the average luminosity of that stimulus (the contrast threshold remains constant when the levels of light increase). This theory is correct for background luminosities of about 10 cd/m2, under which the contrast threshold increases as the luminosity decreases. Additionally, the theory is invalid when dealing with a high average luminosity.

There are many different models that take into consideration this adaptation process. The most commonly used model uses a non-linear compressive structure, usually a log function or a cube root. Daly [DAL 93] brings forth the idea that using a non-linear logarithm over-estimates the visual sensibility in areas of low intensity, and that a cube root is instead, the best possible approximation. Peli [PEL 90] and Lubin [LUB 93] give a definition of contrast based on the ratio between the energy of a spatial frequency band and the local mean luminance obtained by a low-pass filter applied to the image.

The background luminance is not the only value acting upon the visibility of a signal. Other characteristics are also important, such as for example, the spatial frequency content. Contrast sensitivity functions (CSF) are generally used to measure these dependencies. The CSF is defined as the inverse of the contrast threshold. It is a function of the signal spatial frequency to be detected (it must be noted that, often, it is the radial spatial frequency that applies; in fact, it is also a function of the orientation). In a medical context, various more or less complex CSF models translate the relation between the contrast threshold and the stimulus spatial frequency. These models are almost all established from experimental results regarding the detection of sinusoidal signals, using Michelson's definition of contrast. Peli *et al.* [PEL 93] provide a rather complete set of function curves displaying the sensibility to achromatic signals for different stimuli configurations. These results suggest that differences in stimuli greatly influence the shape of the CSF. The assumption that the CSF can be assimilated to a function of a bandpass filter transfer in spatial frequency is only verified for sinusoidal signals limited to a specific spatial extension. It is clear that the gain at spatial frequency (0,0) could not be zero otherwise it means no sensitivity to the general luminance of an image.

5.4.3.3. *Spatio-frequency decomposition*

Various physiological facts reveal that most of the cells in the visual system are responding to particular magnitude types fundamentally determining visual signals such as the color, orientation or spatial frequency. This has been confirmed by results of psycho-physical experiments that suggest the existence of a spatio-frequency deconstruction in visual channels to deal with the information. Thus, Sakrison [SAK 77] notes that when representing a stimulus containing various frequency components, such as for example a sawtooth, the fundamental component is the only one to stabilize the stimulus visibility threshold. Simulating these visual channel phenomena leads to the definition of numerous filters that then characterize the subbands or visual channels. The construction of a bank of filters modeling this decomposition with precision is often criticized. While many authors [DAU 84] [PHI 84] disprove the existence of independent channels in a polar representation, the decomposition characteristics are usually defined in terms of radial or angular selectivity. Regarding the achromatic component, the width of the bands varies, according to the authors, from 1 to 2 octaves in radial frequency sensibility, and 20 to 60 degrees in angular selectivity.

5.4.3.4. *Masking effect*

The contrast masking effect is a phenomenon known to translate the change in visibility of a signal by the presence of another signal. This change may be a reduction of the visibility threshold, in which case we are dealing with the facilitation effect. The visibility threshold may increase, in which case we are talking about the masking effect. The masking effect is a sensitive step in visual quality metrics procedures because bad modeling of the masking effects means that the general appreciation of visible errors will be not good. The main problem is that the implied phenomena are all very complex. There are several experimental conditions and different models available. Experimental results are strongly dependent on the signals used. The masking effects vary according to the frequency content, phase and color orientation of signals, or even according to the degree of familiarity that the observer has with the signals involved.

Traditional experiments measure the capacity to detect sinusoidal signals in the presence of another sinusoidal signal masking the first. Such experiments deal with the masking of signals oriented differently [FOL 94a] or of a different spatial frequency [LEG 80]. Most experiments lead to a model type similar to that defined by Legge and Foley. This model, now widely used for questions relating to the contrast masking effect, predicts the degree with which a second masking sinusoidal signal hides the presence of a target sinusoidal signal. This type of model has been improved over time [FOL 94b] following the introduction of a new spatio-frequency decomposition concept. We can observe that the masking effect depends on the energy present in a channel as well as on the energy in adjacent channels in terms of orientation. Models used to control the increase in contrast have gained in popularity. This is because they are able to predict the visibility thresholds of simple signals equally well. Initiated by Teo and Heeger [TEO 94], these models are constantly being improved. The more specific they become the better they can explain the multiple interactions occurring between visual channels.

Other experiments [LEG 87] [GEG 92] look at the masking effect of a signal by a spectrum noise limited by a frequency band. These experiments deal with masking effects that are considerably more significant than those occurring with simple signals. However, Swift and Smith [SWI 83] have shown that, if the observers are given enough time to familiarize themselves with a noise type masking signal, the results will be similar to those obtained with sinusoidal masking signals. Watson *et al.* [WAT 97] have studied this phenomenon using different masking signals: white noise, colored noise, sinusoidal and natural images. Results have shown that familiarization can work to reduce the masking effect linked to noise down to the effect deriving from a sinusoidal signal. This confirms the results presented by Swift and Smith. Finally, how important the contrast masking effect is, depends on the degree of familiarity that the observer has with the image. In the case of medical images, physicians are usually very familiar with the images that they are given to evaluate.

5.4.3.5. *Visual distortion measures*

The first visual distortion measure with the visual capacity to evaluate the quality of an image was defined by Mannos and Sakrison [MAN 74]. The visual system was then known as a mono-channel (filtered by a contrast sensibility function after a treatment by a non-linear function of adaptation to luminance). This approach has been extended to include the study of masking effects in order to increase the visibility of errors. Various authors ([LUK 82], [XU 94], [HAN 94], [KUS 01]) have suggested such an extension for a mono-channel metric. These mono-channel metrics involving masking effects are a great improvement from the simple mono-channel metrics. However, since masking effects are taken into account, it is not possible to include aspects of the visual system's receptor fields. As a result, the models are under constant evaluation and improvement. They are based on magnitudes that are indirectly representative of those treated by the HVS.

Zetzsche and Hauske were the first to introduce a metric using a multi-channel model [ZET 89] to evaluate the quality of fixed monochrome images. Various authors have taken an interest in what is known as aperiodic masking linked to the structures of natural images (lines and outline), as opposed to the masking phenomenon measured in psychophysics with sinusoidal signals. Since the model is optimized using results of an experiment on the visibility of a line masked by a contour according to the distance between both, the masking effects are kept in a degradation configuration that is far too specific. Comes suggested two additional multi-channel criteria [COM 90] [COM 95] based on the idea that the original image is the actual signal masking the errors in the distorted image. This suggestion may however turn out to be a disaster if distortions are simultaneously masked. The most reasonable measures are those based on a visual representation of a reference image as well as of the distorted one that use psychophysical tests detecting simple signals. The two widely known measures fitting these descriptions are the *Visible Difference Predictor* (VDP) defined by Daly [DAL 93] and the *Visible Discrimination Model* (VDM) defined by Lubin [LUB 95] and considered as one of the standards of quality assessment. The structure of the VDM model is similar to the criterion established by Daly, yet with a few interesting differences. A study [LI 98] is still unable to depict the significant performance differences between both criteria. Nevertheless, Lubin's VDM is most widely known and largely used for medical applications. We will therefore focus in greater detail on the VDM.

The main specificity of the measure holds at its first level, modeling the optical part of the retinal visual system. The optical filtering carried out by the eye takes place at first on the image signal by applying a filter with a two-dimensional Gaussian convolution kernel. It is then sampled a second time in order to model the filtering and the sampling linked to the dispersion of cones in the retina. This approach is very different to Daly's model, as it includes extra foveal vision phenomena. Each resulting signal, coming from both the original and the distorted image, is decomposed according to a Laplacian multi-resolution pyramid, creating a hierarchy of 7 different levels of filtered signals. The central frequencies vary between 32 and 0.5 cycles/degree and are separated from one another by an octave. A 5-level contrast pyramid is then constructed, in which the level k contrast on each site is calculated according to a technique similar to Peli's technique [PEL 90]:

$$C_k = \frac{G_k - G^i_{k+1}}{G^{ii}_{k+2} + \Phi^k} \quad [5.6]$$

where k varies between 0 and 4, and the offset parameters Φ equals 0.025 in order to avoid a division by zero. The G^i_{k+1} image derives from the G_{k+1} image by interpolation with a filter, so that G_k and G^i_{k+1} have the same size and scale. By applying the filter twice on G_{k+2}, we obtain G^{ii}_{k+2}. Using this definition of contrast is another original aspect of the Lubin method compared to the Daly criterion.

The C_k contrast signals are then filtered by four pairs of angular selectivity filters of a 65 degree band-width. Each R_{kl} response corresponding to level k and orientation l, is normalized by the detection threshold Mt (v, L, W), given by Barten's CSF at the central frequency v of each filter, for an image of a mean luminosity L and width W.

$$\frac{1}{M_t(v,L,W)} = a(L,W).v.\exp(.b.v).\sqrt{1+0.6.\exp(b(L).v)} \qquad [5.7]$$

with

$$a(L,W) = \frac{540.(1+0.7L)^{-0.2}}{1+12/[W.(1+v/3)^2]} \qquad [5.8]$$

$$b(L) = 0.3.(1+100/L)^{0.15} \qquad [5.9]$$

The intra channel masking is taken into account, by applying, on each normalized response Rn_{kl}, a non-linearity of sigmoid type:

$$T_{kl} = \frac{(k+2).|Rn_{kl}|^{\alpha}}{k.|Rn_{kl}|^{\alpha-\omega} + |Rn_{kl}|^{\varepsilon} + 1} \qquad [5.10]$$

The parameters have been optimized from the results of psychophysical experiments found in medical literature. Other authors give $\alpha = 0.5$, $\varepsilon = 1.1$, $\omega = 0.068$ and $k = 0.1$. After having gathered all images of similar resolution, we can calculate at each point the figure D_{12} using the Minkowsky summation:

$$D_{12}(m,n) = \left\{ \sum_{k=0}^{4} \sum_{l=0}^{3} \left| T_{kl}^{1}(m,n) - T_{kl}^{2}(m,n) \right|^{\beta} \right\}^{1/\beta} \qquad [5.11]$$

with $\beta = 2.4$ and where T^1_{kl} and T^2_{kl} correspond to the responses of the original and degraded images. The model allows us to obtain distortion maps in the same way as the Daly model provides probability determining how likely we are to detect the differences between two images at each spatial site.

On the basis of this mode, it is possible to engage some improvements both in terms of subband definition and regarding the masking model. Various authors [JAC 96], [JOH 99], [KRU 03], [CHA 04] have suggested using such an approach in medical imaging. The results are often similar to those obtained in ROC studies, even if there is not yet a technique to construct a general quality score.

5.4.4. *Analysis of the modification of quantitative clinical parameters*

Using images is not only about displaying them. Images may be processed in a variety of ways: for improvement, segmentation, three-dimensional visualization, quantification, etc. It is crucial to carefully study the potential influence that compression could have on the image, and the distortions it may engender. Such analyses are about calculating clinical semantic parameters (surface of an organ as shown in Figure 5.12, surface of an anomaly, ejection fraction, degree of stenosis, etc.) and then comparing the results obtained on the original image with those obtained on the same images after compression [KON 97] [CAV 99].

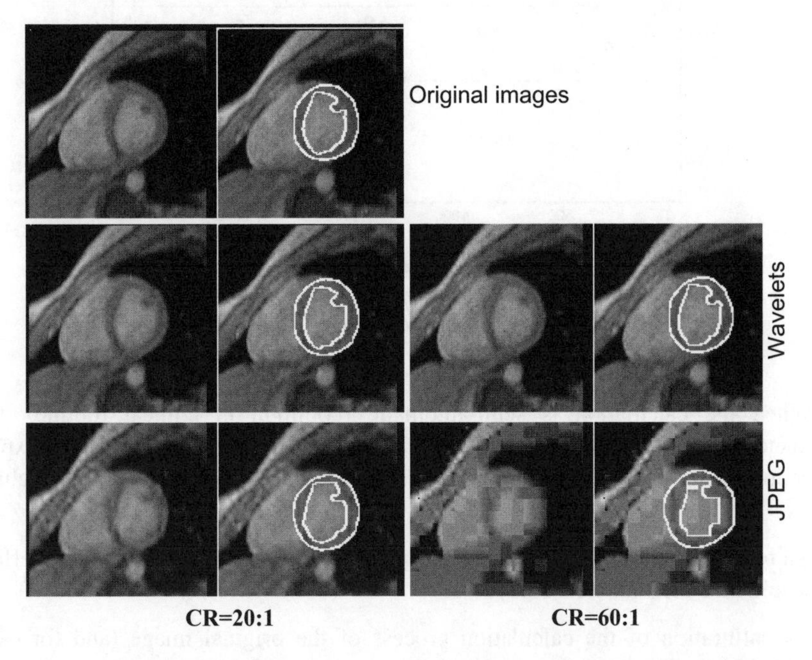

Figure 5.12. *Modification of myocardial surface on cardiac MRI after compression based on wavelets (SPIHT) and after JPEG [CAV 99]*

Descriptive and quantitative statistical methods are used to test whether compressing the image significantly reduces its reliability. Using this approach it is however impossible to compare the measured values with the real unknown values (notion of accuracy). It is possible to check whether compressing the image would cause discordances between different measures (notion of reliability). For this verification to take place where a gold standard is absent, Bland and Altman suggest the use of a graphic method [BLA 86]. This graphic method clearly outlines the narrow concordance between two measures by tracing the differences between a pair of measurements (in this case, the original and the compressed version) according to the average of each one of those pairs as shown in Figure 5.13. The straight line representing the average value of these differences is surrounded by two other straight lines that account for the interval of distortion tolerance. The average of differences indicates the mean bias between the two methods.

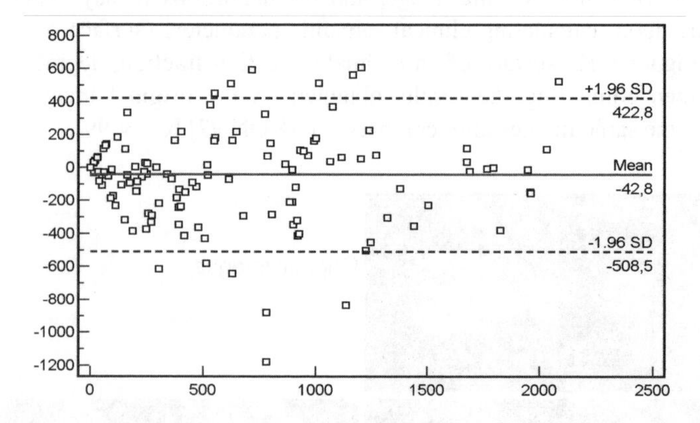

Figure 5.13. *Example of Bland and Altman analysis for the evaluation of degradations caused by a wavelet compression and a JPEG compression on the myocardial surface calculated on cardiac MRIs [CAV 99]*

When the calculation is semi-automatic (common in clinical routine), the parameter value that is being calculated depends on the choice of the expert (selection of points or zones). The measuring process must therefore include [KON 97]:

– a replica of all measures, separated by a few weeks to avoid the learning effect that could bias the statistical study;

– a calibration of the calculation process of the original image (and for each replica). This calibration is then reapplied on the corresponding compressed images. Thus, we can blame clinical parameter errors solely on the compression process.

When analyzing the data we look to answer two main questions:

– does compression reduce the clinical parameters accuracy? We therefore need to analyze whether the values measured on the compressed images are significantly different to the values measured on the original image;

– does compression make measurements of clinical parameters less precise? To answer this question we must find out whether the gap between two replicated measures is significantly different for compressed and original images.

5.5. Conclusion

Quality assessment must demonstrate the fact that losing information is not always linked to a loss of diagnosis quality and reliability. There are numerous new methods for quality image assessment that depict any loss of information. Such assessment methods reduce our apprehensions and help us to point out potential stains and algorithms, and determine which compression processes and compression rates are most appropriate.

In fact, the image quality assessment methods described in this chapter are essential when deciding on which lossy compression method (giving better compression rates compared to lossless compression method) to use in clinical routine. In this case, lossy compression methods could be used during the first diagnostic process. This progress (often referred to as a true revolution) in terms of controlling compression of medical images will enable us to make optimal use of PACS in health services.

5.6. Bibliography

[AME 05] AMERICAN COLLEGE OF RADIOLOGY (ACR), "ACR technical standard for teleradiology", *ACR Practice Guideline*, p. 801-810, October 2005, http://www.acr.org/.

[BER 97] BERETTA P., Compression d'images appliquée aux angiographies cardiaques: aspects algorithmiques, évaluation de la qualité diagnostique, University thesis, INSA Lyon, November 1997.

[BLA 86] BLAND J.M., ALTMAN D.G., "Statistical method for assessing agreement between two methods of clinical measurement", *The Lancet*, vol. 8, p. 307-310, 1986.

[CAV 99] CAVARO-MÉNARD C., LE DUFF A., BALZER P., DENIZOT B., MOREL O., JALLET P., LE JEUNE J.J., "Quality assessment of compressed cardiac MRI. Effect of lossy compression on computerized physiological parameters", *Proceedings of 10th International Conference on Image Analysis and Processing (ICIAP'99)*, Venice, Italy, p. 1034-1037, September 1999.

[CAV 01] CAVARO-MÉNARD C., GOUPIL F., DENIZOT B., TANGUY J.Y., LE JEUNE J.J., CARON-POITREAU C., "Wavelet compression of numerical chest radiograph: quantitative and qualitative evaluation of degradations", *Proceedings of International Conference on Visualization, Imaging and Image Processing (VIIP'01)*, Marbella, Spain, p. 406-410, September 2001.

[CHA 90] CHAKRABORTY D.P., WINTER L., "Free-response methodology: alternate analysis and a new observer-performance experiment", *Radiology*, vol. 174, no. 3, p. 873-881, March 1990.

[CHA 04] CHAKRABORTY D.P., "Predicting detection task performance using a visual discrimination model", *Proc of SPIE Medical Imaging*, vol. 5372, p. 53-61, 2004.

[CHE 99] CHERNICK M.R., *Bootstrap Methods: A Practitioner's Guide*, Wiley Series in Probability and Statistics, October 1999.

[CHE 03] CHEN T.J., CHUANG K.S., WU J., CHEN S.C., HWANG I.M., JAN M.L., "Quality degradation in lossy wavelet image compression", *Journal of Digital Imaging*, vol. 16, no. 2, p. 210-215, June 2003.

[CHE 06] CHEN T.J., CHUANG K.S., CHANG J.H., SHIAO Y.H., CHUANG C.C., "A blurring index for medical images", *Journal of Digital Imaging*, vol. 19, no. 2, p. 118-125, June 2006.

[COH 60] COHEN J., "A coefficient of agreement for nominal scales", *Educational and Psychological Measurement*, vol. 20, p. 27-46, 1960.

[COM 90] COMES S., MACQ B., "Human visual quality criterion", in *SPIE*, vol. 1360, p. 2-13, 1990.

[COM 95] COMES S., BRUYNDONCKX O., MASCQ B., "Image quality criterion based on the cancellation of the masked noise", *ICASSP*, vol. 4, p. 2635-2638, 1995.

[COS 94] COSMAN P.C., DAVIDSON H.C., BERGIN C.J., TSENG C.W., MOSES L.E., RISKIN E.A., OLSHEN R.A., GRAY R.M., " Thoracic CT images: effect of lossy image compression on diagnostic accuracy", *Radiology*, vol. 190, no. 2, p. 517-524, February 1994.

[DAL 93] DALY S., *The visible differences predictor: an algorithm for the assessment of image fidelity*, in *Digital Image and Human Vision*, A.B. Watson (ed.), p.179-206, MIT Press, 1993.

[DAU 84] DAUGMAN J.G., "Spatial visual channels in the fourier plane", *Vision Research*, vol. 24, no. 9, p. 891-910, 1984.

[ERI 02] ERICKSON B.J., "Irreversible compression of medical images", *Journal of Digital Imaging*, vol. 15, no. 1, p. 5-14, March 2002.

[FOL 94a] FOLEY J.M., "Human luminance pattern-vision mechanisms: masking experiments require a new model", *Journal of the Optical Society of America*, vol. 11, no. 6, p. 1710-1719, 1994.

[FOL 94b] FOLEY J.M., BOYTON G.M., "A new model of human luminance pattern vision mechanisms: Analysis of the effects of pattern orientation, spatial phase and temporal frequency", in *SPIE*, vol. 2054, p. 32-42, 1994.

[GEG 92] GEGENFURTNER K.R., KIPER D.C., "Contrast detection in luminance and chromatic noise", *Journal of the Optical Society of America*, vol. 9, no. 11, p. 1880-1888, 1992.

[HAN 94] HANGAI S., SUZUKI K., MIHAUCHI K., "Advanced Wsnr for coded monochrome picture evaluation using fractal dimension", *PCS*, p. 92.95, 1994.

[HIS 02] HISTACE A., CAVARO-MENARD C., "Wavelet compression of numerical chest radiographs: a quantitative evaluations of the degradations of reconstructed images", *Proceedings of 16th International Eurasip Conference on Analysis of Biomedical Signals and images (BioSignal 2002)*, Brno, Czech Republic, p. 313-315, June 2002.

[JAC 96] JACKSON W.B., BEEBEE P., JARED D.A., BIEGELSEN D.K., LARIMER J.O., LUBIN J., GILLE J., "X-ray image system design using a human visual model", *Proc. SPIE 2708*, p. 29-40, 1996.

[JOH 99] JOHNSON J.P., LUBIN J., KRUPINSKI E.A., PETERSON H.A., ROEHRIG H., BAYSINGER A., "Visual discrimination model for digital mammography", *Proc. SPIE 3663*, p. 253-263, 1999.

[KON 97] KONING G., BERETTA P.V., ZWART P., HEKKING E., REIBER J.H.C., "Effect of lossy data compression on quantitative coronary measurements", *International Journal of Cardiac Imaging*, vol. 13, p. 261-270, 1997.

[KRU 03] KRUPINSKI E., JOHNSON J., ROEHRIG H., LUBIN J., "Using a human visual system model to optimize soft-copy performance with digital mammography display: influence of display phosphor", *Academic Radiology*, vol. 10, p. 161-166, 2003.

[KUS 01] KUSAYAMA T., HAMAMOTO T., HANGAI S., "A proposal of objective measure considering human visual characteristics", *PCS*, p. 132.135, 2001.

[LEG 80] LEGGE G.E., FOLEY J.M., "Contrast masking in human vision", *Journal of the Optical Society of America*, vol. 70, no. 12, p. 1458-1471, 1980.

[LEG 87] LEGGE G.E., KERSTEN D., BURGESS A.E., "Contrast discrimination in noise", *Journal of the Optical Society of America*, vol. 4, no. 2, p. 391-403, 1987.

[LI 98] LI B., MEYER G.W., KLASSEN R.V., "A comparison of two quality image models", *Human Vision and Electronics Imaging, SPIE*, vol. 3299, p. 98-109, 1998.

[LIN 89] LIN L.I., "A concordance correlation coefficient to evaluate reproducibility", *Biometrics*, vol. 45, p. 255-268, 1989.

[LUB 93] LUBIN J., "The use of psychophysical data models in the analysis of display system performance", in *Digital Images and Human Vision*, A.B. Watson (ed.), p. 163-178. MIT Press, 1993.

[LUB 95] LUBIN J., "A visual discrimination model for imaging system design and evaluation", in *Vision Models for Target Detection and Recognition*, E. Peli (ed.), p. 245-283, World Scientific Publishing, 1995.

[LUK 82] LUKAS F.X.J., BUDRIKIS Z.L., "Picture quality prediction based on a visual model", *IEEE Transactions on Communications*, vol. 30, no. 7, p. 1679-1692, 1982.

[MAN 74] MANNOS J., SAKRISON D., "The effects of visual fidelity criterion on the encoding of images", *IEEE Transactions on Informatics Theory*, vol. 20, no. 4, p. 525-536, 1974.

[MET 89] METZ C.E., "Some practical issues of experimental design and data analysis in radiological ROC studies", *Investigate Radiology*, vol. 24, no. 3, p. 234-245, March 1989.

[MIY 98] MIYAHARA M., KOTANI K., ALGAZI V.R., "Objective picture quality scale (PQS) for image coding", *IEEE Transactions on Communications*, vol. 46, no. 9, p. 1215-1226, September 1998.

[PEL 90] PELI E., "Contrast in complex images", *Journal of Optical Society of America*, vol. 7, no. 10, p. 2032-2040, 1990.

[PEL 93] PELI E., AREND L.E., YOUNG G.M., GOLDSTEIN R.B., "Contrast sensitivity to patch stimuli: Effects of spatial bandwidth and temporal presentation", *Spatial Vision*, vol. 7, no. 1, p. 1-14, 1993.

[PEN 05] PENEDO M., SOUTO M., TAHOCES P.G., CARREIRA J.M., VILLALON J., PORTO G., SEOANE C., VIDAL J.J., BERBAUM K.S., CHAKRABORTY D.P., FAJARDO L.L., "Free-response receiver operating characteristic evaluation of lossy JPEG2000 and object-based set partitioning in hierarchical trees compression of digitized mammograms", *Radiology*, vol. 237, no. 2, p. 450-457, November 2005.

[PHI 84] PHILIPS G.C., WILSON H.R. "Orientation bandwidths of spatial mechanisms measured by masking", *Journal of the Optical Society of America*, vol. 1, no. 2, p. 226-232, 1984.

[PRZ 04] PRZELASKOWSKI A., "Vector quality measure of lossy compressed medical images", *Computers in Biology and Medicine*, vol. 34, p. 193-207, 2004.

[SAK 77] SAKRISON D.J. "On the role of the observer and a distortion measure in image transmission", *IEEE Transactions on Communications*, vol. COM-25, no. 11, p. 1251-1267, 1977.

[SUN 02] SUNG M.M., KIM H.J., YOO S.K., CHOI B.W., NAM J.E., KIM H.S., LEE J.H., YOO H.S., "Clinical evaluation of compression ratios using JPEG2000 on computed radiography chest images", *Journal of Digital Imaging*, vol. 15, no. 2, p. 78-83, June 2002.

[SWI 83] SWIFT D.J., SMITH R.A., "Spatial frequency masking and Weber's Law", *Vision Research*, vol. 23, p. 495.506, 1983.

[TEO 94] TEO P.C., HEEGER D.J., "Perceptual image distortion", *Proceedings ICIP*, p. 982-986, 1994.

[WAT 97] WATSON A.B., BORTHWICK R., TAYLOR M., "Image quality and entropy masking", in *SPIE*, vol. 3016, 1997.

[XU 94] XU W., HAUSKE G., "Picture quality evaluation based on error segmentation", in *SPIE*, vol. 2308, p. 1454-1465, 1994.

[YAM 01] YAMAMOTO S., JOHKOH T., MIHARA N., UMEDA T., AZUMA M., NAKANISHI S., NARUMI Y., NAITO H., NAKAMURA H., "Evaluation of compressed lung CT image quality using quantitative analysis", *Radiation Medicine*, vol. 19, no. 6, p. 321-329, 2001.

[ZET 89] ZETZSHE C., HAUSKE G., "Multiple channel model for the prediction of subjective image quality", in *SPIE*, vol. 1077, p. 209-216, 1989.

[ZHI 05] ZHIGANG L., KUNCHENG L.I., JINGHONG Z., SHULIANG L., "The study of diagnostic accuracy of chest nodules by using different compression methods", *European Journal of Radiology*, vol. 55, no. 2, p. 255-257, August 2005.

Chapter 6

Compression of Physiological Signals

6.1. Introduction

The aim of this chapter is to provide the reader with a general overview of the compression of physiological signals. The specificities of these signals have been covered in Chapter 3; whereas, in this chapter, the EEG compression is discussed and special attention will be given to the ECG signal. This is explained by the fact that the ECG is somehow, more concerned with compression, especially when used for monitoring purposes. Moreover, a huge number of research publications are dedicated to the compression of this type of signal. In fact, some specific and various requirements, i.e. transmission by Internet, wireless transmission, long-term storage on Holter monitors (i.e. for 24 hours or more), make the compression of the ECG an important tool.

For the reasons outlined above, this chapter is organized as follows: in section 6.2, the main standards used for coding the physiological signals are presented. Section 6.3 is dedicated to EEG compression, while in section 6.4, various ECG compression techniques are described.

Chapter written by Amine NAÏT-ALI.

6.2. Standards for coding physiological signals

Unlike the DICOM standard, devoted to medical images, as discussed in Chapter 4, it is important to specify that an exclusive norm or standard does not exist for coding physiological signals. In other words, the few available norms are not systematically accepted by both the European and American communities. Nevertheless, it seems to us to be essential to look at the main existing norms, which are summarized below.

6.2.1. *CEN/ENV 1064 Norm*

This European standard, also known as "SCP-ECG", has been specially developed to code the ECG. Using this norm, coding the signals can be achieved either in compressed mode or non-compressed mode [ENV 96]. It includes the information related to the sampling frequency, filtering as well as other useful specifications. This norm is usually suggested for use with ECG databases. It allows the user to share the same utility software for reading and analyzing the data.

6.2.2. *ASTM 1467 Norm*

This norm is mainly appropriate to neuro-physiological signals such as the electroencephalogram (EEG), evoked potentials (EP), electromyogram (EMG) [ASM 94]. It is also used for monitoring using ECG, gastrointestinal signal, etc. In addition, this norm specifically includes a set of useful clinical information such as:

– sampling frequency;

– channel identification;

– filter parameters;

– electrode positions;

– stimulation parameters;

– types of drugs used during the acquisition process, etc.

6.2.3. *EDF norm*

The EDF (European Data Format) norm was introduced in the last decade by a team of engineers actively working in the field of biomedical engineering [KEM 92]. The format which has been used allows an important flexibility for both exchange and storage of multichannel physiological signals. As in the previous norm, it includes clinical information related to both the patient and the acquisition protocol. This norm is commonly used in Europe and it was extended in 2002 with the proposed format EDF+.

6.2.4. *Other norms*

Finally, we can cite other norms such as the CEN-TC251/FEF [CEN 95], the EBS (extensible biosignal format) for EEGs and EMGs, the SIGIF for neurophysiological signals including the compression option, the MIT arrhythmia database format and finally the DICOM supplement 30, proposed by the DICOM committee following clinician recommendations.

The compression of physiological signals has not been systematically included in the codecs mentioned above. However, it seems obvious that the standardization of compression has now become an important target to be attained.

6.3. EEG compression

6.3.1. *Time-domain EEG compression*

Generally, most of the techniques proposed in the literature devoted to EEG compression are mainly prediction based. This can be explained by the fact that the EEG is a low-frequency signal, which is characterized by a high temporal correlation. Some of these techniques are in fact a direct application of classical digital signal processing methods. For instance, we can point out the Linear Prediction Coding (LPC), the Markovian Prediction, the Adaptive Linear Prediction and Neural Network Prediction based methods. On the other hand, some approaches include the information related to the long-term temporal correlation of the samples. In fact, if we analyze the correlation function of an EEG segment, we will note that spaced samples present a non-neglected correlation that should be taken into account during processing. This information might be integrated into various dedicated codecs. Finally, we can also evoke the techniques which consist of correcting the errors of the prediction using information intrinsic to the EEG. For more details, the reader can refer to the following reference [ANT 97].

6.3.2. *Frequency-domain EEG compression*

The compression of the EEG in the frequency domain did not come from classical techniques such as Karhunen-Loève Transform (KLT) or the Discrete Cosine Transform (DCT). As has already been mentioned, the EEG signal is dominated by low frequencies, mainly lower than 20 Hz. In fact, it is considered that the main energy is located around the α rhythm (between 8 Hz and 13 Hz).

6.3.3. *Time-frequency EEG compression*

Among the time-frequency techniques, the wavelet transform has been commonly used to compress the EEG [CAR 04]. In this technique, the signal is segmented and decomposed using Wavelet Packets. The coefficients are coded afterwards. Other algorithms such as the well known EZW (Embedded Zerotree Wavelet) have also been successfully applied to compress the EEG signal [LU 04]. Even if the obtained results seem significant, we think that the various codecs can be improved by pre-processing the EEG signal by reducing or eliminating the artefacts, which contaminate the EEG.

6.3.4. *Spatio-temporal compression of the EEG*

These approaches have the advantage of combining two aspects. The first aspect consists of taking into consideration the temporal correlation using the techniques pointed out previously, whereas the second aspect includes the spatial correlation due to a multichannel record [ANT 97]. In this method, a lossless compression technique is used.

6.3.5. *Compression of the EEG by parameter extraction*

This last approach is different from the techniques introduced previously. In fact, the EEG is compressed using an uncommon method in the sense that only the main parameters which allow an objective diagnostic are extracted. They can either be stored or transmitted but cannot under any circumstances be used to reconstruct the temporal signal [AGA 01]. However, this approach involves three stages:

– segmentation: this consists of isolating the stationary EEG segment of interest;

– feature extraction: each EEG segment is modeled as a statistical process (AR, MA, ARMA, etc.);

– classification: the analysis of the extracted parameters allows the identification of the different phases of anomalies.

6.4. ECG compression

As pointed out in the introduction to this chapter, the EEG compression field has been somehow less critical than the ECG compression. However, in this section, some ECG compression techniques will be described. Some of them are appropriate for real time transmission, whereas others are more suitable for storage, basically when Holter monitors are used.

6.4.1. *State of the art*

For purposes of clarity, we have gathered in Table 6.1 the most recent research pertaining to the ECG compression field. This has been highlighted, on one hand by specifying the country of each concerned research team working in this field and on the other hand the corresponding methods used. In fact, if we consider the number of articles published over the past six years, we will observe that not less that 20 papers have been published in international journals. This clearly demonstrates the importance of this field of research.

The ECG compression techniques can be classified into three broad categories: direct methods, transform-based methods and parameter extraction methods. In the direct methods category, the original samples are compressed directly. In the transformation methods category, the original samples are transformed and the compression is performed in the new domain. Among the algorithms which employ the transforms, we can hold up several algorithms based on the discrete cosine transform, and the wavelet transform. In the category of the methods using the extraction of parameters, the features of the processed signal are extracted and then used *a posteriori* for the reconstruction.

In this chapter, the scheme (i.e. of three categories) presented above will not be taken as reference. In fact, this section is structured so that the ECG compression techniques dedicated to real time transmission are first presented in section 6.4.4, whereas in section 6.4.5, the techniques designed mainly for storage purpose are described. Both of these techniques will be preceded by two sections describing, on the one hand, the evaluation of the performances (section 6.4.2), and on the other hand, the pre-processing techniques of the ECG signal (see section 6.4.3).

Research team (countries)	Approaches	Year of publication	References
USA	Wavelet packet	2001	[HAN 01]
Jordan	Wavelet transform of the prediction error	2001	[AHM 01]
Jordan	Wavelet transform	2001	[ALS 01]
Brazil	Optimal quantization of the DCT	2001	[BAT 01]
Norway	Minimization of the distortion rate	2001	[NYG 01]
Taiwan	Vectorial quantization	2001	[MIA 01]
Taiwan	SVD	2001	[WEI 01]
Finland	R-R lossless compression	2002	[GIU 02]
Jordan	Wavelet transform	2002	[RAJ 02]
Taiwan	Vectorial quantization	2002	[MIA 02]
USA	JPEG 2000	2003	[BIL 03]
Taiwan	Shape adaptation	2004	[CHE 04]
Spain	Max-Lloyd quantization	2004	[ROD 04]
Finland	"Review"	2004	[KOS 04]
Taiwan	Vectorial quantization-wavelets	2005	[MIA 05]
France	Neural networks, polynomial projection, Hilbert transform, Lorentzian modeling, radon transform, interleaving.	2005-2007	[CHA 05] [BOR 05] [NUN 05] [OUA 07] [NAI 07a] [NAI 07b]

Table 6.1. *Recent research work related to ECG compression*

6.4.2. *Evaluation of the performances of ECG compression methods*

The performance of the proposed algorithm is evaluated using the Compression Ratio (CR) and the Percent Root-Mean-Square Difference (PRD) in % which is commonly used to measure the distortion resulting from ECG compression.

These two definitions are given by equations [6.1] and [6.2] respectively.

$$CR = \frac{N_x}{N_{\hat{x}}} \qquad [6.1]$$

where:

N_x denotes the number of bits used to code the original signal;

$N_{\hat{x}}$ denotes the number of bits used to code the reconstructed signal.

$$PRD = \sqrt{\frac{\sum_{n=1}^{N}(x(n)-\hat{x}(n))^2}{\sum_{n=1}^{N}x^2(n)}}.100 \qquad [6.2]$$

where:

$x(n)$ is the original signal to be compressed, recorded on N samples;

$\hat{x}(n)$ represents the reconstructed signal, recorded on N samples.

It is also important to point out that the PRD does not provide a significant evaluation, especially when the DC-component is included in the calculation; for more information, see [ALS 03]. In addition, even if the PRD and CR are considered as two important criteria for the evaluation of a given ECG compression technique, it is important to take into consideration other significant parameters, i.e. the calculation complexity for both coding and decoding as well as the robustness of the technique with respect to noise. In addition, we must specify whether the technique is more appropriate for real-time transmission or for storage.

6.4.3. *ECG pre-processing*

Based on the above comments, the first ECG pre-processing consists of suppressing the DC-component. This component can be added at the reception during the reconstruction process. A second important pre-processing technique consists of segmenting the ECG in order to isolate the different beats before the coding phase. In this case, the ECG signal is processed segment by segment. However, it is also important to know that if the segmentation is performed by a non-supervised approach, a border effect can be observed after the reconstruction phase.

Several ECG segmentation techniques have been studied in the literature for purposes of either compression or classification. The reader can refer for instance to various techniques using hidden Markov chains, wavelets [CLA 02] or genetic algorithms [GAC 03].

6.4.4. *ECG compression for real-time transmission*

In this section we present a certain number of techniques, classified according to their domain of processing, i.e. the time-domain or the frequency domain. These approaches are mainly based on parametrical modeling. The quantization and entropic coding will not be detailed in this chapter since these classical functionalities should be integrated into the final compression scheme. However, the time-domain approaches are described in section 6.4.4.1, while the frequency-domain approaches are presented in section 6.4.4.2.

6.4.4.1. *Time domain ECG compression*

6.4.4.1.1. Gaussian modeling of the ECG beat

If we consider $x(1)$, $x(2),\ldots,$ $x(N)$, N samples of a measured ECG beat, denoted by x(n); modeling this signal using a sum of M Gaussians consists of approximating it by a set of Gaussians allowing the best fitting in the sense of least squares.

The model is defined as follows:

$$\hat{x}(n)=\sum_{i=1}^{M} A_i.exp\left[\frac{-\left(n-\mu_i\right)^2}{\sigma_i^2}\right] \qquad i=1\ldots M \text{ and } n=1\ldots N \qquad [6.3]$$

where A_i is the amplitude, μ_i (points along a temporal scale) is the mean value and σ_i is the standard deviation of Gaussian i.

Since, it is not evident to explicitly calculate the parameters of the model, this could be fitted using a non-linear least squares optimization technique.

It is clear that the choice of the Gaussian model is intuitive in the sense that the shape of each wave constituting a normal ECG beat can be approximated by a Gaussian. Therefore, in order to identify the parameters of equation 6.3, several optimization techniques can be used. For example, we can use any classical optimization technique suited to non-convex criteria, including the metaheuristic approaches. Of course, it is well known that these methods are time consuming, but if we consider that the ECG is a low frequency signal (i.e. one beat per second on average) and that the compression of the kth beat is achieved during the acquisition of the $(k+1)^{th}$ beat, the processing time generally becomes sufficient. In fact, some specific processors dedicated to Digital Signal Processing, like DSPs (Digital Signal Processors) or FPGAs (Field Programmable Gate Fields), can meet the real time requirements.

Let us reconsider our equation 6.3 for which we have to identify the parameters described above. These parameters can be determined simply by minimizing the following criterion:

$$J = \sum_{n=1}^{N} \left\{ \left[x(n) - \hat{x}(n) \right]^2 \right\} \quad [6.4]$$

This can also be expressed by:

$$J = \sum_{n=1}^{N} \left\{ \left[x(n) - \sum_{i=1}^{M} A_i . exp \left[\frac{-(n-\mu_i)^2}{\sigma_i^2} \right] \right]^2 \right\} \quad [6.5]$$

It is also evident that the number of Gaussians (*M*) to be used is of great importance. If *M* is under-estimated, the quality of the reconstructed signal decreases, whereas if the *M* is over estimated, the calculation increases. Thus, we can for example use some information criteria to identify the optimal order.

In practice, an empirical approach can be used to determine the most appropriate order. For example, experiences on normal and abnormal (for instance PVCs) beats show that an order of 5 or 6 can be suitable for modeling each ECG beat. It is also obvious that if the recorded ECG contains some specific significant high frequencies, we have to increase the order to fit the signal properly.

Figures 6.1 and 6.2 represent two typical ECG beats (normal/PVC) as well as their corresponding parametrical models. The PVC beat is reconstructed using 5 Gaussians, whereas the normal ECG beat is obtained using 6 Gaussians.

Using these approaches, the compression ratio depends of course on the following parameters:

– *M*: number of Gaussians (model order) used to reconstruct an ECG beat;

– N_e: number of samples of a single ECG beat (this depends on the frequency sampling);

– B_e: number of bits used for coding each ECG beat;

– B_p: number of bits used for coding the parameters (A_i, μ_i , σ_i) of the *M* Gaussians.

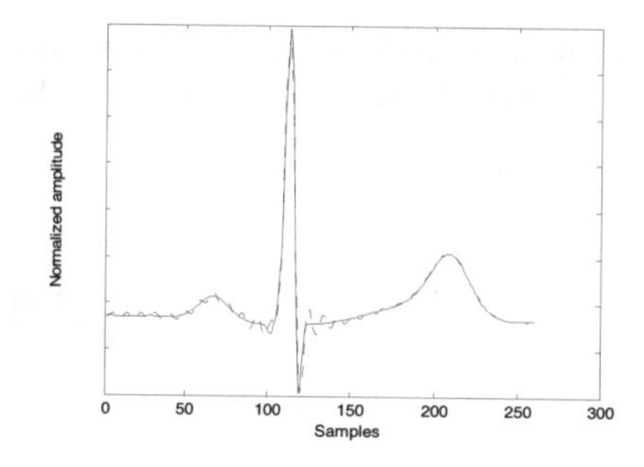

Figure 6.1. *[-] normal ECG beat (original signal); [..] normal ECG beat reconstructed using 6 Gaussians*

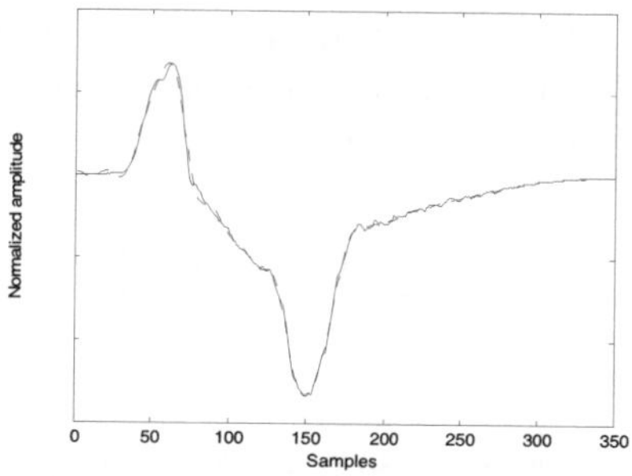

Figure 6.2. *[-] PVC (original signal); [..] PVC reconstructed using 5 Gaussians*

The compression ratio (without taking into account the quantization and the entropic coding) is given by:

$$CR = \frac{N_e B_e}{3.M.B_p} \qquad [6.6]$$

When the model order is invariant (i.e. does not vary dynamically from one beat to another), the transmission mode is performed in a fixed bitrate. In addition, using this mode of transmission, we should transmit parameter N_e in each frame. This is dependent of course on the duration of each segment and the duration depends on the cardiac rhythm, (i.e. long durations for bradycardia and short durations for tachycardia). Furthermore, the parameters in each frame are considered to be coded using a fixed number of bits. For a static model, we have to note that parameters M and B_p should be included only in the first frame.

By compressing the ECG beats without taking into account the redundancy intra-beats, experiments show that on average, a compression ratio of 15 can be achieved. In fact, this performance can be improved if we include the information related to the redundancy. For instance, this can occur when an arrhythmia signal is recorded.

The validation of the ECG compression techniques is generally achieved using international databases, such as the following:

– MIT-BIH Arrhythmia Database;

– MIT-BIH Atrial Fibrillation;

– MIT-BIH Long-Term Database;

– MIT-BIH Noise Stress Test Database.

If we explore the articles dedicated to ECG compression, in most of cases, the MIT-BIH Arrhytmia Database is the most widely-used. This database contains 234 signals where each one is identified by a single reference.

6.4.4.1.2. ECG compression using the deconvolution principle

In this approach, we consider that each ECG beat is the impulsional response of a linear system (see Figure 6.3). When the ECG beats are stationary (i.e. no modification of the shape), the deconvolution process provides an impulsional signal. In such a situation, we have to transmit only the position and the amplitude of each impulse. Therefore, in order to reconstruct the ECG signal at reception, a simple convolution is performed. This problem is considered as an inverse problem for which several approaches can be used to solve it (see Figure 6.4).

Generally, the ECG beats are not stationary and the recorded signal is not periodic. In addition, the shape of each beat can change with time. Since the proposed approach is still under consideration, we will attempt, in this chapter to present the principles of this technique as well as some preliminary results.

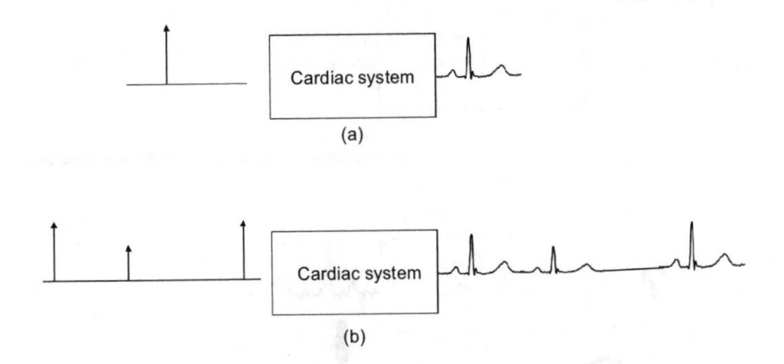

Figure 6.3. *(a) Model to generate an ECG beat; (b) generating an ECG signal using the same principle*

Let us consider that $x_1(n), x_2(n)...x_N(n)$ represent M ECG beats of an ECG signal. Each ith beat is expressed by a vector $\mathbf{x}_i$.

The first beat is considered as a reference. It is denoted $\mathbf{x}_m$:

$$\mathbf{x}_m = \mathbf{x}_1 \qquad [6.7]$$

Thus, each beat $\mathbf{x}_i$ i=2,...M is considered obtained from the convolution between an impulsional signal $\mathbf{z}_i$ and the reference signal $\mathbf{x}_m$.

This can be expressed as follows:

$$\mathbf{x}_i = \mathbf{X}_m \mathbf{z}_i + \mathbf{b}_i \qquad [6.8]$$

where $\mathbf{X}_m$ is a matrix having a Toeplitz structure, obtained from $\mathbf{x}_m$ and $\mathbf{b}_i$ is a zero-mean Gaussian noise vector.

The problem then consists of identifying $\mathbf{z}_i$. In fact, the more significant the correlation is between $\mathbf{x}_i$ and $\mathbf{x}_m$, the more $\mathbf{z}_i$ converges to an impulsional signal. The idea of the compression then consists of concentrating the information of a given ECG beat in a pseudo-impulse.

From equation [6.8], $\mathbf{z}_i$ can be estimated using the least squares by minimizing the Euclidian distance. In this case, the solution is given by:

$$\hat{\mathbf{z}}_i = \arg\min \left\{ \left\| \mathbf{x}_i - \mathbf{X}_m \mathbf{z}_i \right\|^2 \right\} \qquad [6.9]$$

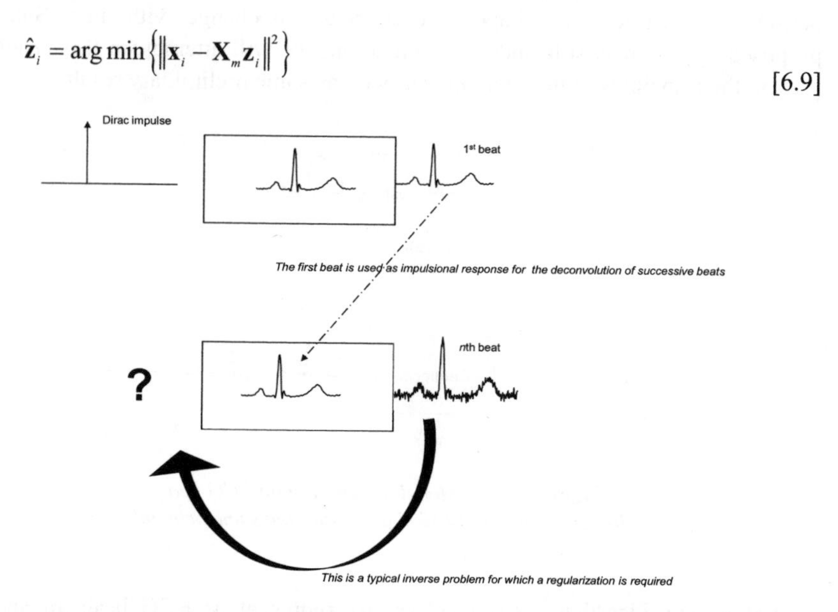

Figure 6.4. *Using the first ECG beat as an impulse response of the cardiac system, each recorded beat is deconvolved using the reference beat in order to estimate the pseudo-impulse (this is a typical inverse problem which necessitates regularization)*

Figure 6.5 represents an original ECG signal. Using the first beat as a reference, the deconvolution of this signal, represented in Figure 6.6, clearly shows the impulsional aspect. In the next step, a simple thresholding of the low amplitudes is achieved as shown in Figure 6.7. This allows us to increase the number of zero values. For transmission purposes, only the non-zero samples are coded.

On reception, the ECG signal is reconstructed using a simple convolution with the reference beat (see Figure 6.8). It is obvious that for such a transmission mode, the reference signal should be transmitted before transmitting the impulsional signal.

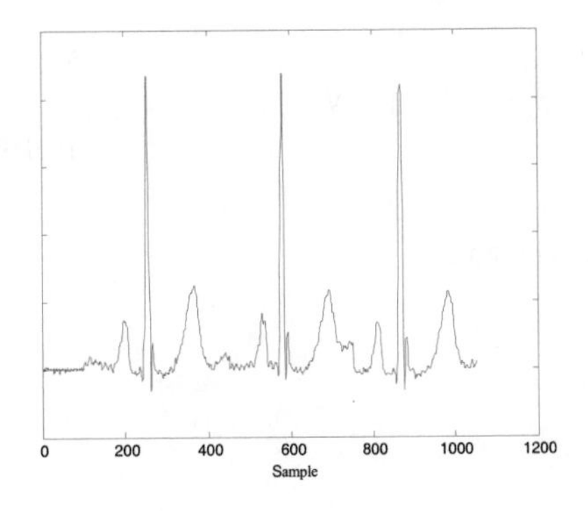

Figure 6.5. *Original ECG signal*

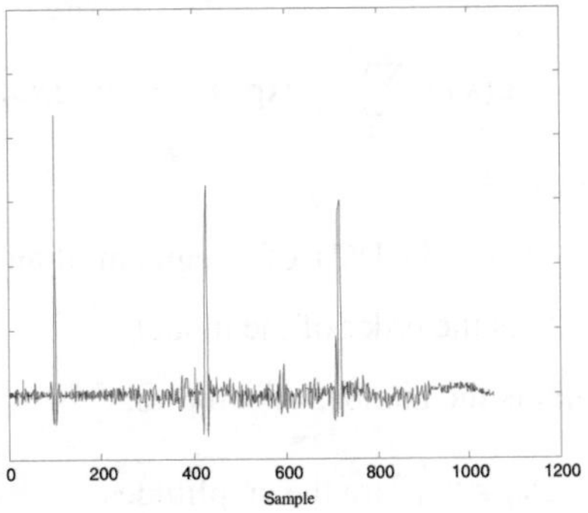

Figure 6.6. *Signal obtained after deconvolution using the first beat*

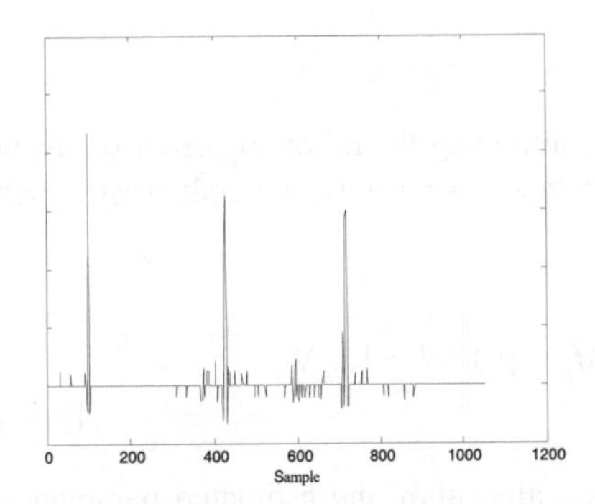

Figure 6.7. *Thresholding of the deconvolved signal*

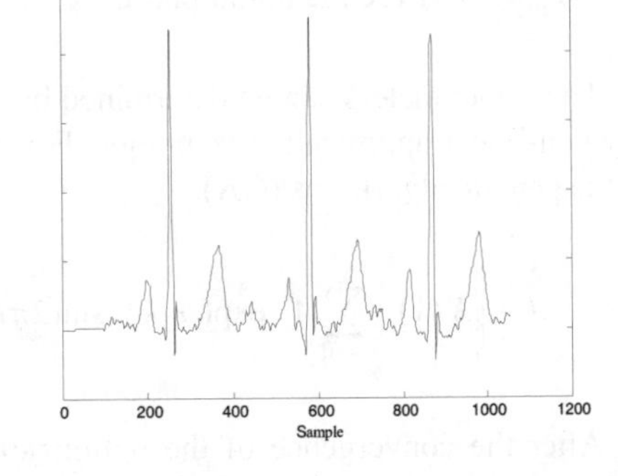

Figure 6.8. *Reconstruction of the ECG signal using the thresholded impulsional signal*

6.4.4.2. *Compression of the ECG in the frequency domain*

As presented in section 6.4.4.1, the proposed approaches are based on the parametrical modeling of the ECG signal in the time domain. In this section, the idea consists of modeling the transform of the ECG signal such as the DCT or the Fourier transform.

In Figures 6.9 and 6.10, we represent respectively the DCT of a normal ECG beat and the DCT of the PVC. We can observe that the obtained curves have a damped sinusoid aspect. Thus, the idea consists of modeling them mathematically as follows:

$$X(k) = \sum_{m=1}^{M} A_m \exp(\alpha_m k) \sin(2\pi k f_m + \phi_m) \qquad k = 1,...N \qquad [6.10]$$

where:

– X(k) is the DCT of a segment of an ECG signal;

– M is the order of the model;

– N is the number of samples;

– $A_m \in \Re_+^*$ are the amplitudes;

– α_m are the damping factors;

– f_m are the frequencies;

– $\phi_m \in [-\pi,\pi]$ are the initial phases;

These parameters can be determined by minimizing the following criterion using any non-linear optimization technique. For instance, we can use a metaheuristic such as the genetic algorithms (GA).

$$J = \left\| X(k) - \sum_{m=1}^{M} A_m \exp(\alpha_m k) \sin(2\pi k f_m + \phi_m) \right\|^2 \quad k = 1,...N \qquad [6.11]$$

After the convergence of the optimization algorithm, the estimated parameters are used to reconstruct the DCT model (Figure 6.11). Using the inverse Discrete Cosine Transform, we obtain the temporal ECG signal as shown in Figure 6.12.

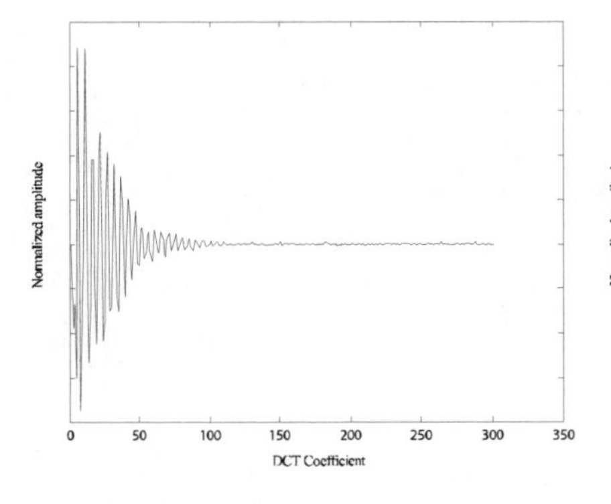

Figure 6.9. *DCT of a normal ECG beat*

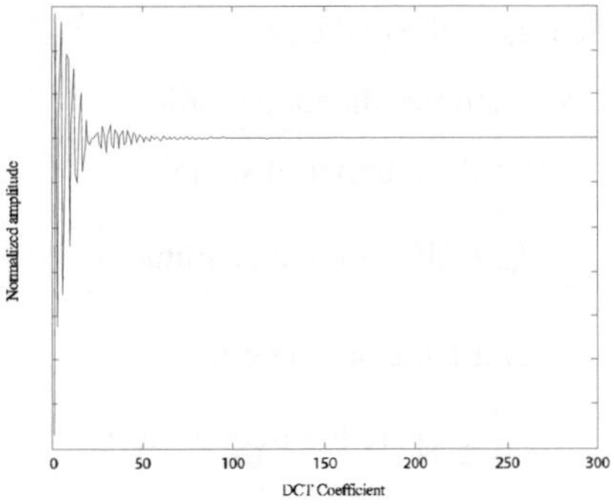

Figure 6.10. *DCT of a PVC*

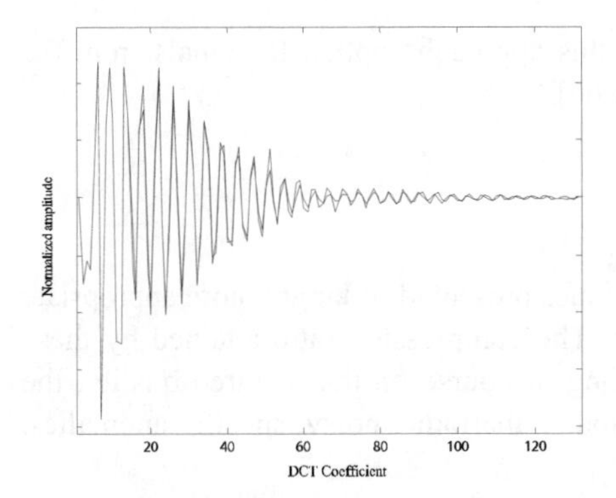

Figure 6.11. *[-] original DCT; [..] reconstructed DCT using 12 damped sinusoids*

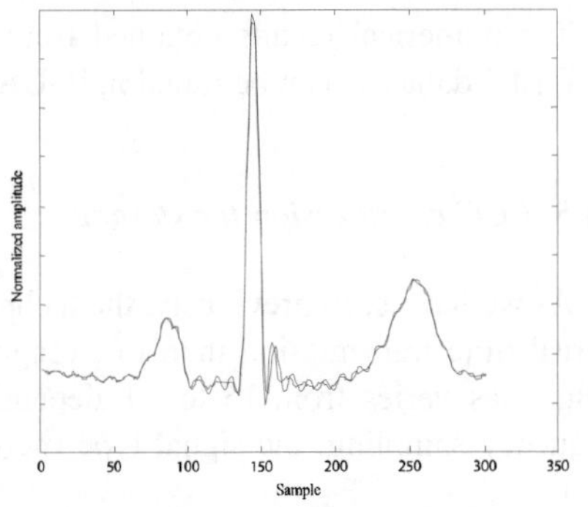

Figure 6.12. *[-] original ECG beat; [..] reconstructed ECG beat*

On the other hand, when calculating the Fourier transform of an ECG beat, we show that the curves corresponding to both the real part and the imaginary one, present as previously a damped sinusoid aspect. In this case, the problem is processed in the complex-domain without using any global optimization technique. In fact, some methods such as Prony are very appropriate for such situation.

The Discrete Fourier Transform of a given ECG signal is modeled as follows:

$$X(k) = \sum_{m=1}^{M} A_m \exp(\mathrm{j}\phi_{\mathrm{m}}) \exp(\alpha_{\mathrm{m}} k + j2\pi k f_m) \qquad k = 1,...N \qquad [6.12]$$

where as in the real case:

– M denotes the model order;

– N is the number of samples;

– $A_m \in \Re_+^*$ are the amplitudes;

– α_m are the damping factors;

– $f_m \in \left[-\frac{1}{2}, \frac{1}{2}\right]$ are the frequencies;

– $\phi_m \in [-\pi, \pi]$ are the initial phases.

The problem consists of identifying the set of parameters $\left\{A_m, f_m \alpha_m, \phi_m,\right\}_{m=1}^{M}$.

The numerical results obtained from this approach applied to signals from the MIT-BIH database can be found in [OUA 06].

6.4.5. *ECG compression for storage*

As we have seen previously, the techniques presented so far are more appropriate to real time transmission than to storage. The compression ratio attained by these techniques varies from 15 to 20, depending of course on the required quality, the frequency sampling, the signal type (periodic, aperiodic, noisy, specific anomalies, etc.).

In this section, the presented approaches have been developed specifically for ECG storage. Thus, we will show how we can achieve very impressive compression ratios (i.e. 100 or more).

The first approach is presented in the section 6.4.5.1 and is based on the synchronization and the polynomial modeling of the dynamic of the ECG beats, whereas the second approach is based on the principle of the synchronization and interleaving (see section 6.4.5.2). Finally, an ECG compression technique that uses the standard JPEG 2000 will be presented in section 6.4.5.3.

When using these three techniques, the first step requires the separation of the ECG beats using any appropriate segmentation algorithm. Since the ECG signal is basically not periodic, the different segmented beats do not necessarily have the same duration. Therefore, in order to gather the set of ECG beats in a same matrix $\mathbf{X}$ so that each line contains one ECG beat, it is essential to perform an extrapolation of the different segments in order to equalize the durations. The variable N denotes here the size of the larger segment.

$$\mathbf{X} = \begin{bmatrix} x_1(0) & x_1(1) & \cdots & x_1(N-1) \\ x_2(0) & x_2(1) & \cdots & x_2(N-1) \\ \vdots & \vdots & \ddots & \vdots \\ x_M(0) & x_M(1) & \cdots & x_M(N-1) \end{bmatrix} = \begin{bmatrix} \mathbf{x}_1^t \\ \mathbf{x}_2^t \\ \vdots \\ \mathbf{x}_M^t \end{bmatrix} \qquad [6.13]$$

It is also important to be aware that the beats included in matrix $\mathbf{X}$ are highly correlated to each other. However, in order to minimize the fast transitions (i.e. high frequencies), the beats should be aligned (i.e. synchronized). In other words, we have to reduce the high frequencies which can occur between successive beats as shown respectively in Figures 6.13 and 6.14 representing one column of matrix $\mathbf{X}$).

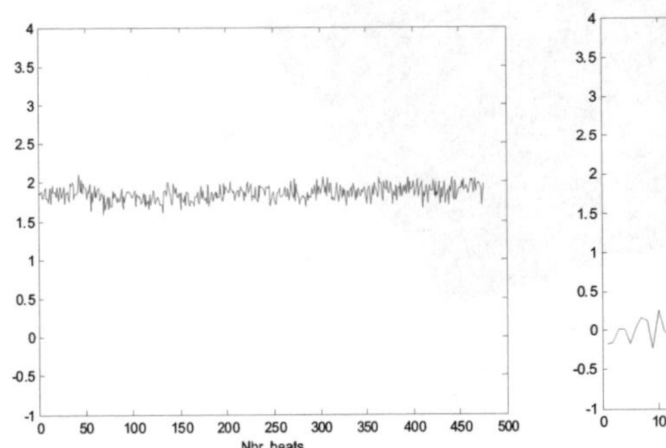

Figure 6.13. *Low frequencies after synchronizing the ECG beats*

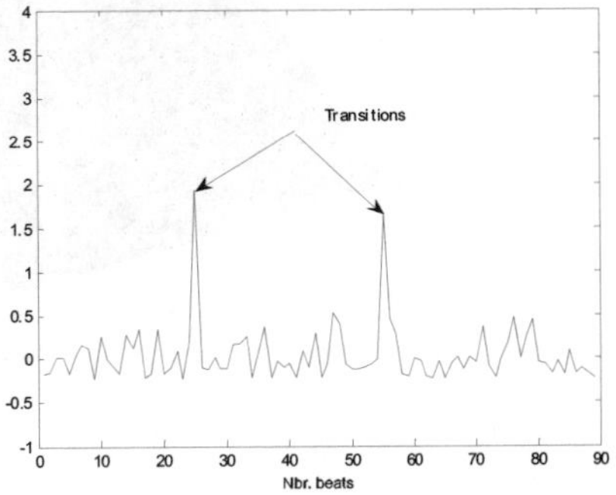

Figure 6.14. *High frequencies for non-synchronized ECG beats*

Synchronization can be obtained easily using some basic correlation techniques.

6.4.5.1. *Synchronization and polynomial modeling*

The transform which leads to matrix $\mathbf{X}$ allows us to represent the ECG signal as an image. The non-stationarity of the ECG generally due to arrhythmias leads to a desynchronized surface as depicted in Figure 6.15. Since for an abnormal ECG signal, the shape of some beats might be time-variant, it is then essential to include a pre-processing phase which consists of gathering similar shapes in the same matrix. For example, suppose an ECG signal contains both normal beats and PVCs. In such a situation, the original matrix should be decomposed into two matrices so that each matrix contains only one specific beat type. The beats in each matrix should be synchronized; see Figure 6.16 for normal beats and Figure 6.17 for PVCs.

Now, since each row of the obtained matrixes contain a single beat, the processing technique consists of projecting each column on a polynomial basis as follows:

$$\hat{x}(m,n) = a(n)_0 + a(n)_1.m + a(n)_2.m^2 + ...a(n)_p.m^p \quad [6.14]$$

$$m = 1,...M \quad n = 0,1,...N-1$$

where $a(n)_i$ represents the ith coefficient of the p order polynomial at instant n.

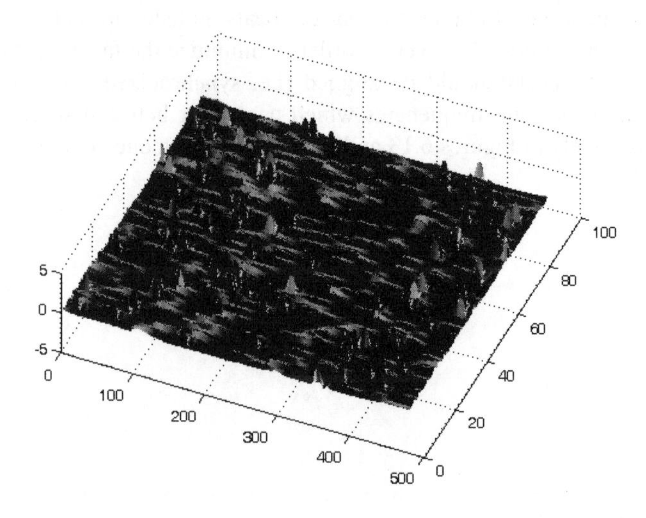

Figure 6.15. *Image representation of the ECG signal. Each row contains one ECG beat. Since the signal is not periodic, the obtained surface is desynchronized*

These coefficients can be easily estimated using a least squares criterion:

$$J = \left\| x(m,n) - \hat{x}(m,n) \right\|^2 \quad [6.15]$$

Using a vectorial representation, the coefficients of each polynomial corresponding to the nth column are estimated by:

$$\hat{\mathbf{a}}_n = \arg\min \left\{ \left\| \mathbf{x}_n - \hat{\mathbf{x}}_n \right\|^2 \right\} \quad [6.16]$$

where:

$$\hat{\mathbf{a}}_n = \begin{bmatrix} a_0 & a_2 & \cdots & a_{p-1} & a_p \end{bmatrix}^t$$

$$\mathbf{x}_n = \begin{bmatrix} x(1,n) & x(2,n) & \cdots & x(M,n) \end{bmatrix}^t$$

$$\widehat{\mathbf{x}}_n = \left[\widehat{x}(1,n) \quad \widehat{x}(2,n) \quad \cdots \quad \widehat{x}(M,n)\right]^t$$

The final solution is given by:

$$\hat{\mathbf{a}}_n = \left[\mathbf{C}^t\mathbf{C}\right]^{-1}\mathbf{C}^t\mathbf{x}_n \qquad [6.17]$$

where $\mathbf{C}$ is the classical Vandermonde matrix, expressed as follows:

$$\mathbf{C} = \begin{bmatrix} 1 & 0 & \cdots & 0 \\ 1 & 1 & \cdots & 1^p \\ \vdots & \vdots & \ddots & \vdots \\ 1 & N-1 & \cdots & (N-1)^p \end{bmatrix}$$

The polynomial $\hat{\mathbf{x}}_n$ is calculated using the following equation:

$$\hat{\mathbf{x}}_n = \mathbf{C}\,\hat{\mathbf{a}}_n \qquad [6.18]$$

By projecting matrix $\mathbf{X}$ on the same polynomial basis, another matrix of coefficients, denoted $\mathbf{A}$ is derived. This is given by:

$$\mathbf{A} = \begin{bmatrix} a_{10} & a_{11} & \cdots & a_{1(N-1)} \\ a_{20} & a_{21} & \cdots & a_{2(N-1)} \\ \vdots & \vdots & \ddots & \vdots \\ a_{M0} & a_{M1} & \cdots & a_{M(N-1)} \end{bmatrix} = \left[\mathbf{a}_1 \mid \mathbf{a}_2 \mid \mathbf{a}_3 \mid \ldots \mid \mathbf{a}_N\right] \qquad [6.19]$$

From this equation, we can note that the size of the matrix is $\mathbf{A}$ $p \times N$ whereas it is $M \times N$ for matrix $\mathbf{X}$. Therefore, we can only store matrix $\mathbf{A}$ with the possibility of estimating the elements of matrix $\mathbf{X}$ allowing of course the reconstruction of the ECG signal.

The compression ratio obtained by this technique depends of course on the polynomial order as well as the number of samples of the longest segment.

As mentioned previously, when the ECG beats are synchronized, the variance in each column is considerably small. This also leads to a reduced polynomial order. In Figure 6.18, we show one column of a matrix composed of 480 ECG beats, projected on a six order polynomial basis. In fact, these six coefficients can reproduce up to "infinity" the tendency of whatever the number of ECG beats.

Consequently, high compression ratios can be achieved with this technique. On the other hand, when the shape of the beats change, the approach becomes less interesting as shown in Figure 6.19. In fact, the reduced polynomial order cannot reproduce some abrupt changes. Therefore, a classification (as mentioned previously) of beats becomes in this case the most appropriate solution in order to overcome this disadvantage.

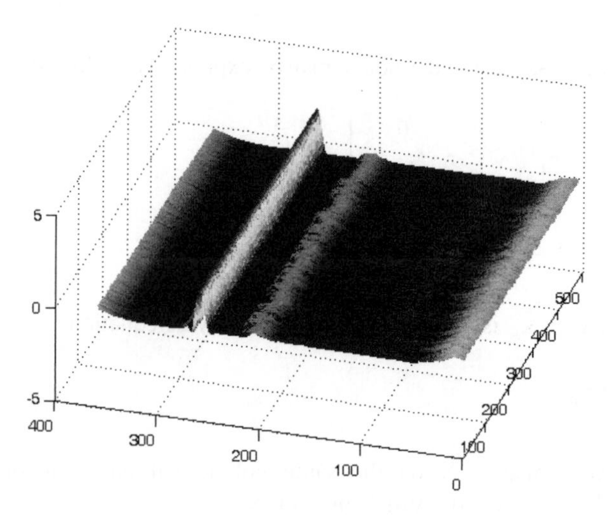

Figure 6.16. *The alignment of the ECG beats reduces the high frequencies*

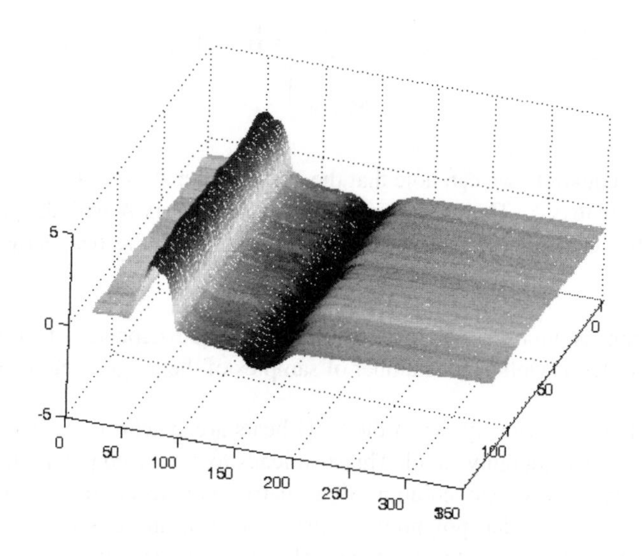

Figure 6.17. *Alignment of PVCs*

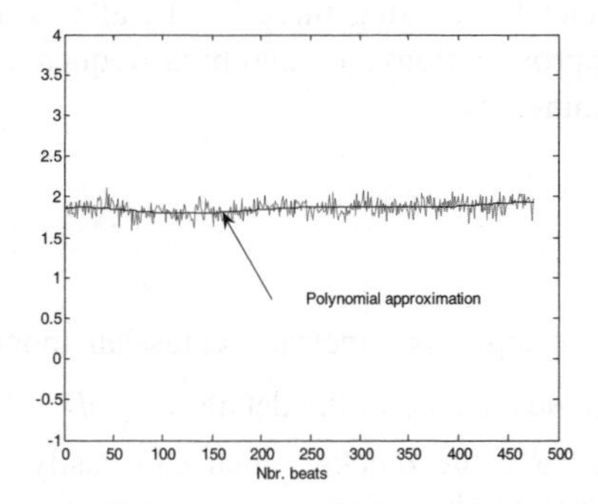

Figure 6.18. *Polynomial modeling of the low frequencies after ECG beat synchronization*

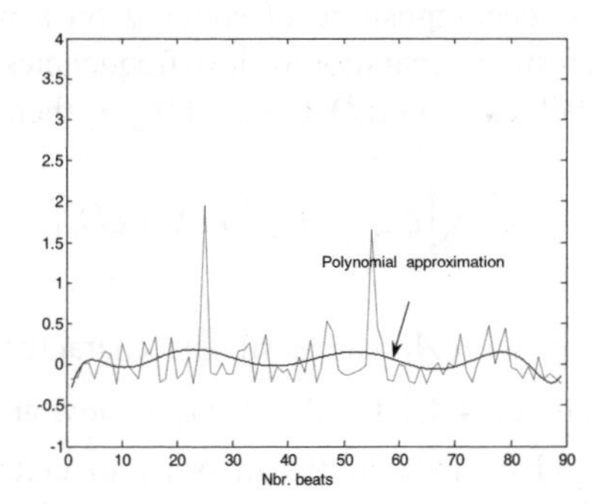

Figure 6.19. *Polynomial modeling with a reduced order (non-synchronized case)*

6.4.5.2. *Synchronization and interleaving*

This technique is simple in the sense that it uses matrix **X** [6.13] after synchronizing the ECG beats (in each row). However, an interleaving of these rows is then achieved in order to obtain a single signal denoted **z**, having the following structure:

$$\mathbf{z}=\left[x_1(0)\mid x_2(0)\mid\cdots\mid x_M(0)\mid\quad\mid x_1(N-1)\mid x_2(N-1)\mid\cdots\mid x_M(N-1)\right]^t \qquad [6.20]$$

Since the ECG beats have correlated shapes, the resulting signal **z** has a $M\times N$ size. Therefore, the shape of the obtained signal becomes very close to that of a single beat (Figure 6.20). Signal **z** seems to be noisy due to the variations between each beat. At this stage of processing, no compression is performed since it is only a transformation matrix-vector.

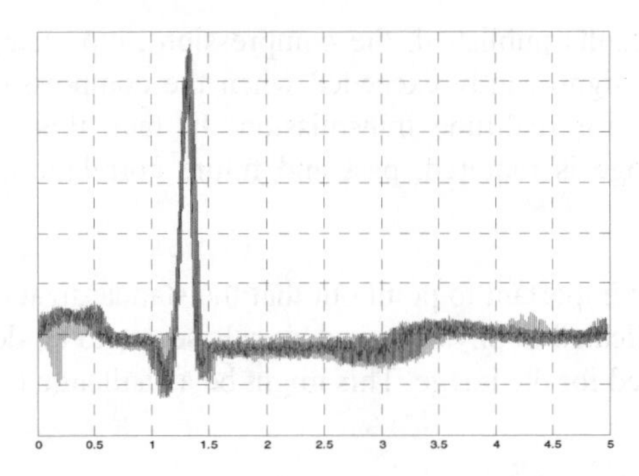

Figure 6.20. *Signal obtained after ECG beat interleaving*

A decomposition of signal $\mathbf{z}$ on a wavelet basis using three levels, allows an interesting separation of low frequencies (approximations cA_3) and high frequencies (details cD_3, cD_2, cD_1). Signal $\mathbf{z}_w$ is then obtained by:

$$\mathbf{z}_w = \left[cA_3 \mid cD_3 \mid cD_2 \mid cD_1 \right]^t \qquad [6.21]$$

Signal cA_3 can easily be characterized using a parametrical Gaussian model (section 6.4.4.1.1). Moreover, the low amplitudes related to the details cD_3, cD_2, cD_1 should be thresholded in order to create zero value blocks which can easily be coded. The numerical results obtained by applying those approaches on real signals are presented in [NAI 06].

6.4.5.3. *Compression of the ECG signal using the JPEG 2000 standard*

The most recent techniques for compressing the ECG signal consist of transforming the signal to an image before the coding operation. However, the standard JPEG 2000, initially dedicated to compressing images, has been successfully used to compress ECG signals. As in the previous technique, a preprocessing technique is required which consists of synchronizing the beats initially stored in a matrix **X**. Decoding the data requires of course the use of the parameters (delays) used for synchronization purposes; for more details about this technique, see [BIL 03].

6.5. Conclusion

Throughout the course of this chapter, several recent compression techniques basically dedicated to the ECG signal have been presented. The choice of method of course depends on the condition of recording.

Based on the results published, the compression ratio changes from 15 to 20. These rates can be significantly exceeded when the compression is performed for storage rather than for real-time transmission. In fact, this seems to be logical because when storage is required, past and future correlations can be taken into account.

Finally, it seems important to point out that the standardization of the techniques of compression dedicated to physiological signals seems to be delayed in relation to the progress achieved for the image. This might be a challenge for the future.

6.6. Bibliography

[AGA 01] AGARVAL R., GOTMAN J., "Long-term EEG compression for intensive-care setting", *IEEE Eng. in Med. and Biol.*, vol. 9, p. 23-29, 2001.

[AHM 01] AHMEDA S., ABO-ZAHHAD M., "A new hybrid algorithm for ECG signal compression based on the wavelet transformation of the linearly predicted error", *Med. Eng. Phys.*, vol. 23, p. 117-26, 2001.

[ALS 01] ALSHAMALI A., AL-SMADI A., "Combined coding and wavelet transform for ECG compression", *J. Med. Eng. Technol.*, vol. 25, p. 212-216, 2001.

[ALS 03] ALSHAMALI A., ALFAHOUM A., "Comments on an efficient coding algorithm for the compression of ECG signals using the wavelet transform", *IEEE Trans. Biomed. Eng.*, vol. 50, p. 1034-1037, 2003.

[ANT 97] ANTONIOL G., TONNELA P., "EEG data compression techniques", *IEEE Eng. in Med. and Biol.*, vol. 44, 1997.

[ASM 94] ASTM E1467-94, American Society for Testing and Materials, STM, (www.astm.org), 1916 Race St., Philadelphia, PA 19103, USA, 1994.

[BAT 01] BATISTA L., MELCHER E., CARVALHO L., "Compression of ECG signals by optimized quantization of discrete cosine transform coefficients", *Med. Eng. Phys.*, vol. 23, p. 127-34, 2001.

[BIL 03] BILGIN A., MARCELLIN M., ALTBACH M., "Compression of ECG Signals using JPEG2000", *IEEE Trans. on Cons. Electr.*, vol. 49, p. 833-840, 2003.

[BOR 05] BORSALI R., NAIT-ALI A., "ECG compression using an ensemble polynomial modeling: comparison with the wavelet based technique", *Biomed. Eng.*, Springer-Verlag, vol. 39, no. 3, p. 138-142, 2005.

[CAR 04] CARDENAS-BARRERA J., LORENZO-GINORI J., RODRIGUEZ-VALDIVIA E., "A wavelet-packets based algorithm for EEG signal compression", *Med. Inform. Intern. Med.*, vol. 29, p. 15-27, 2004.

[CEN 95] File Exchange Format for Vital Signs, Interim Report, Revision 2, TC251 Secretariat, Stockholm, Sweden, CEN/TC251/PT-40, 2000.

[CHA 05] CHATTERJEE A., NAIT-ALI A., SIARRY P., "An Input-Delay Neural Network Based Approach For Piecewise ECG signal compression", *IEEE Trans. Biom. Eng.*, vol. 52, p. 945-947, 2005.

[CHE 04] CHEN W., HSIEH L., YUAN S., "High performance data compression method with pattern matching for biomedical ECG and arterial pulse waveforms", *Comp. Method. Prog. Biomed.*, vol. 74, 2004.

[CLA 02] CLAVIER L., BOUCHER J.-M., LEPAGE R., BLANC J.-J., CORNILY J.-C., "Automatic P-wave analysis of patients prone to atrial fibrillation", *Med. Biol. Eng. Comput.*, vol. 40, p. 63-71, 2002.

[ENV 96] ENV 1064 Standard communications protocol for computer-assisted electrocardiography, European Committee for Standardisation (CEN), Brussels, Belgium, 1996.

[GAC 03] GACEK A., PEDRYCZ W., "A genetic segmentation of ECG signals", *IEEE Trans. Biomed. Eng.*, vol. 50, p. 1203-1208, 2003.

[GIU 02] GIURCANEANU C., TABUS I., MEREUTA S., "Using contexts and R-R interval estimation in lossless ECG compression", *Comput. Meth. Prog. Biomed.*, vol. 67, p. 177-86, 2002.

[HAN 01] HANG X., GREENBERG N., QIN J., THOMAS J., "Compression of echocardiographic scan line data using wavelet packet transform", *Comput. Cardiol.,* vol. 28, p. 425-7, 2001.

[KEM 92] KEMP B., VÄRRI A., ROSA A.-C., NIELSEN K.-D., GADE J., "A simple format for exchange of digitized polygraphic recordings", *Electroencephalogr. Clin. Neurophysiol.*, vol. 82, p. 391-393, 1992.

[KOS 04] KOSKI A., TOSSAVAINEN T., JUHOLA M., "On lossy transform compression of ECG signals with reference to deformation of their parameter values", *J. Med. Eng. Technol.*, vol. 28, p. 61-66, 2004.

[LU 04] LU M., ZHOU W., "An EEG compression algorithm based on embedded zerotree wavelet (EZW)", *Space Med. Eng.*, vol. 17, p. 232-234, 2004.

[MIA 01] MIAOU S. YEN H., "Multichannel ECG compression using multichannel adaptive vector quantization", *IEEE Trans Biomed. Eng.*, vol. 48, p. 1203-1209, 2001.

[MIA 02] MIAOU S., LIN C., "A quality-on-demand algorithm for wavelet-based compression of electrocardiogram signals", *IEEE Trans Biomed. Eng.*, vol. 49, p. 233-9, 2002.

[MIA 05] MIAOU S., CHAO S., "Wavelet-based lossy-to-lossless ECG compression in a unified vector quantization framework", *IEEE Trans Biomed. Eng.*, vol. 52, p. 539-543, 2005.

[NAI 07a] NAÏT-ALI A., BORSALI R., KHALED W., LEMOINE J., "Time division multiplexing based-method for compressing ECG signals: application for normal and abnormal cases", *Journal of Med. Eng. and Tech.*, Taylor & Francis, vol. 31, no. 5, p. 324-331, 2007.

[NAI 07b] NAÏT-ALI A., "A New Technique for Progressive ECG Transmission using Discrete Radon Transform", *Int. Jour. Biomedical Sciences*, vol. 2, pp. 27-32, 2007.

[NUN 05] NUNES J.-C., NAIT-ALI A., "ECG compression by modeling the instantaneous module/phase of its DCT", *Journal of Clinical Monitoring and Computing, Springer*, vol. 19, no. 3, p. 207-214, 2005.

[NYG 01] NYGAARD R., MELNIKOV G., KATSAGGELOS A., "A rate distortion optimal ECG coding algorithm", *IEEE Trans Biomed. Eng.*, vol. 48, p. 28-40, 2001.

[OUA 07] OUAMRI A., NAIT-ALI A., "ECG compression method using Lorentzian functions Model", *Digital Signal Processing*, vol. 17, p. 319-326, 2007.

[RAJ 02] RAJOUB B., "An efficient coding algorithm for the compression of ECG signals using the wavelet transform", *IEEE Trans Biomed. Eng.*, vol. 49, p. 355-362, 2002.

[ROD 04] RODRIGUEZ M., AYALA A., RODRIGUEZ S., DIAZ-GONZALEZ M., "Application of the Max-Lloyd quantizer for ECG compression in diving mammals", *Comp. Meth. Prog. Biomed.*, vol. 73, p. 13-21, 2004.

[WEI 01] WEI J., CHANG C., CHOU N., JAN G., "ECG data compression using truncated singular value decomposition", *IEEE Trans Inf. Tech. Biomed.*, vol. 5, p. 290-299, 2001.

Chapter 7

Compression of 2D Biomedical Images

7.1. Introduction

On a daily basis, large amounts of medical images are acquired using 2D acquisition imaging systems (e.g., vertebra and lung digital X rays, mammography). Moreover, it is possible to compress temporal sequences (i.e. 2D+t), volume sequences (i.e. 3D) or even spatio-temporal sequences (i.e. 3D+t) by encoding each image separately and independently of all others (i.e. in clinical routine, physicians do not always keep all images but instead select the most relevant and accurate ones). Thus, 2D compression is widely applied to medical images. It is also included in the DICOM format (described in Chapter 4), within various PACS.

This chapter is a review of some basic 2D compression methods which are frequently applied to medical images. Although the common compression techniques and the traditional standards of compression do apply to medical images, some specific methods applied to specific images have been specially developed in order to optimize both the compression rates and the quality of the re-constructed image.

This chapter is made up of three main parts. Section 7.2 will look at the compression of medical images using reversible methods (i.e. lossless). This will be followed by an examination of lossy techniques in section 7.3, and finally, progressive compression methods will be described in section 7.4. Thus, we will show that this type of compression is highly appropriate to the transmission of medical information.

Chapter written by Christine CAVARO-MÉNARD, Amine NAÏT-ALI, Olivier DEFORGES and Marie BABEL.

7.2. Reversible compression of medical images

As is well known, reversible compression methods produce an exact duplicate of the original image and are often incorporated by constructors within some acquisition systems. In the medical field, the lossless nature of these methods is of paramount importance for ethical reasons.

The general scheme of reversible compression methods occurs in two stages: a transformation in order to reduce the inter-pixel correlation, and an entropic coding (e.g. Huffman or arithmetic encoder). The transformation must result in integer values for the encoder to function. It can be a transformation by block, by filter banks or a predictive transformation.

7.2.1. *Lossless compression by standard methods*

A 1998 study [KIV 98] has drawn a detailed comparison of different reversible compression methods using a medical image database made up of 3,000 images from 125 patients. In this study, 10 different gathering techniques have been used: X-ray (lungs), CT (abdomen and head), MRI (abdomen and spine), SPECT (head, heart and total body) and ultrasound. The methods that were tested are the following:

– general compression software such as GZIP, PKZIP, JAR, RAR, YAC, etc.;

– specific methods for the compression of monochrome images:

- methods that are based on the coding algorithm LZW (derived from the inventors' names Lempel, Ziv and Welch), the PNG standard (*Portable Network Graphics*) and GIF standard (*Graphics Interchange Format*),

- a method based on a contextual model and RICE coding: the FELICS algorithm (*Fast, Efficient, Lossless Image Compression System*) [HOW 93],

- methods based on predictive encoding with a fixed predicator: the LJPEG standard (*Lossless JPEG*), the CLIC algorithm (*Context-based Lossless Image Compression*) [TIS 93],

- methods based on adaptive predictive encoding: the LOCO-I standard (*Low Complexity Lossless Compression for Images*) also known as JPEG-LS [WEI 00], the CALIC algorithm (*Context-based Adaptive Lossless Image Compression*) [WU 97], and the TMW algorithm (a reference derived once again from the authors' names) [MEY 97],

- methods based on hierarchical predictive encoding: the CPM algorithm (*Central Prediction Method*) [HUA 91], and the BTPC algorithm (*Binary Tree Predictive Coding*) [ROB 97],

- methods based on integer-to-integer wavelet transform (S transform): the SPIHT algorithm (*Set Partitioning in Hierarchical Trees*) [SAI 96a].

To conclude, Kivijärvi *et al.* have suggested that the BTCP and GIF standards are not well-suited to medical images because they only apply to images encoded on 8 bits and because compression rates are directly related to the image type. The best compression rates are obtained by methods specific to images and in particular, using the CALIC algorithm for most of the images, TMW for CT (average compression rates lying between 2.7:1 and 3.8:1), SPIHT for MRI (between 2:1 and 3:1) and JPEG-LS for SPECT (between 2.7:1 and 5:1). Overall, when it comes to medical imaging, the compression rates achieved using reversible compression techniques vary between 2:1 and 5:1. In some cases, they can reach the compression rate 10:1 on specific images that represent large homogenous areas. In terms of execution time, the TMW method was found to be rather time consuming and the JPEG-LS method (lasting about 0.2 to 14 seconds per image) is faster than the CALIC method (lasting between 0.3 and 60 seconds per image) and the SPIHT method (between 0.4 and 90 seconds per image). Of course the execution time depends on the calculation platform.

Similar conclusions have been reached by Adamson following his study on various cerebral CT [ADA 02].

Recently, Clunie [CLU 06] has evaluated the latest versions of the JPEG-LS standard (ISO/IEC 14495-1) (this version links the RLE encoding to the LOCO-I standard) and the lossless mode of JPEG 2000 (ISO/IEC CD15444-1) (using the integer-to-integer 5.3 wavelet transform) which may be incorporated within the DICOM, on 3,679 images of different organs and obtained using different gathering techniques. The average compression rates obtained for these methods are similar and of the order of 3.8:1. Depending on the gathering technique, JPEG-LS gives on average the following compression rates: 4:1 for CT images, 3.6:1 for MRI images, 6:1 for SPECT and 3.4:1 for ultrasound.

7.2.2. *Specific methods of lossless compression*

In order to optimize the compression rates it is possible to make use of the similitude between two T1-weighted MRIs of the abdomen and two CTs of the thorax. The method defined hereafter will be advantageous for the type of image given: the same modality, acquisition parameters, organ and using similar acquisition systems. For predictive encoding methods it is possible to proceed as follows:

– define specific fixed predictors adapted either by adaptable algorithm LMS (Least Mean Squares) [NIJ 96] or by pseudo-linear recursive regression [CHE 99];

– define for a given predictor, a statistical model of the prediction error on different regions of the image (such as the image background and the studied area). Thus, a specific codebook will be associated with each region of the image during the statistical encoding [MID 99].

7.2.3. *Compression based on the region of interest*

Each region of the image may have more or less importance in the diagnostic process. For example, a brain slice can be separated into two distinct regions: the brain area that is useful to establish a medical diagnosis and the image background which provides no useful information. In order to improve the global compression rates of the image, methods based on regions of interest coding (*ROI coding*) adjust the encoding accuracy to fit the diagnosis information present in the image (i.e. the data in ROI is subject to reversible encoding while other areas are coded using lossy techniques). Regions of interest are either defined manually or after segmentation.

Halpern *et al.* have manually depicted the ROI of 75 CT images of the abdomen [HAL 90]. The data present in the ROI is encoded by a quad-tree reversible process and external data is subject to lossy compression method at different compression rates. Whatever the compression rate (i.e. lower than 50:1) applied on the outside of the ROI, the sensitivity (as defined in Chapter 5) is almost always satisfactory (over 90%) and close to that obtained by reversible compression (96%). On this type of image, compression based on ROI allows us to reach compression rates of up to 28:1, (with a maximum compression rate outside the ROI) when the average compression rate for most reversible compression processes is of 3:1.

The latest JPEG 2000 version allows a manual determination of various ROI of circular shape and allows us to perform a reversible compression of data contained within the ROI (Chapter 2).

The different regions to be coded can be defined after a segmentation phase. Then the coding strategy for each region can be adapted to the content of the information included in this type of region. This coding method works by separating the image into two parts (its contours and its texture) and represents an image symbolically in the form of a mosaic of adjacent regions with continuous variations in their internal pixel amplitudes. The borders of the regions represent the image's contours and can be coded by a reversible process known as the Freeman differential coding method or by a lossy process which consists of approximating its borders by straight lines for example. The luminance signal in each region corresponds to the texture of each object within the image and can be encoded using various methods with compression rates the levels of which vary according to the information contained in that object.

In medical imaging, a reversible compression of images that hold fine diagnosis information is required. Improving the global compression rate of the image can then be obtained without coding the image's background [CAV 96] (Figure 7.1).

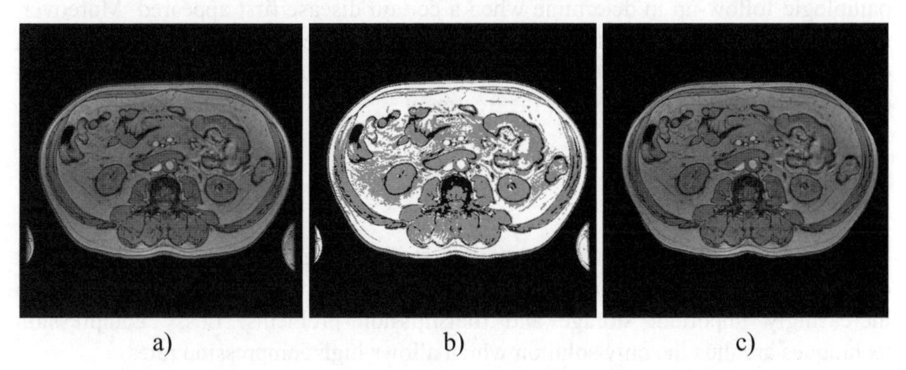

Figure 7.1. *a) Original abdominal MRI (512x512x16 bits); b) fuzzy classification into 3 sets based on pixel intensity (the background set of the image in black holding 56% of pixels);c) reconstructed image after a predictive reversible compression of the pixels in all the other sets TC=7.4*

Adaptive or optimal predictive encoding on each region to be coded can be performed according to a specific exploration adapted to any kind of region's shape (Figure 7.2).

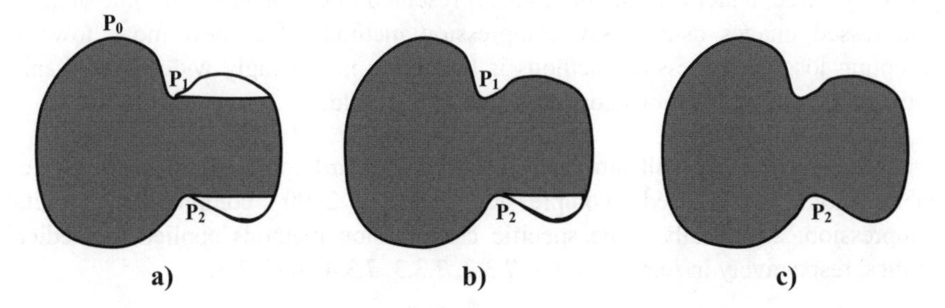

Figure 7.2. *Exploration of a region with a shape: a) exploration from top to bottom starting at point P_0; b) exploration from bottom to top starting at P_1; c) exploration from top to bottom starting at P_2; from [CAV 96]*

The benefits in terms of compression rate largely depend on the image being studied (i.e. the amount of non-coded data), but according to [CAV 96], coding methods based on the contour-texture approach resulted in improvements of 10 to 30% in compression rates. Such methods are truly promising and can be further improved by re-adapting the process to new diagnostic signal models.

Nevertheless, when a storage process is required, it is impossible to determine which details will be useful in the future [KIV 98]. An initially unimportant detail during the acquisition process may become of great informative value for a pathologic follow-up to determine when a certain disease first appeared. Moreover, many pathologies can spread out over the entire image studied and a single detail during the acquisition process may be an important element to take into consideration for the diagnosis or the treatment.

7.2.4. *Conclusion*

As we have seen in the above section, strictly reversible systems have limited efficiency in terms of compression rates and do not provide a long-term solution to increasingly important storage and transmission problems. Lossy compression techniques are thus the only solution which allows high compression rates.

7.3. Lossy compression of medical images

Although reversible compression of medical images provides physicians with resulting images of high analytical quality, the compression rates obtained through these methods are low in comparison with the compression rates obtained using irreversible methods, also known as lossy compression methods. As mentioned in Chapter 1, mindsets have changed over the last few years, and physicians, now generally agree, under certain conditions (presented in Chapters 1 and 5), to analyze compressed images using lossy compression methods. This new move towards accepting lossy compression methods is becoming increasingly widespread thanks to the numerous publications now available in this field.

In this section, we will introduce different standard compression methods, i.e., quantization, DCT-based compression, JPEG 2000 compression, fractal compression, and finally some specific compression methods applied to medical images, respectively in sections 7.3.1, 7.3.2, 7.3.3, 7.3.4 and 7.3.5.

7.3.1. *Quantization of medical images*

As pointed out in Chapter 2, there are two main forms of quantization: scalar quantization often used at the end of an encoding scheme, and vector quantization (VQ). These methods are described in further detail in [BAR 02], Chapter 2. This section will only address vector quantization, a method which might be appropriate to some specific types of images including medical images.

7.3.1.1. *Principles of vector quantization*

Vector quantization is a lossy compression technique which consists of allocating a code from a specific dictionary to each block or pixel vector constituting the original image. Using this technique, a given block in the image is compared to a set of dictionary codes. Consequently, the code chosen will be that which minimizes a distance with respect to the original block of the image to be coded. As a result, the image can be stored or transmitted very simply by storing and transmitting the index of each code. To reconstruct an image, the decoder uses the indices from which it extracts the appropriate dictionary codes (Figure 7.3). It seems obvious that the performances of this method depend on the number of codes contained in the dictionary. On the other hand, the complexity in terms of code-search could be increased significantly.

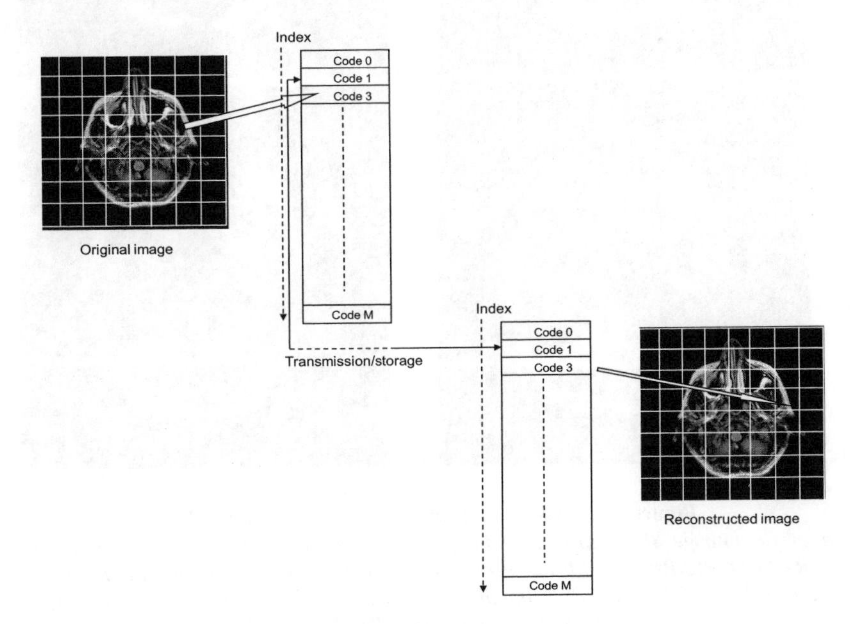

Figure 7.3. *Vector quantification scheme*

7.3.1.2. *A few illustrations*

Figure 7.4 shows various results obtained from a vector quantization of a brain MRI image, having a size of 256x256 (Figure 7.4a). Each pixel in this image is coded on 8 bits. In this example, three dictionaries have been tested. The vectors used are in fact, 4x4 or 2x2 matrices. Figures 7.4b, 7.4c and 7.4d show images quantized at a fixed rate. Although the results obtained at 0.3 bpp (CR=25:1) are obviously unacceptable, the coding 2 bpp (CR=4:1), or even 0.75 bpp (CR=10.5:1) offer an acceptable visual quality.

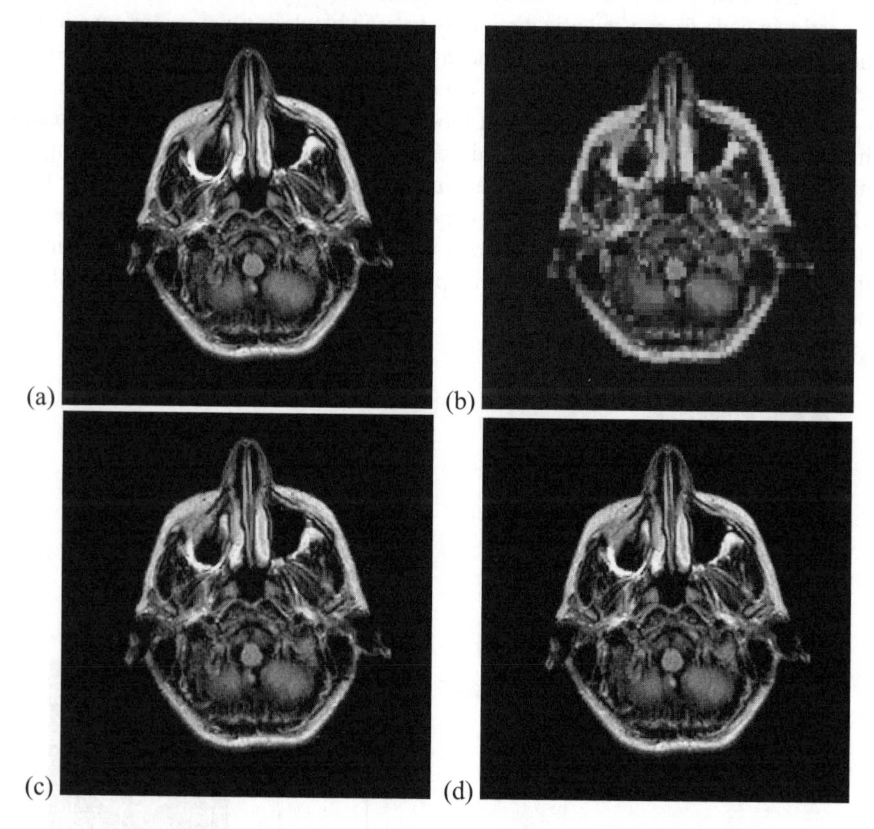

Figure 7.4. *Vector quantization applied to a brain MRI: (a) original image; (b) reconstructed image after vector quantization (2x2), at 0.31 bpp (CR=25:1), PSNR= 24.2 dB; (c) reconstructed image after vector quantization (4x4), at 0.75 bpp (CR= 10.5:1), PSNR= 30.0 dB; (d) reconstructed image after vector quantization (2x2), at 2 bpp (CR= 4:1), PSNR= 33.6 dB*

As mentioned previously, one of the disadvantages of the vector quantization is the significant calculation time required for such schemes. More specifically, while searching for a code in the dictionary, the encoder carries out a polling process. Search procedures related to the compression of medical images using rich dictionaries, have often included alternative methods such as the well known tree-structured vector quantization. This can be either balanced or unbalanced.

7.3.1.3. *Balanced tree-structured vector quantization*

As mentioned above, vector quantization involves complex calculations (in order to access codes). This complexity increases in proportion to *M*, where *M* stands for the number of codes used in the dictionary. To overcome this problem, tree-structured vector quantization (*TSVS*) has been developed in [BUS 80] and effectively used in [COS 93] to quantify X-ray and MRI images [RIS 90]. This method enables them to access codes whose complexity increase in proportion to the dictionary size logarithm (Figure 7.5a).

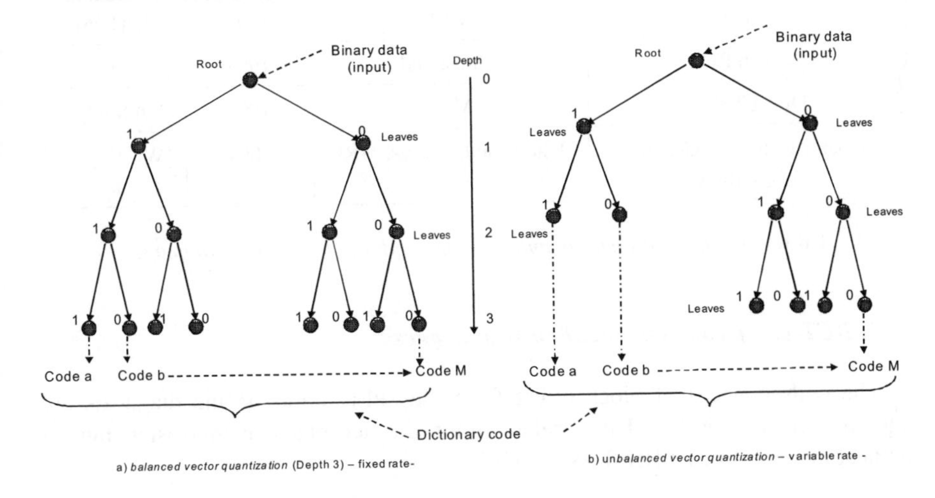

Figure 7.5. *Comparison of (a) balanced vector quantization and (b) unbalanced vector quantization*

7.3.1.4. *Pruned tree-structured vector quantization*

During vector quantization of medical images, different regions of the image must sometimes be encoded by a higher or lower number of bits, depending on how useful the information is in each of those regions. This is performed by placing the tree-structured leaves at different depths. This method is known as pruned tree-structure vector quantization or PTSVQ (Figure 7.5b). The PTSVQ always results in variable rates. For more information, the reader can refer to other alternatives, such as EPTSVQ and ECVQ [RIS 90].

7.3.1.5. *Other vector quantization methods applied to medical images*

Table 7.1 shows a set of different approaches applied to medical images and provides a general overview of the different techniques that are developed and evaluated in this context. In this table, the second column indicates the type of medical images used during the evaluation of the given algorithms. The performances that are recorded in this table will not remain consistent once these techniques are applied to other types of images. In fact, the performance of an encoder is directly related to the specificities of the images being processed.

Methods/approaches	Image Type	Year	Reference
PTSVQ, EPTSVQ, ECVQ	MRI	1990	[RIS 90]
TSVQ	X-rays	1993	[COS 93]
SVQ	MRI	1996	[MOH 96]
VQ on ROI	Ultrasound	1998	[CZI 98]
DI-VLTSVQ	MRI	1999	[HAN 99]
Quantification by Genetic Algorithms	Ultrasound, X-rays, MRI	2004	[WU 04]

Table 7.1. *Different vector quantization methods applied to medical images*

7.3.2. *DCT-based compression of medical images*

The consequences of block effect from the JPEG norm on the diagnosis are explained in section 5.2. This artefact limits the acceptable compression rate in medical imaging to 10:1 [BER 94] [CAV 99].

Based on the DCT transformation of the entire image, Full-Frame DCT compression avoids creating any such artefact. For this reason, it has been widely used in the medical imaging field [CHA 89] [LO 91] [BER 94]. However, it is necessary to adapt the quantization of the DCT coefficients to the specificities of the image being coded. For example, Béretta [BER 94] has segmented the DCT plan into circular frequency zones according to the distribution of frequency components of cardiac angiograms on the DCT plane. The circular zones have been separated according to their direction (horizontal or vertical) (Figure 7.6).

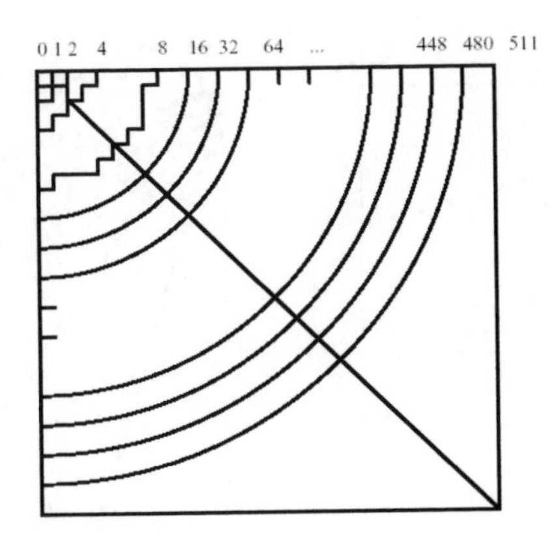

Figure 7.6. *Splitting of the DCT map into circular zones with vertical direction (above the diagonal) and horizontal direction (below the diagonal)*

Within circular zones, the DCT coefficients generally have low variances. Thus, the same number of bits can be allocated to all coefficients of a circular zone, and the same quantization step can be applied. A truncated Laplacian analytical model can be used to design an optimal midtread uniform quantizer. For an *a priori* number of bits per coefficient and with the observed dynamics and variance of a zone, the optimal quantization step can be calculated, with minimum quantization error. This evaluation can be used in an integer bit allocation algorithm based on the theory of marginal analysis.

On digital cardiac imaging systems, images are enhanced in order to outline the edges of vessels and ventricles. Edge enhancement filters emphasize the visibility of diagnostic information as well as the possible compression artefacts in the image. Therefore, Béretta [BER 95] has suggested incorporating unsharp masking in the DCT domain before performing the quantization (i.e. pre-enhancement) or during the quantization process (i.e. post-enhancement). The table representing the bit allocation will differ if the image edges are enhanced either before or after the quantization process (Figure 7.7).

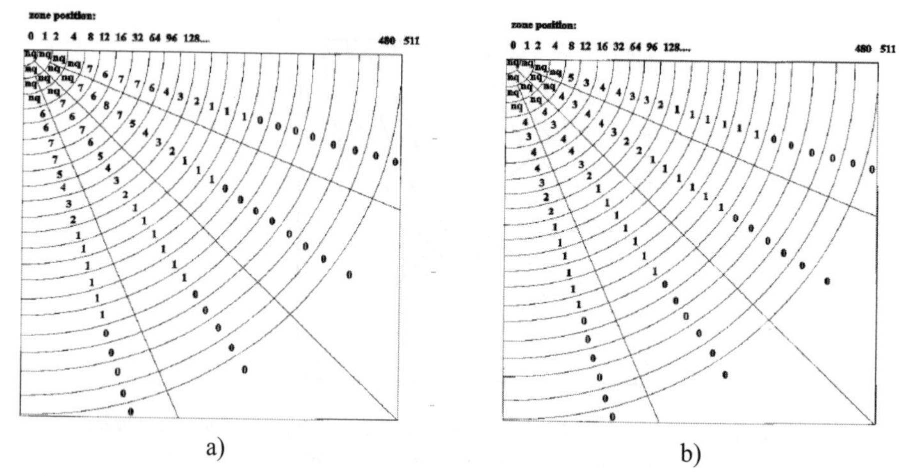

Figure 7.7. *a) Bit allocation of the post-enhanced coronary image at CR=12.6:1; b) bit allocation of the pre-enhanced coronary image at CR=12.8:1; from [BER 95]*

An inverse filter method has been introduced for calculating a de-enhanced image from the enhanced-compressed one. In fact, images which have not been enhanced are needed for other types of treatments, such as for quantitative measures for example. A regularized restoration can be used to improve the quality of the decompressed de-enhanced image. The results show a significant improvement in the quality of the enhanced image compressed with a Full-Frame DCT transformation, whereas the JPEG coded image shows a blocking effect.

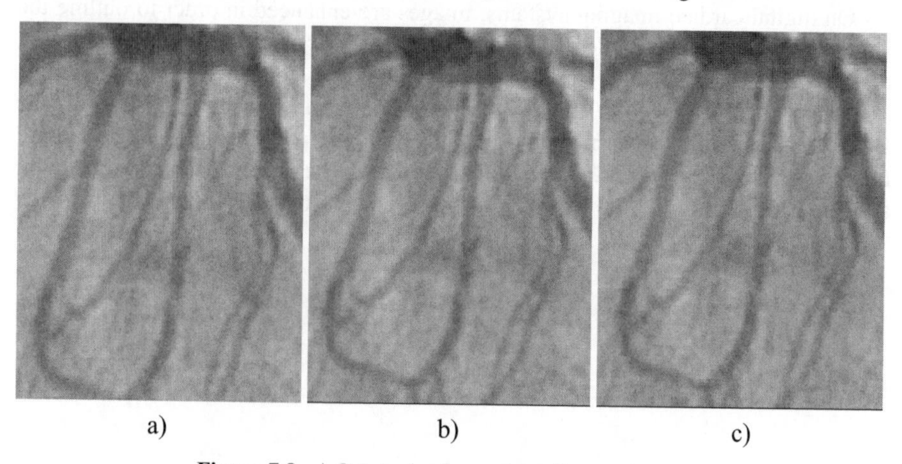

Figure 7.8. *a) Original enhanced cardiac angiogram; b) pre-enhanced Full-Frame DCT coded image (CR=12.8:1); c) enhanced JPEG coded image (CR=12.8:1); from [BER 95]*

7.3.3. *JPEG 2000 lossy compression of medical images*

Recent works have studied the use of JPEG 2000 compression on a variety of medical images. Table 7.2 shows the range of acceptable compression rates defined after analyzing the accuracy of the diagnosis. Section 5.2 outlines the effects of the smoothing effect caused by the JPEG 2000.

Image Type	Acceptable compression rate	Reference
Digital chest radiograph	20:1 (so that lesions can still be detected)	[SUN 02] [CAV 01]
Mammography	20:1 (detecting lesions)	[SUR 04]
Lung CT image	10:1 (so that the volume of nodules can still be measured)	[KO 05]
Ultrasound	12:1	[CHE 05]
Coronary angiogram	30:1 (after optimizing JPEG 2000 options)	[ZHA 04]

Table 7.2. *Applying JPEG 2000 compression to different medical images*

7.3.3.1. *Optimizing the JPEG 2000 parameters for the compression of medical images*

In this section we will be looking at a compression approach which has been applied to medical images. This technique is based on the optimization of the options of the JPEG 2000 standard (Table 7.3) with respect to a given objective function. Zhang *et al.* [ZHA 04] applied this approach to angiograms. In their study, Zhang *et al.* have determined the best set of parameters (JPEG 2000 options), which guarantee optimal compression rates while allowing, at the same time, the detection of useful information from the considered image, initially compressed using a lossy technique (Figure 7.9).

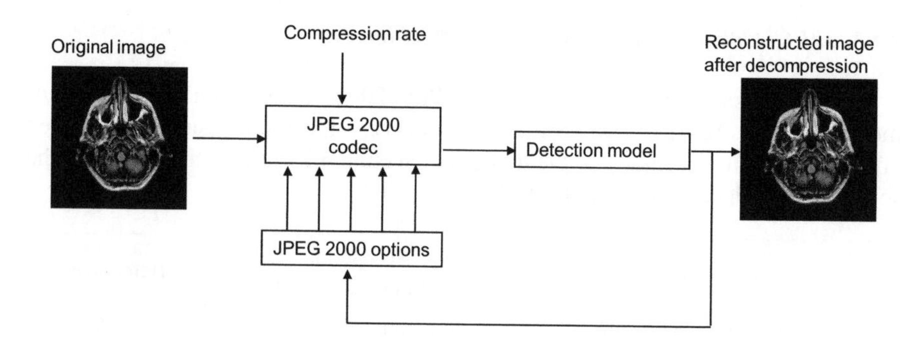

Figure 7.9. *Generalized block diagram illustrating the approach proposed in [ZHA 04]*

Genes	Coder options	Range
Gene 1	Size: tile	32, 64, [88-92], [108-114], 128, [148-152], [216-240], 256, [436-484], 512 (by default)
Gene 2	Resolution	2, 3, 4, 5, 6 (by default), 7, 8
Gene 3	Mode	Int (by default), real
Gene 4	Size: codeblocks	32x64, 32x32, 64x32, 64x64 (by default), 256x16, 16x256, 16x128, 128x16, 128x32, 32x128, 128x8, 8, 256x8, 8x256
Gene 5	Size: Precinct	256x256, 128x128, 512x256, 64x64, 32x32, 256x512 (by default)

Table 7.3. *JPEG 2000 options to be optimized*

It is important to note that with such an approach, the type of anomaly to be identified, as well as its approximate location within the image, are assumed to be known *a priori* (Figure 7.10).

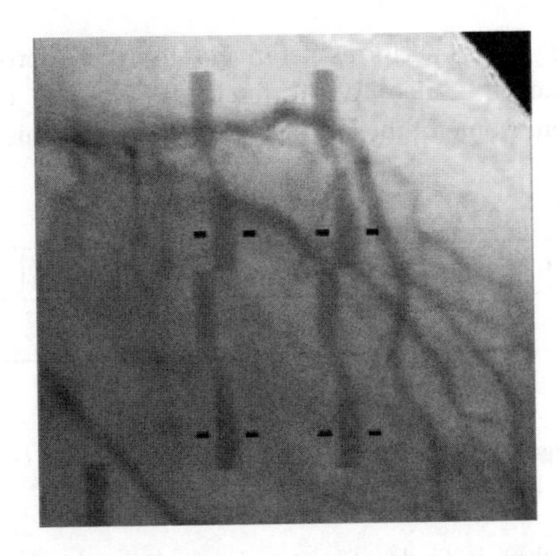

Figure 7.10. *Angiogram: the type of anomalies to be detected as well as their location is supposedly already known (between the dark lines, according to [ZHA 04])*

The detection model used is known as the NPWE (*None-Pre-whitening matched filter With an Eye filter*) model. It is based on the performance of the human visual system, and acts as a meta-heuristic optimizing technique. This model applies genetic algorithms (GA), widely used to minimize (or maximize) non-convex criteria; for further information on how this method works, see [SIA 03]. It is worth noting that the GA uses a stochastic search technique based on the principle of crossover, mutation and reproduction. Figure 7.11 shows a diagram that evaluates the objective function for a fixed *a priori* compression rate.

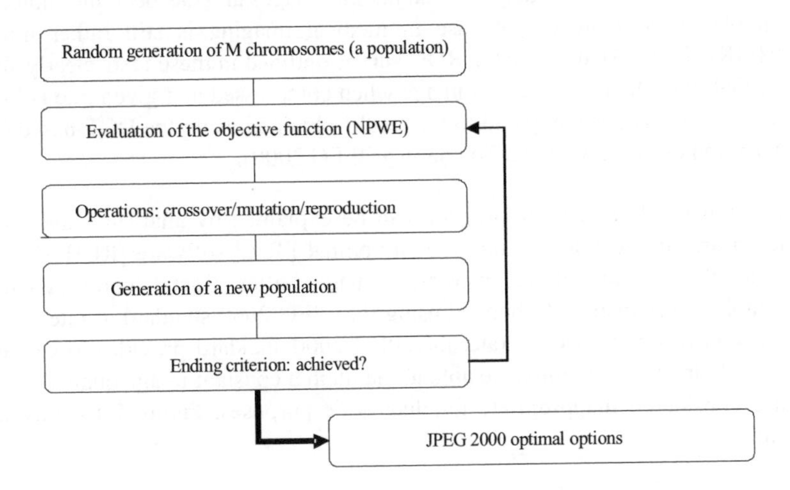

Figure 7.11. *Using the GA to optimize JPEG 2000 options*

This particular algorithm performs iteratively until the appropriate conditions of convergence are reached. Since five different JPEG 2000 parameters must be optimized, each chromosome of the GA should include five genes (Figure 7.12).

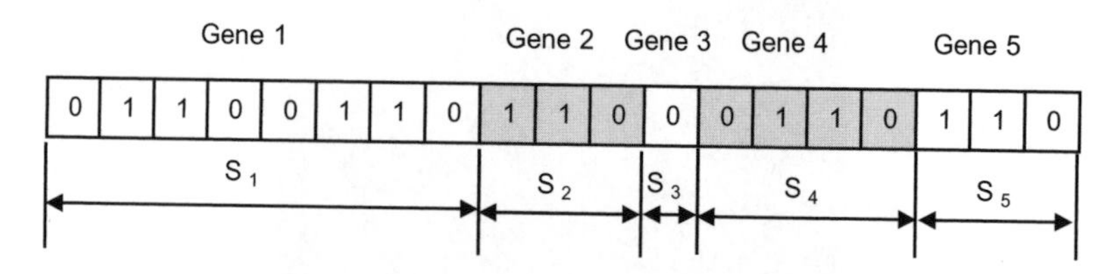

Figure 7.12. *Structure of a chromosome used in GA*

Although this method is fairly efficient, due to its optimal mode, we must take into consideration the time required by the GA to converge. This process often requires the use of a powerful computer (e.g. multiprocessor system) to overcome the problem of calculation complexity. Nevertheless, this technique can be used to identify an optimal range of options that fits a given type of image for a given detection model. In other words, a learning phase can be used in order to approximate the optimal values of JPEG 2000 options which can be used *a posteriori* to compress any image of the same type and under the same conditions.

7.3.4. *Fractal compression*

Fractal compression is widely used on natural images and has been the subject of many publications. However, its use in medical imaging is still rather limited [RIC 98] [KOT 03]. Analyses using ROC curves outlined in these texts clearly show that the quality of the reconstructed image when compressed at a given rate is lower for fractal compressions compared to the results obtained using the DCT-based (e.g. JPEG) method or the wavelet-based one (e.g. JPEG 2000).

The problem of fractal compression can be explained by analyzing the results obtained from cardiac MRIs using the non-optimal FRAP software [RUH 97]. It is clear that the quality of reconstructed images after fractal compression is considerably lower than that obtained using the JPEG 2000 standard at rates below 40:1. Beyond this compression rate, the JPEG 2000 standard provides low quality images while fractal algorithms are able to maintain a constant image quality even if it is also considered inappropriate for diagnostic purposes. Figure 7.14 illustrates this point.

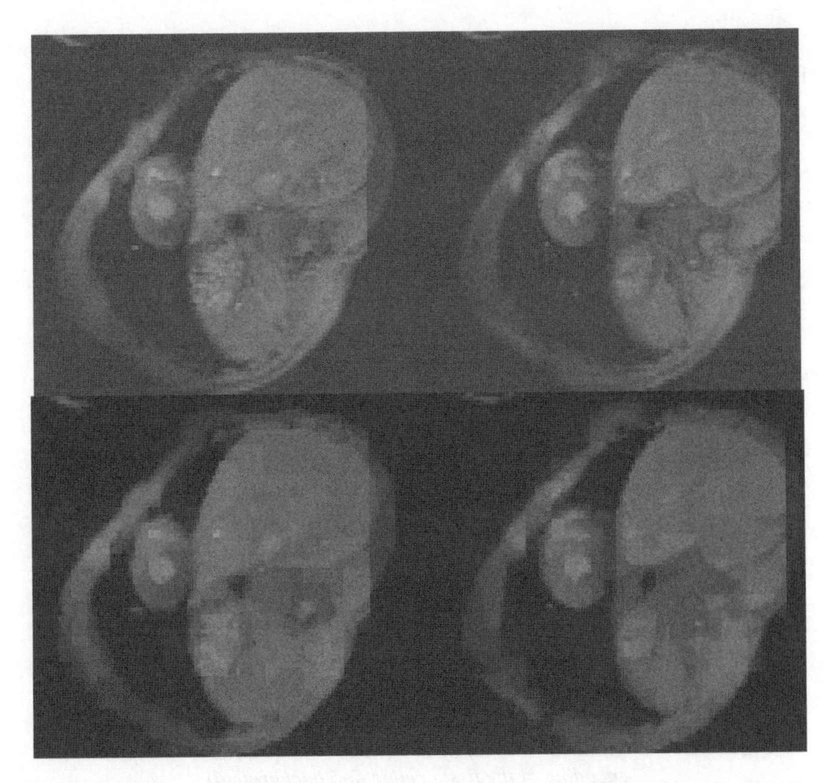

Figure 7.14. *Fractal compression of 2 successive phases on a similar cardiac MRI slice. Above, original images. Below, reconstructed images after fractal compression at a compression rate of 60:1*

7.3.5. *Some specific compression methods*

7.3.5.1. *Compression of mammography images*

The encoding method based on regions of interest is widely used for the lossy compression of digital mammography. As an example to illustrate this case, we may refer to the works of Penedo *et al.* [PEN 03], in which they test and compare two techniques with traditional compression methods. One of those techniques is the OB-SPIHT (*Object-Based extension of the Set Partitioning In Hierarchical Trees*) and the other is the OB-SPECK technique (*Object-Based extension of the Set Partitioning Embedded block Coding algorithm*). Since digital mammography is often set against a black background, a segmentation phase is obviously necessary to isolate the tissue from the backdrop as shown in Figure 7.13. The various processing steps following segmentation are:

– encoding of the edges: the edge is often encoded either with or without losses. For further information see [KAN 85] and [EDE 85];

– decomposing the region of interest by wavelets [BAR 94];

– encoding phase using OB-SPIHT or OB-SPECK [SAI 96b], [ISL 99].

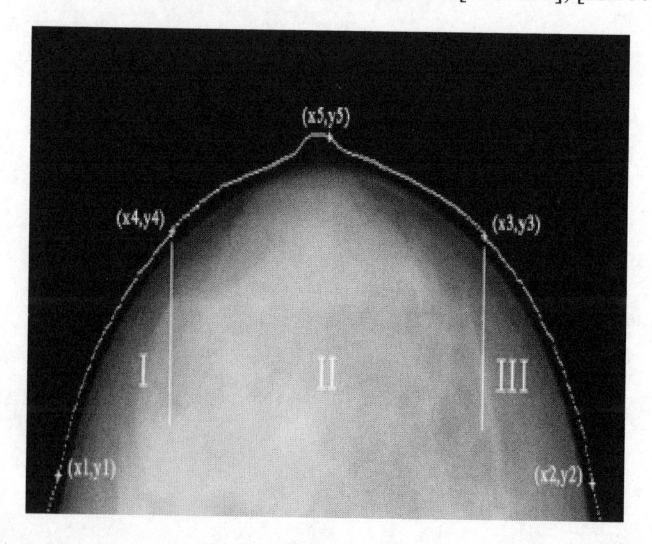

Figure 7.13. *Digital mammography in Penedo et al. [PEN 03] (independent encoding of the tissue, edge and background)*

7.3.5.2. *Compression of ultrasound images*

Ideally, it would be best to find a way of combining compression techniques with methods reducing the speckle noise which is specific to ultrasound images. One of the recent works published on this topic deals with ultrasound images using the following distinct steps [GUP 04]:

– calculating the logarithm of the original image (while taking the speckle noise into consideration);

– decomposing the obtained image using wavelets;

– establishing a threshold (after estimation) for the coefficients that correspond to the speckle noise and that are located in different subbands;

– classification of the coefficients organized into subbands in different classes; each class is modeled by a generalized Gaussian;

– carrying out an adaptive quantization of each class;

– carrying out an entropic coding.

For further references see [CHE 05] in which the edges of ultrasound images are restored *a posteriori* by a post-processing technique using morphologic filters.

7.4. Progressive compression of medical images

The design of new encoding techniques is no longer aimed solely at improving performances in terms of compression rates. The development of digital image transmission solutions using different communication media naturally implies new services requiring the adaptation of the coding process to this purpose. Scalable encoding allows the user to download the images progressively; from low quality up to the desired more refined quality. The loading time required for a given image depends on its size, on the transmission rate, on the number of users sharing the same network, etc.

Said has summarized in [SAI 99] the ideal properties of a still image encoding scheme. Among them, we can specifically mention: compression efficiency, scalability, good quality at low bitrates, flexibility and adaptability, rate and quality control, algorithm uniqueness (with/without losses), reduced complexity, error robustness (for instance in a wireless transmission context) and region of interest (ROI) decoding at decoder level.

In the medical imaging context, the use of progressive encoding methods increases network fluidity especially when transmitting over PACS networks or over the Internet. It would be far too ambitious for us to list all the works that have been published in this domain within a single chapter. Nevertheless, section 7.4.1. presents some of the more recent major works relative to this topic. As an example, section 7.4.2 exposes the LAR (*Locally Adaptive Resolution*) progressive compression technique principles. This particular technique has been developed by a French research group.

7.4.1. *State-of-the-art progressive medical image compression techniques*

As we have seen in Chapter 2, the JPEG 2000 standard was specified in such a way that it includes a scalable description of the image. This feature is obtained thanks to on one hand the particular properties of wavelets which allow the decomposition of an image at different resolutions and, on the other hand, the use of *codestreams* in order to organize the transmitted information.

Before wavelets were incorporated into the JPEG 2000 standard, many researchers were dealing with the problem of progressive compression, defining methods as *lossy to lossless* or even, *lossy to near losseless* compression techniques.

From all the progressive compression solutions that incorporate wavelets, we can cite the one designed by Cristea *et al.* [CRI 98] and applied to MRI images and the one outlined in Munteano *et al.* [MUN 99] and applied to coronary angiograms. These authors propose an approach based on an integer wavelet decomposition scheme (designed through a lifting scheme) and based on intra-band exploration (or non-inter-band) of transformed coefficients. In their respective works, the results have been compared to those obtained by JPEG, SPHIT, and CALIC standards, etc.

In [RAM 06], the SPHIT encoder has been tested on X-ray images, for a progressive transmission on low bitrate networks. Data is first stored in the DICOM format (Chapter 4). This method involves the modification of the TSUID (*Transfer Syntax Unique Identification*) field inserted in DICOM files header, so as to determine which type of encoder has to be applied on the image. During transmission, the header is first transmitted, followed by the SPHIT compressed image. In this particular study, evaluation is achieved using the following references: JPEG, JPEG-LS and JPEG 2000. The authors' conclusions tend to highlight the advantages of the SPHIT compression method; however, the results obtained in this study cannot be generalized to apply to all types of medical images.

In addition to wavelet-based solutions, Grüter *et al.* designed subband decomposition schemes based on non-linear polynomial prediction models [GRU 00]. Evaluations of this particular ROPD (*Rank-Order Polynomial Decomposition*) technique have been performed on MRI, X-ray and ultrasound images. Comparisons have been established using the JPEG methods, SPHIT, WTCQ, etc.

7.4.2. *LAR progressive compression of medical images*

In this section we describe the LAR method, an encoding technique that tends to fit all the aforementioned characteristics. Section 7.4.2.1 presents the main principles of LAR encoding scheme, as a basis of advanced algorithms. The rest of this presentation deals with the process of encoding medical images only. Section 7.4.2.2 outlines the global progressive scheme and the LAR applications for lossless coding purposes. Section 7.4.2.3 briefly describes the principles of ROI encoding process and its use on medical images. More detailed descriptions, particularly for LAR low bitrate encoding scheme, are available in [DEF 04] and [BAB 05a].

7.4.2.1. *Characteristics of the LAR encoding method*

Classically, an image can be described as the superposition of a local texture (fine details) over some low bitrate global image information (coarse details). The LAR compression method is based on this concept which results in the successful processing of these two types of data. Thus, the overall scheme of this approach consists of two scalable layers (Figure 7.15): an initial one to encode an image at low bitrates, and a second one for visual quality enhancement at medium/high bitrates.

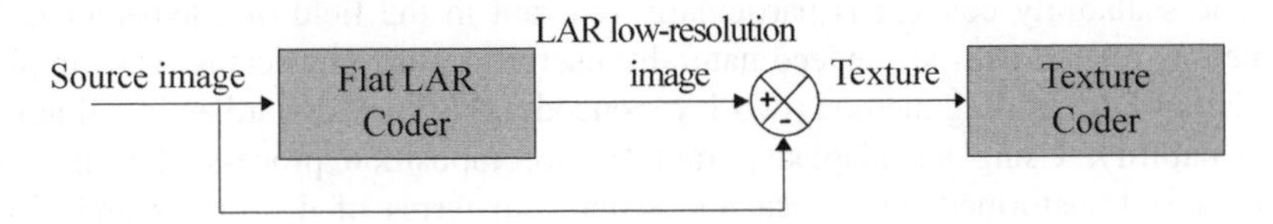

Figure 7.15. *General scheme of 2-layer LAR encoding*

The first layer of the LAR scheme, called the flat LAR coder, provides a low-resolution image of high visual quality. This image relies on the definition of a specific "quadtree" partitioning process. The size of each block is estimated using a specific criterion measuring local activity, so that smaller blocks are located on the image's contours, and larger blocks are located in homogenous areas. Figure 7.16 shows the resulting variable-size block representation. The LAR low-resolution image is obtained when filling each block by its mean luminance value. To enhance the visual quality, and taking into consideration the model used to describe block content, a low-complexity post-processing adapted to the variable size-block representation is applied to smooth uniform zones while not damaging contours.

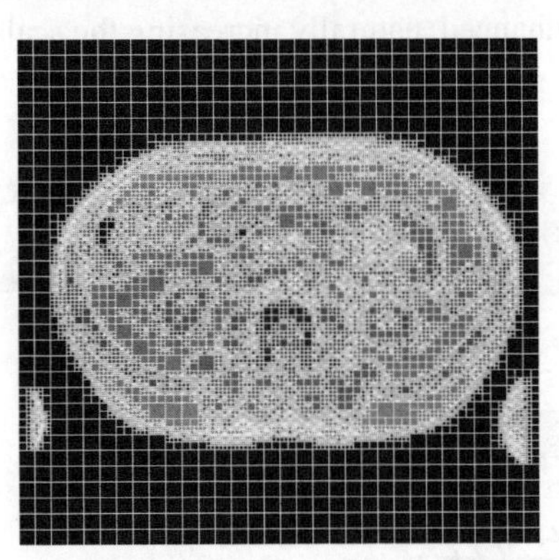

Figure 7.16. *Visual representation of the quadtree partitioning proper to the flat LAR coder*

The LAR flat coder produces a low bitrate image. This being so, the main feature of the FLAT coder consists of preserving contours while smoothing homogenous parts of the image. Adapting block sizes to the image content enables high compression ratios together with a satisfactory visual result. Enhancing the low-resolution image is realized using the second layer encoding process (texture layer).

When no quantification is applied, it is thus possible to losslessly encode the image. This feature is particularly relevant when dealing with medical images.

7.4.2.2. *Progressive LAR encoding*

The scalability concept is particularly relevant in the field of telemedicine; its numerous related uses are indeed naturally useful to most physicians. The simplest version of LAR coding method (two-layer encoding) brings forward an initial notion of scalability. Using an adapted pyramidal decomposition process, the encoding method is transformed into a scalable scheme in terms of distortions and visual quality. The aim is to design a unified coding system able to efficiently address low bitrates up to lossless compression. For that purpose, three methods have been developed: LAR-APP (Pyramidal LAR approach) [BAB 03], Interleaved S+P [BAB 05b], RWHaT+P (*Reversible Walsh-Hadamard Transform + Prediction*) [DEF 06]. The overall approach used in these three techniques is identical.

To fit the Quadtree partition, dyadic decomposition is carried out. The first and second layers of the flat LAR are replaced by two successive pyramidal descent processes, but the image representation content is preserved: the first decomposition reconstructs the low-resolution image (LAR-image) while the second one processes the local texture information. As shown in Figure 7.17, the first pyramid pass performs a conditional decomposition in accordance with the quadtree partition depicted in Figure 7.16. Consequently the local content-adapted resolution of the image is successively enhanced, naturally increasing the scalability of the method.

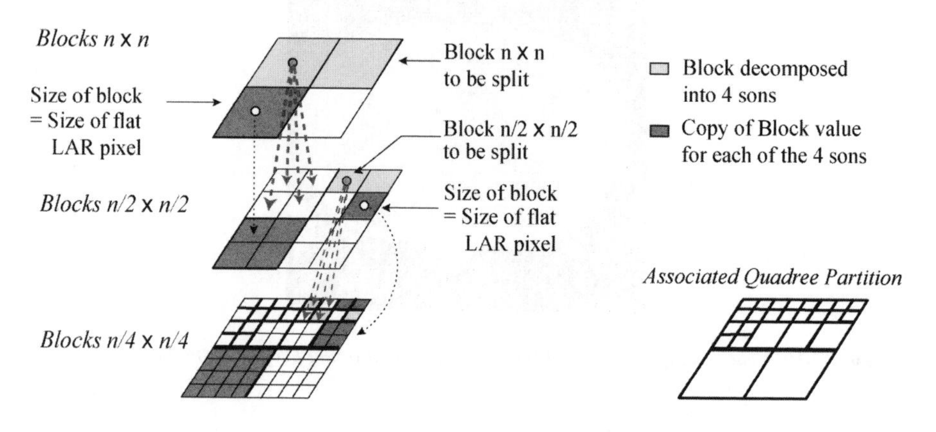

Figure 7.17. *Conditional pyramidal decomposition according to the quadtree flat LAR partition*

The second pyramidal decomposition, as a dual process, makes it possible to recover the local texture of the image. Blocks that are not decomposed during the first pass are processed during the second one.

As a consequence, pyramidal methods, exploiting the properties of the LAR coding scheme, can be seen as highly scalable methods, in the sense that progressive transmission of data according to resolution or image quality is made feasible. Figure 7.18 shows a typical multi-resolution representation.

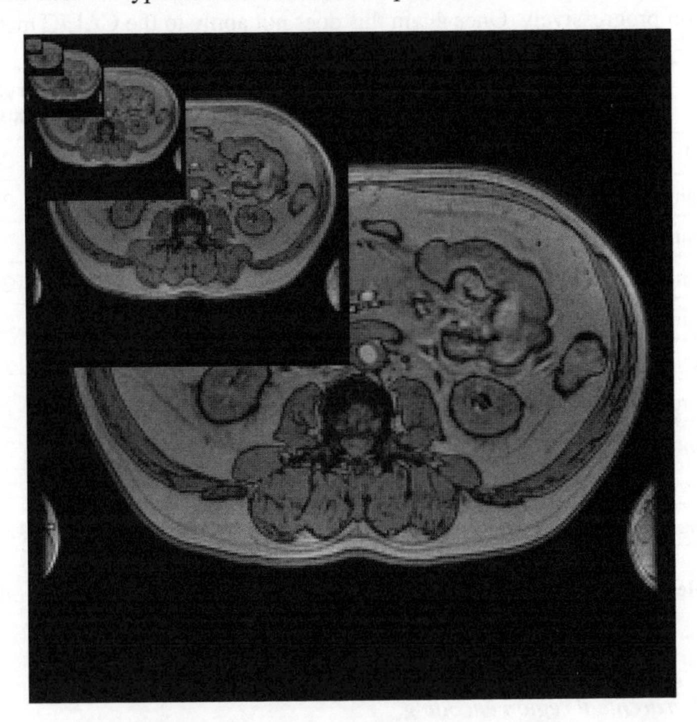

Figure 7.18. *Multi-resolution representation*

Although the three scalable LAR algorithms are based on the same principles, they differ in terms of decomposition stages and pixel encoding processes. The LAR-APP works in the spatial domain using an enriched inter and intra-level prediction. The Interleaved S+P and RWHaT+P methods perform transformation (respectively 1D S-Interleaved transform and 2D Walsh-Hadamard transform) before encoding transformed coefficients using a prediction step.

Performances obtained using the Interleaved S+P method on a large range of medical images are recorded in [BAB 05a]. This particular method largely outperforms the CALIC method in terms of zeroth-order entropy [WU 95]. Nevertheless, the RWHaT+P algorithm produces better results, in comparison to those obtained by the Interleaved S+P method.

Table 7.4 displays several compression results from the RWHaT+P and Interleaved S+P methods: when compared to CALIC, these results clearly indicate the efficiency of the proposed scheme. Moreover, besides the quantitative aspect of this analysis, the RWHaT+P method offers a scalable solution able to transmit information progressively. Once again this does not apply to the CALIC method.

	Original Entropy (bpp)	**Entropy (bpp) CALIC**	**Entropy (bpp) Interleaved S+P**	**Entropy (bpp) RWHaT+P**
Abdomen1	5.30	3.39	3.11	2.92
MR2DBrain100001	3.55	3.33	2.84	2.76
MR2DBrain100002	3.50	3.30	2.83	2.74
MR2DBrain100003	3.48	3.31	2.83	2.73
MR2DBrain100004	3.53	3.27	2.80	2.69
MR2DBrain100005	3.58	3.29	2.82	2.71
MR2DBrain100006	3.63	3.33	2.88	2.77
XR2DLung1	7.14	2.37	2.45	2.39
XR2DLung2	7.05	2.45	2.51	2.44
XR2DLung3	7.21	2.42	2.49	2.43

Table 7.4. *Results of the RWHaT+P and the Interleaved S+P method application on medical images (zero-order entropies – bits per pixel)*

7.4.2.3. *Hierarchical region encoding*

Image encoding methods that operate using region representation are highly usefully in the sense that they associate both compression and fine descriptions of images. Nevertheless, designing such approaches encounters two recurrent difficulties:

– shape description, using polygons, produces an information overhead, which can be fairly significant at low bitrates. To reduce this overhead, we need to limit the number of regions to obtain rudimentary simplified regions;

– region-based methods mainly preserve the “shape” component and often neglect the “content” component. Consequently, for a given representation, an encoded shape becomes independent of its content.

To gradually enhance the quality of reconstructed images while using scalable coding, the idea is to insert a segmentation stage calculated at both the coder and the decoder. This stage uses only first-layer rebuilt images and is efficient because the low bitrate LAR images keep their global content, in particular object contours. It leads to a fine hierarchical region representation at no cost, as no further information is transmitted to describe region shapes.

In fact, the flat layer can be interpreted as a pre-segmentation phase in a split-and-merge segmentation scheme, where small blocks are located on the contours, and larger blocks represent smooth areas. Figure 7.19 illustrates the general LAR region encoding scheme. The choice of resolution level used in the flat LAR coder determines how fine the segmentation will be.

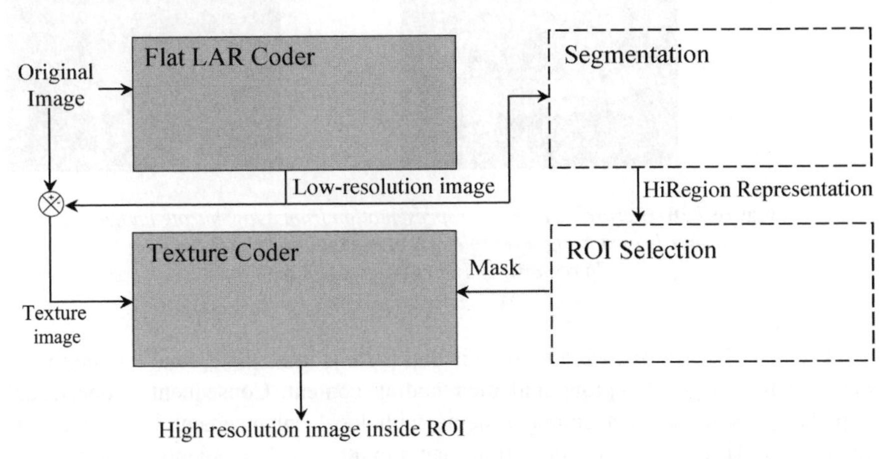

Figure 7.19. *General scheme of LAR region encoding*

The segmentation algorithm is built upon the intrinsic properties of the LAR low bitrate image. It involves an iterating, merging process of regions, the initial regions being the blocks from the low-resolution image. Merging is controlled by a single threshold parameter. By iteratively increasing thresholds, a hierarchical segmentation is obtained and enables an efficient description of the image content from finest to coarsest scale. The resulting representation tree can then be scanned globally by level, or locally by region. Figure 7.20 illustrates this type of hierarchical representation obtained at different levels of that tree.

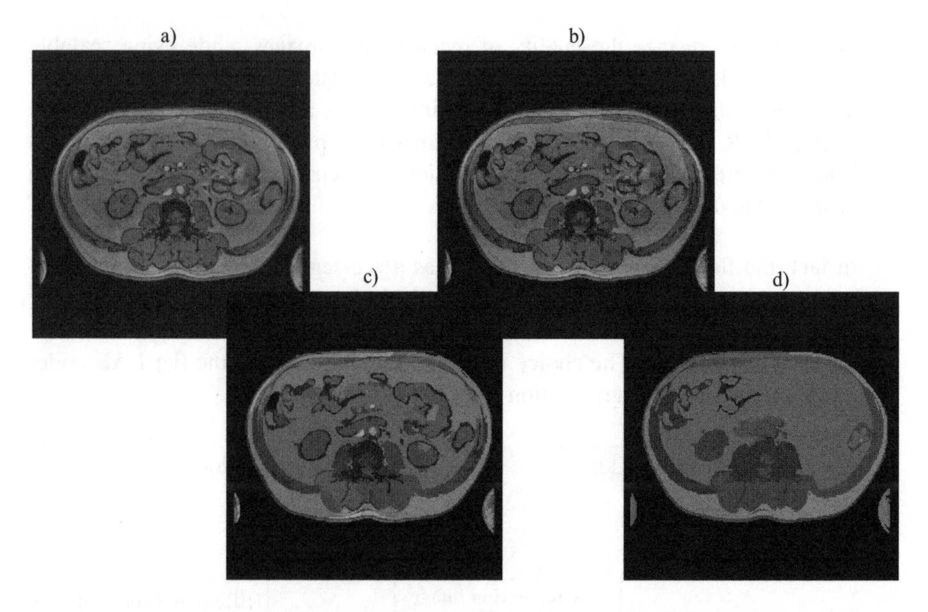

Figure 7.20. *Hierarchical region representation from a low bitrate image of level L1: a) starting level (block representation): 16,202 regions; b) 876 regions; c) 124 regions; d) 18 regions*

The main benefit of such a method is that there is immediate, total compatibility between the shape of regions and their coding content. Consequently, one direct application is found in a coding scheme with local enhancement in Regions Of Interest (ROI). From the segmentation map simultaneously available in both coder and decoder, either device can define its own ROI as a set of regions. Thus, an ROI will simply be specified by the labels of its regions.

The method provides a *semi-automatic tool for ROI selection*. Each region, and consequently each ROI, consists of a set of blocks defined in the initial partition. Then *the enhancement of an ROI is straightforward as it merely requires execution of the texture codec for the validated blocks, i.e. those inside the ROI*: the ROI acts simply as a direct on/off control for block-level enhancement. This type of application can easily be implemented using a simple graphic interface that enables us to select the concerned regions. Figure 7.21 illustrates this procedure.

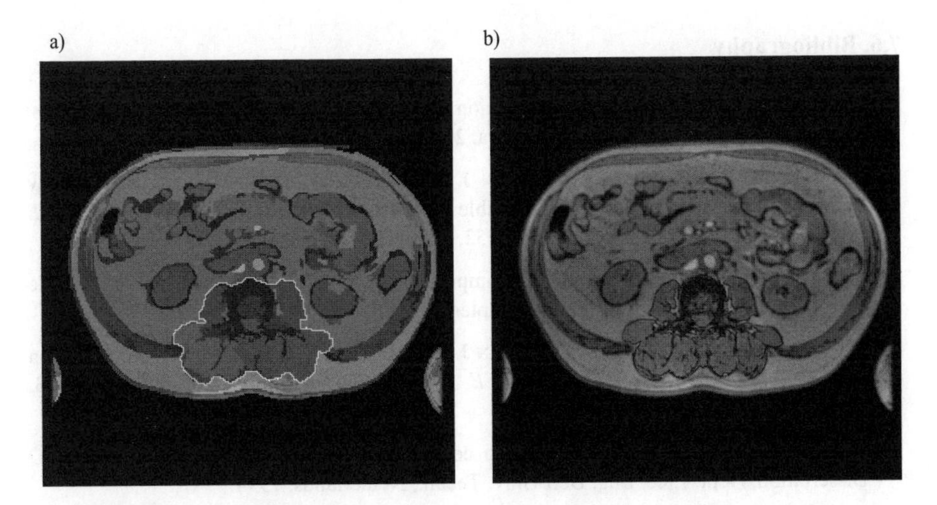

Figure 7.21. *ROI encoding: a) 124 regions representation and ROI selection; b) low bitrate encoding of the image background (first pass, 0.55 bpp), ROI losslessly coded*

7.5. Conclusion

Numerous image compression methods have been developed. They may be defined either as reversible methods (offering low compression ratios but guaranteeing an exact or near-lossless reconstruction of the image), irreversible methods (designed for higher compression ratios at the cost of a quality loss that must be controlled and characterized) or scalable methods (fully adapted to data transmission purposes and enabling lossy to lossless reconstructions). Choosing one method mainly depends on the use of images. In the case of the needs of a first diagnosis, a reversible compression would be most suitable. However, if compressed data has to be stored on low-capacity data supports, an irreversible compression would be necessary. Finally, scalable techniques clearly suit data transmission.

All compression solutions presented in this chapter have been applied, or even adapted for the purposes of medical images. The expression "medical images" represents images of various modalities (X-ray, MRI, ultrasound images, etc.) despite the fact that, as we have seen in Chapter 3, the image specificities and properties differ according to these modalities. As a consequence, the following question remains: why should they all be compressed using the same algorithm? The term "medical image" alone may not and should not justify the use of a particular method. At this stage, we are convinced that it is most pertinent to apply algorithms that are adapted to each type of image. Unsurprisingly, this discussion still remains open.

7.6. Bibliography

[ADA 02] ADAMSON C., "Lossless compression of magnetic resonance imaging data", Thesis, Monash University, Melbourne, Australia, 2002.

[BAB 03] BABEL M., DEFORGES O., ROSIN J. "Lossless and Lossy Minimal Redundancy Pyramidal Decomposition for Scalable Image Compression Technique", *IEEE ICASSP'03*, Hong Kong, vol. 3, p. 249-252, September 2005.

[BAB 05a] BABEL M., "Représentation et compression de séquences d'images par la méthode LAR", PhD Thesis, INSA de Rennes, September 2005.

[BAB 05b] BABEL M., DEFORGES O., RONSIN J., "Interleaved S+P Pyramidal Decomposition with Refined Prediction Model", *IEEE ICIP'05, Gênes*, Italy, vol. 2, p. 750-753, September 2005.

[BAR 94] BARNARD H., "Image and video coding using a wavelet decomposition", PhD Dissertation, Dept. Elec. Eng. Delf Univ. Techn., Netherlands, 1994.

[BAR 02] BARLAUD M., LABIT C., *Compression et codage des images et des vidéos*, Traité IC2, Série traitement du signal et de l'image, Hermes, Paris, 2002.

[BER 94] BERETTA P., PROST R., AMIEL M., "Optimal bit allocation for Full-Frame DCT coding scheme – Application to cardiac angiography", *SPIE Image Capture, Formatting and Display*, vol. 2164, p. 291-311, 1994.

[BER 95] BERETTA P., PROST R., "Unsharp masking and its inverse processing integrated in a compression/decompression scheme – Application to cardiac angiograms", *SPIE Medical Imaging*, vol. 2431, p. 233-244, 1995.

[BUS 80] BUZOI A, GRAY A., GRAY R., MARKEL J., "Speech coding based upon vector quantization", *IEEE Trans. Acous. Speech and Signal Processing,* vol. 28, p. 562-574, 1980.

[CAV 96] CAVARO-MÉNARD C., "A coding system based on a contour-texture model for medical image compression", *Innovation et Technologie en Biologie et Médecine*, vol. 17, no. 5, p. 359-371, 1996.

[CAV 99] CAVARO-MÉNARD C., LE DUFF A., BALZER P., DENIZOT B., MOREL O., JALLET P., LE JEUNE J.J., "Quality assessment of compressed cardiac MRI. Effect of lossy compression on computerized physiological parameters", *Proceedings of 10th International Conference on Image Analysis and Processing (ICIAP'99)*, Venice, Italy, p. 1034-1037, September 1999.

[CAV 01] CAVARO-MÉNARD C., GOUPIL F., DENIZOT B., TANGUY J.Y., LE JEUNE J.J., CARON-POITREAU C., "Wavelet compression of numerical chest radiograph: quantitative et qualitative evaluation of degradations", *Proceedings of International Conference on Visualization, Imaging and Image Processing (VIIP'01)*, Marbella, Spain, p. 406-410, September 2001.

[CHA 89] CHAN K.K., LOU S.L., HUANG H.K., "Radiological image compression using Full-Frame Cosine transform with adaptive bit-allocation", *Computerized Medical Imaging and Graphics*, vol. 13, no. 2, p. 153-159, 1989.

[CHE 99] CHEN Z.D., CHANG R.F., KUO W.J., "Adaptive predictive multiplicative autoregressive model for medical image compression", *IEEE Trans. on Medical Imaging*, vol. 18, no. 2, p. 181-184, 1999.

[CHE 05] CHEN Y.Y., TAI S.-C., "Enhancing ultrasound by morphology filter and eliminating ringing effect", *European Journal of Radiology*, vol. 53, p. 293-305, 2005.

[CLU 06] CLUNIE D.A., "Lossless compression of greyscale medical images – Effectiveness of traditional and state of the art approaches", *SPIE MI 2006, Proceedings of SPIE, The International Society for Optical Engineering,* vol. 3980, p. 74-84, 2006.

[COS 93] COSMAN P., TSENG C., GRAY R., OLSHEN R., MOSES E., DAVIDSON H., BERGIN C AND RISKIN E, "Tree-structured vector quantization of CT chest scans: Image quality and Diagnostic accuracy", *IEEE Trans. on Med. Imag.*, vol. 12, p. 727-739, 1993.

[CRI 98] CRISTEA P., CORNEALIS J., MUNTEANU A., "Progressive lossless coding medical images", *Future Generation Computer Systems*, vol. 14, p. 23-32, 1998.

[CZI 98] CZIHO A., CAZUGUEL G., SOLAIMAN B., ROUX C., "Medical image compression using region of interest vector quantization", *Proc. of the 20th Int. Conf. of the IEEE EMBS*, vol. 20, no. 3, 1998

[DEF 04] DEFORGES O., "Codage d'images par la méthode LAR et méthodologie Adéquation Algorithme Architecture. De la définition des algorithmes de compression au prototypage rapide sur architectures parallèles hétérogènes", Authorization to conduct research, (HDR), University of Rennes I, November 2004.

[DEF 06] DEFORGES O., BABEL M., MOTSCH J., "The RWHT+P for an improved lossless multiresolution coding", *EUSIPCO'06*, Florence, Italy, September 2006.

[EDE 85] EDEN M., KOCHER M., "On the performance of a contour coding algorithm in the context of image coding. Part I: contour segment coding", *Signal Processing,* vol. 8, p. 381-386, 1985.

[GUP 04] GUPTA N., SWAMY N., PLOTKIN E., "Despeckling of medical ultrasound images using data and rate adaptive lossy compression", *IEEE Trans. on Med. Imag.*, vol. 24, p. 743-454, 2005.

[GRU 00] GRÜTER R., EGGER O., VESIN J.M., "Rank-order polynomial subband decomposition for medical image compression", *IEEE Transactions on Medical Imaging*, vol. 19, p. 1044-1052, October 2000.

[HAL 90] HALPERN E.J., PREMKUMAR A., MULLEN D.J., NG C.C., LEVY H.M, NEWHOUSE J.H., AMIS E.S., SANDERS L.M., MUN I.K., "Application of region of interest definition to quadtree-based compression of CT images", *Invest. Radiology*, vol. 25, no. 6, p. 703-707, June 1990.

[HAN 99] HAN J., KIM H., "A differential index assignment scheme for tree-structured vector quantization", *IEEE Trans. on Medical Imaging*, vol. 18, p. 442-447, 1999.

[HOW 93] HOWARD P.G., VITTER J.S., "Fast and efficient lossless image compression", *Proceedings of the IEEE Data Compression Conference 1993*, p. 351-361, 1993.

[HUA 91] HUANG L., BIJAOUI A., "An efficient image compression algorithm without distortion", *Pattern Recognition Letters*, vol. 12, p. 69-72, 1991.

[ISL 99] ISLAM A., PEARLMAN W., "An embedded and efficient low-complexity hierarchical image coder", *Proc., SPIE,* vol. 3653, p. 294-305, 1999.

[KAN 85] KANEKO T., OKUDAIRA M., "Encoding of arbitrary curves based on the chain code representation", *IEEE Trans. Comm.*, vol. 33, p. 697-707, 1985.

[KIV 98] KIVIJÄRVI J., OJALA T., KAUKORANTA T., KUBA A., NYUL L., NEVALAINEN O., "A comparison of lossless compression methods for medical images", *Computerized Medical Imaging and Graphics*, vol. 22, p. 323-339, August 1998.

[KO 05] KO J.P., CHANG J., BOMSZTYK E., BABB J.S., NAIDICH D.P., RUSINEK H., "Effect of CT image compression on computer-assisted lung nodule volume measurement", *Radiology*, vol. 237, no. 1, p. 83-88, 2005.

[KOT 03] KOTTER E., ROESNER A., TORSTEN WINTERER J., GHANEM N., EINERT A., JAEGER D., UHRMEISTER P., LANGER M., "Evaluation of lossy data compression of chest X-rays: a receiver operating characteristic study", *Investigative Radiology*, vol. 38, no. 5, p. 243-249, 2003.

[LO 91] LO S. C., HUANG H. K., "Full-Frame entropy coding for radiological image compression", *SPIE Image Capture, Formatting and Display*, vol. 1444, p. 265-271, 1991.

[MEY 97] MEYER B., TISCHER P., "TMW a new method for lossless image compression", *Proceedings of International Picture Coding Symposium*, Berlin, p. 217-220, 1997.

[MID 99] MIDTVIK M., HOVIG I., "Reversible compression of MR images", *IEEE Trans. on Medical Imaging*, vol. 18, no. 9, p. 795-800, 1999.

[MOH 96] MOHSENIAN N., SHAHRI H., NASRABADI N., "Scalar-vector quantization of medical images", *IEEE Trans. on Image Processing.*, vol. 12, p. 727-739, 1993.

[MUN 99] MUNTEANU A., CORNELIS J., "Wavelet-based lossless compression of coronary angiographic images", *IEEE Transactions on Medical Imaging*, vol. 18, p. 272-281, 1999.

[NIJ 96] NIJIM Y.W., STEARNS S.D., MIKHAEL W.B., "Differentiation applied to lossless compression of medical images", *IEEE Trans. on Medical Imaging*, vol. 15, no. 4, p. 555-559, 1996.

[PEN 03] PENEDO M., PEARLMAN W., TAHOCES P., SOUTO M., VIDAL J., "Region-based wavelet coding methods for digital mammography", *IEEE Trans. on Med. Imag.*, vol. 22, no. 10, p. 1288-1296, 2003.

[PEN 05] PENEDO M., SOUTO M., TAHOCES P.G., CARREIRA J.M., VILLALON J., PORTO G., SEOANE C., VIDAL J.J., BERBAUM K.S., CHAKRABORTY D.P., FAJARDO L.L., "Free-response receiver operating characteristic evaluation of lossy JPEG 2000 and object-based set partitioning in hierarchical trees compression of digitized mammograms", *Radiology*, vol. 237, no. 2, p. 450-457, November 2005.

[RAM 06] RAMAKRISHMAN B., SRIRAAM N., "Internet transmission of DICOM images with effective low bandwidth utilization", *Digital Signal Processing*, vol. 16, no. 6, p. 824-831, 2006

[RIC 98] RICKE J., MAASS P., LOPEZ HANNINEN E., LIEBIG T., AMTHAUER H., STROSZCZYNSKI C., SCHAUER W., BOSKAMP T., WOLF M. "Wavelet versus JPEG (Joint Photographic Expert Group) and fractal compression. Impact on the detection of low-contrast details in computed radiographs", *Investigative Radiology*, vol. 33, no. 8, p. 456-463, 1998.

[RIS 90] RISKIN E, LOOKKABAUGH, CHOU P., GRAY R., "Variable rate vector quantization for medical image compression", *IEEE Trans. on Medical Imaging*, vol. 9, p. 290-298, 1990.

[ROB 97] ROBINSON J.A., "Efficient general-purpose image compression with binary tree predictive coding", *IEEE Transactions on Image Processing*, vol. 6, no. 4, p. 601-608, 1997.

[RUH 97] RUHL M., HARTENSTEIN H., SAUPE D. "Adaptive partitioning for fractal image compression", *IEEE International Conference on Image Processing ICIP'97*, Santa Barbara, October 1997.

[SAI 96a] SAID A., PEARLMAN W.A., "Image multi-resolution representation for lossless and lossy compression", *IEEE Transactions on Image Processing*, vol. 5, no. 9, p. 1303-1310, 1996.

[SAI 96b] SAID A., PEARLMAN W., "A new fast and efficient image codec based on set partitioning in hierarchical trees", *IEEE Trans. Circuits, Syst. Video Techno.*, vol. 6, p. 243-250, 1996.

[SAI 99] SAID A., "Wavelet-Based Image Compression", Imaging Technology Department, Hewlett Packard Laboratories, 1999.

[SIA 03] SIARRY P., DREO J., PETROWSKI A., TAILLARD E., "*Métaheuristiques pour l'optimisation difficile*", Edition Eyrolles, 2003.

[SUN 02] SUNG M.M., KIM H.J., YOO S.K., CHOI B.W., NAM J.E., KIM H.S., LEE J.H., YOO H.S., "Clinical evaluation of compression ratios using JPEG2000 on computed radiography chest images", *Journal of Digital Imaging*, vol. 15, no. 2, p. 78-83, June 2002.

[SUR 04] SURYANARAYANAN S., KARELLAS A., VEDANTHAM S., WALDROP S.M., D'ORSI C.J., "A perceptual evaluation of JPEG2000 image compression for digital mammography: contrast-detail characteristics", *Journal of Digital Imaging*, vol. 17, no. 1, p. 64-70, 2004.

[TIS 93] TISCHER P.E., WORLEY R.T., MAEDER A.J., GOODWIN M., "Context-based lossless image compression", *The Computer Journal,* vol. 36, no. 1, p. 68-77, 1993.

[WEI 00] WEINBERGER M. J., SEROUSSI G., SAPIRO G., "The LOCO-I lossless image compression algorithm: Principles and standardization into JPEG-LS", *IEEE Transactions on Image processing*, vol. 9, p. 1309-1324, 2000.

[WU 95] WU X., MEMON N., "A Context-based, Adaptive, Lossless/Nearly-Lossless Coding Scheme for Continuous-Tone Images (CALIC)", International Standards Organization working document, ISO/IEC SC29/WG 1/N256, vol. 202, 1995.

[WU 97] WU X., MEMON N.D., "Context-based adaptive lossless image compression", *IEEE Transactions on Information Theory*, vol. 23, p. 337-343, 1997.

[WU 04] WU Y., "GA-based DCT quantisation table design procedure for medical images", *IEEE Proc. Vis. Image Signal Processing*, vol. 151, no. 5, p. 353-359, 2004.

[ZHA 04] ZHANG Y., PHAM B., ECKSTEIN M., "Automated optimization of JPEG2000 Encoder options based on model observer Performance for detecting variable signals in X-Ray coronary angiograms", *IEEE Trans. on Med. Imag.*, vol. 23, p. 459-474, 2004.

Chapter 8

Compression of Dynamic and Volumetric Medical Sequences

8.1. Introduction

Most of the current medical imaging techniques produce three-dimensional (3D) data distributions. Some of them are intrinsically volumetric and represented as a set of two-dimensional (2D) slices, such as magnetic resonance imaging (MRI), computerized tomography (CT), positron emission tomography (PET) and 3D ultrasound, while others (such as angiography and echography) describe the temporal evolution of a dynamic phenomenon as a time sequence of 2D static images (frames), and thus are more correctly labelled as 2D+t. When displayed in rapid succession, these frames are perceived as continuous motion by the human eye.

The most commonly used digital modalities of medical volumetric data generate multiple slices in a single examination. One slice is normally a cross-section of the body part. Its adjacent slices are cross-sections parallel to the slice under consideration. Multiple slices generated this way are normally anatomically or physiologically correlated to each other (Figure 8.1). In other words, there are some image structural similarities between adjacent slices. Although it is possible to compress an image set slice by slice, more efficient compression can be achieved by exploring the correlation between slices.

Chapter written by Azza OULED ZAID, Christian OLIVIER and Amine NAÏT-ALI.

Medical 2D+t and 3D images have had a great impact on the diagnosis of diseases and surgical planning. The limitations in storage space and transmission bandwidth on the one hand, and the growing size of medical image data sets on the other, have pushed the design of ad-hoc tools for their manipulation. The increasing demand for efficiently storing and transmitting digital medical image data sets has triggered a vast investigation of volumetric and dynamic image compression. More importantly, compression may help to postpone the acquisition of new storage devices or networks when these reach their maximal capacity.

A common characteristic of digital images is that neighboring pixels have a high degree of correlation. To enhance the coding performance, data compression aims to reduce the spatial or temporal redundancy by first decreasing the amount of correlation in the data and then encoding the resulting data.

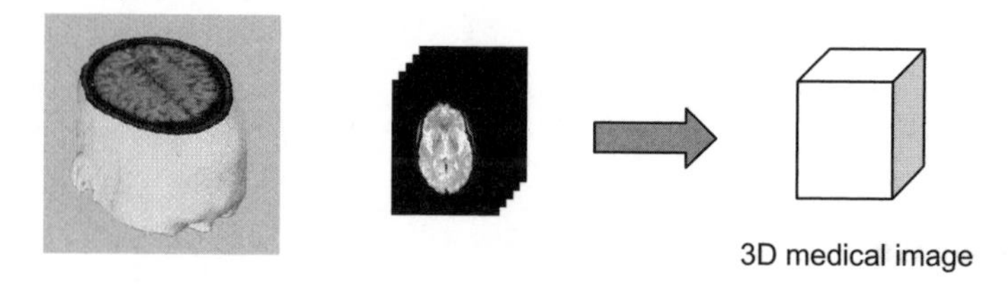

Figure 8.1. *Example of volumetric medical image data set*

The two major types of medical image data set compression are lossless and lossy. Lossless techniques allow an exact reconstruction of the original image. Unfortunately, the tight constraints imposed by lossless compression usually limit the compression ratio to about 2 or 4:1. Lossy techniques can reduce images by arbitrarily large ratios but do not perfectly reproduce the original image. However, the reproduction may be good enough that no image degradation is perceptible and diagnostic value is not compromised.

When lossless compression is used, 3D image data can be represented as multiple two-dimensional (2D) slices, it is possible to code these 2D images independently on a slice-by-slice basis. Several methods have been proposed for lossless (medical) 2D image compression, such as predictive coding like Differential Pulse Code Modulation (DPCM), Context-based Adaptive Lossless Image Coding (CALIC), Hierarchical INTerpolation (HINT) or simply a variable length coder, for example arithmetic coding, Huffman and Lempel-Ziv. However, such 2D methods do not exploit the data redundancy in the third dimension, i.e. the property that pixels in the same position in neighboring image frames are usually very similar. As pixels are correlated in all three dimensions, a better approach is to consider the whole set of slices as a single 3D data set. Although dynamic medical sequences differ in nature to volumetric ones, it is difficult to distinguish between lossless compression techniques applied to the two families. This aspect will be considered in detail in section 8.2.

While further research into lossless methods may produce some modest improvements, significantly higher compression ratios, reaching 60:1, can be achieved using lossy approaches. Unfortunately, lossy compression schemes may only achieve modest compression before significant information is lost (see Chapter 5). The choice of the compression rate is consequently adjusted in a way which will reduce the volume of data without discarding visually or diagnostically important information.

The development of a lossy compression algorithm adapted to (2D+t) sequences is a critical problem. For example, the ultrasound images are corrupted by a random granular pattern, called a "speckle" pattern. This noise is generated by physical phenomena related to the acquisition technique and can be considered as a texture representing information about the observed medium. Furthermore, in the case of angiogram sequences, images obtained from projection radiography may reveal lesions by image details that are extremely sensitive to lossy compression since they are small or have poorly defined borders (e.g. the edge of coronary arteries), and are only distinguishable by subtle variations in contrast. The lossy (irreversible) compression approaches to 2D+t medical images will be considered in section 8.3.

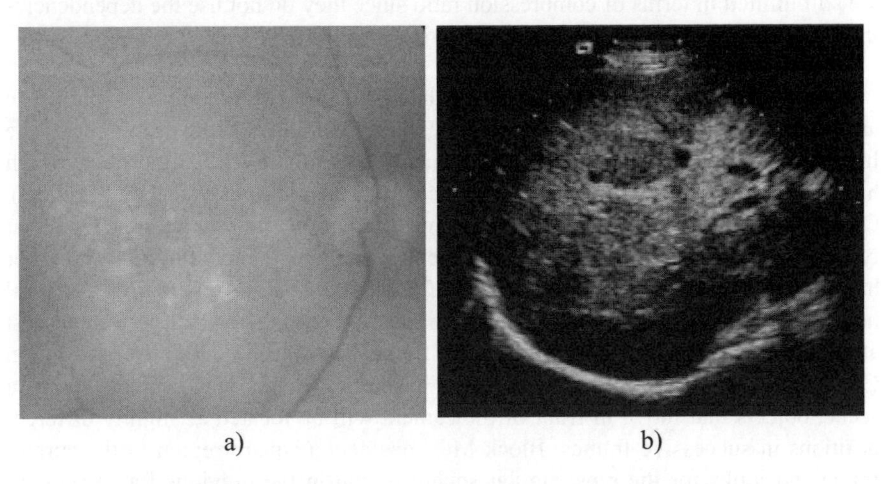

a) b)

Figure 8.2. *Example of dynamic medical images, from left to right: angiographic image; echographic (ultrasonic) image*

Section 8.4 deals with the irreversible compression methods of the 3D medical data sets. As will be discussed later, the direct use of dynamic data compression approaches is not appropriate for volumetric medical images. These coding techniques are usually based on object motion estimation approaches. When applied to volumetric images, the inter-image correlation is considered as a temporal one. The assumption of object motion is not realistic in 3D image sets, where image frames are not snap shots of moving objects but instead plane slices of 3D objects. It should be noted that fully 3D wavelet-based coding systems are very promising techniques in the field of volumetric medical data compression.

8.2. Reversible compression of (2D+t) and 3D medical data sets

As mentioned earlier, in order to efficiently compress volumetric or dynamic medical sequence, it is important to use spatial and temporal redundancy. Image sequence coding can use redundancies at both intra-frame (within a frame) and inter-frame (between frames) levels. At the inter-frame level the redundancy between successive images across time or along the third spatial dimension, is used. As *2D+t* and *3D* data can be represented as sets of 2D images, it is possible to code these 2D images independently on an image-by-image basis. Several excellent 2D lossless compression algorithms are in existence, which are presented in Chapters 2 and 7. Some of these reversible 2D compression techniques, such as the JPEG-LS or CALIC algorithm, are applied independently on each 2D image of a dynamic or volumetric data set. However, as mentioned by Clunie [CLU 00], such 2D methods are still limited in terms of compression ratio since they do not use the dependencies that exist among pixel values across the third scale.

Lossless video coding techniques have been investigated to compress volumetric and dynamic medical image sequences. Those that have been proposed in the literature are often essentially 2D techniques, such as the CALIC algorithm, which have been modified to use some information from the previously encoded frame(s). Other approaches combine the intra-frame prediction and modeling steps that typically occur in 2D techniques with inter-frame context modeling. On the other hand, some techniques adapt predictive 2D image coders by providing several possible predictors and encoding each pixel using the predictor that performed best at the same spatial position in the previous frame. These techniques also use Motion Compensation (MC). The principle of MC is simple: when a video camera shoots a scene, objects that move in front of the camera will be located at slightly different positions in successive frames. Block MC considers a square region in the current frame and looks for the most similar square region in the previous frame (motion estimation). The pixels in that region are then used to predict the pixels in the current frame.

Existing 2D lossless coders [ASS 98] have been extended in order to exploit the inter-image correlation in 3D sequences. These extended coders use an intra-frame predictor, but they incorporate inter-frame information in the context model. A simplified description of this technique is shown in Figure 8.3; for each pixel a fully intra-frame prediction is made. Then, after subtracting this prediction from the original pixel value, a residual is obtained (and stored for later use in the compression of the next frame). In the context modeling step, the context parameter for encoding the current pixel is a quantized version of the intra-frame prediction error in the previous frame at the same position. The underlying idea is that in smooth regions, where intra-frame prediction works well, the context parameter will be low. On the other hand, near moving or static edges, the context parameter will be high. Thus, the context model effectively forces the use of different probability tables in both of these region types. It was experimentally shown that this inter-frame context modeling leads to an additional 10% increase in compression ratio compared to 2D fully predictive techniques.

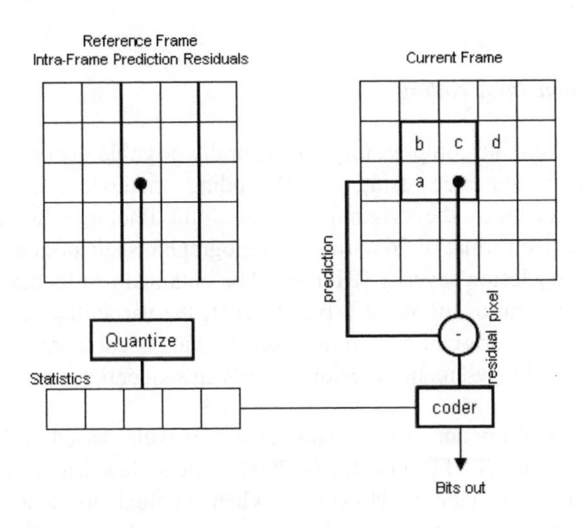

Figure 8.3. *Combination of intra-frame prediction and inter-frame context modeling*

The GRINT (*Generalized Recursive Interpolation*) coding approach [AIA 96], which is an extension of HINT algorithm [AIA 96], is a progressive inter-frame reversible compression technique of tomographic sections that typically occur in the medical field. An image sequence is decimated by a factor of 2, first along rows only, then along columns only, and possibly along slices only, recursively in a sequel, thus creating a gray level hyper-pyramid whose number of voxels halves at every level. The top of the pyramid (root) is stored and then directionally interpolated by means of a 1D kernel. Interpolation errors with the underlying equally-sized hyper-layer are stored as well. The same procedure is repeated, until the image sequence is completely decomposed. The advantage of the novel scheme with respect to other non-causal DPCM schemes is twofold: firstly, interpolation is performed from all error-free values, thereby reducing the variance of residuals; secondly, different correlation values along rows, columns and sections can be used for a better decorrelation.

8.3. Irreversible compression of (2D+t) medical sequences

Due to the limited performance of lossless compression techniques, in terms of compression ratio, there has been significant interest in developing efficient lossy (irreversible) image compression techniques that achieve much higher compression ratios without affecting clinical decision making. Similar to reversible compression schemes, the developed irreversible compression algorithms are also classified as either intra or inter-frame.

8.3.1. *Intra-frame lossy coding*

Similar to reversible compression, it is naturally possible to compress each frame of the dynamic sequence using a 2D coding method. For example, three angiographers reviewed the original and the Joint Photographic Experts Group (JPEG) compressed format of 96 coronary angiographic sequences in a blind fashion to assess coronary lesion severity [RIG 96]. The obtained results have shown that at small compression ratios, of about a 10:1 to 20:1, the variability in assessing lesion severity between the original and compressed formats is comparable to the reported variability in visual assessment of lesion severity in sequential analysis of cine film.

The high performance that characterizes wavelet-based 2D compression algorithms, such as SPIHT and JPEG 2000 coders, has triggered their use on dynamic medical sequences. However, when applied to video angiographic sequences the efficiency of wavelet-based encoding schemes is hampered by the comparatively small valued transform coefficients in the high frequency subbands. The high frequency subbands can be broadly subdivided into *regions* containing diagnostically important information, and regions which do not. Regions which contain diagnostically important information tend to contain far more structure in the high frequency subbands detailing, for example, the precise position of an artery's boundary. In contrast, regions not containing diagnostically important information tend to be noise-like in texture [GIB 01b]. Significant amounts of noise-like texture reduce the effectiveness of the wavelet coding approaches, especially at the relatively high bitrates required for diagnostically lossless images. This higher bitrate results in a much larger number of *significant* wavelet coefficients in the high frequency subbands which in turn must be encoded in the final bit stream. This is inconvenient as it weakens the key quality of energy compaction, requiring the transmission of significant amounts of information to identify the locations of *significant* wavelet coefficients. To overcome this problem, region-based approaches have been investigated to distinguish between the two different types of regions.

Despite the improvement in efficacy of object-based frameworks, which integrate ROI-based functionalities [RAB 02], to assign high priority to the semantically relevant object, to be represented with up to lossless quality and lower priority to the background, these frameworks are not adapted to angiographic images. In fact, the ROI is generally assimilated to square blocks of fixed size that do not necessary fit to medical image structures. In order to improve the diagnosis quality of reconstructed angiographic sequences, the method developed in [GIB 01a] proposes making a distinction between the two different types of regions, with the regions containing diagnostically important information encoded using a standard wavelet encoding algorithm, and the remaining area encoded more efficiently using a wavelet parameter texture modeling approach. This procedure is applied to the two highest frequency subbands, with the remaining levels of the wavelet pyramid encoded in their entirety using a modified version of the SPIHT coder (see Figure 8.4). Despite its compression efficiency, as well as the main concern with important diagnostic information, this method has the disadvantage of not exploiting the temporal correlations between successive frames.

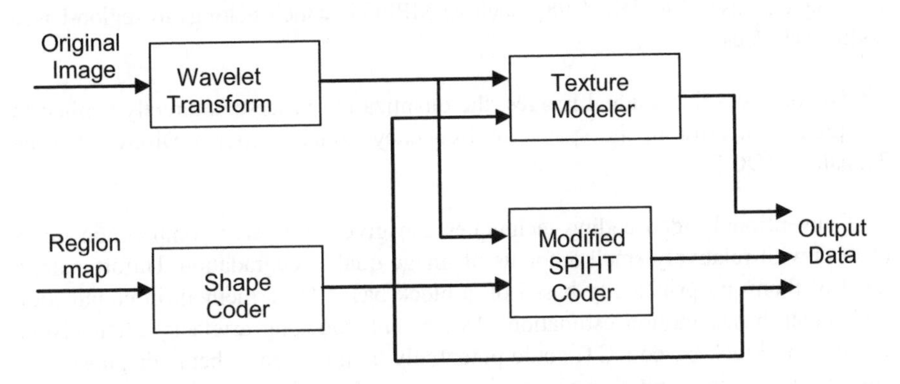

Figure 8.4. *Alternation between MC and intra-image prediction*

Recently, compression systems that support ROI coding using a pre-segmentation approach have been investigated in many image applications. However, segmentation is one of the most significant and difficult tasks in the area of image processing. If a segmentation algorithm fails to detect the correct ROIs then the result would be disastrous since the important diagnostic data in the original image would risk being lost. To solve the problems associated with the pre-segmentation approach, a region scalable coding scheme, based on a post-segmentation approach, has been developed [YOO 04] for interactive medical applications such as telemedicine. Its principle can be described as follows: lossless compression is applied to the whole image and stored in the server. By applying special treatment (set partitioning and rearrangement) to the transformed data, small regions of an image are identified and compressed separately. This will allow a client to access a specific region of interest efficiently by specifying the ROI.

8.3.2. *Inter-frame lossy coding*

8.3.2.1. *Conventional video coding techniques*

The conventional video coding methods have much in common, usually being centered on a combined motion estimation-compensation strategy and followed by an often frequency-based, residual coding method in order to carry out a good prediction which takes into account the motion present between consecutive images. Several approaches have been suggested in order to take this movement into account:

– block matching MC, which is adopted in MPEG1 [GAL 91], MPEG2, H.261, H.263 and H.264 coding schemes;

– region-based MC [KAT 98], such as MPEG4, which belongs to region-based coding schemes.

To compress the residual images, the quantization process is directly applied to the prediction error in the spatial or frequency domain, after a Discrete Cosine Transform (DCT).

Conventional video coding techniques can give impressive compression ratios (100:1) with relatively small amounts of image quality degradation. Unfortunately, most of them are principally based on a block-based DCT method, in conjunction with block-based motion estimation. As a result, blocking artefacts, often visible when using block-based DCT, could potentially be mistaken as being diagnostically significant or interpreted erroneously as pathological signs. Furthermore, the effectiveness assessment results of MC, applied to dynamic medical sequences have shown that conventional MC strategies are not optimal for this type of data, specifically angiogram video sequences. This is largely due to the particular structure of this type of data, i.e. the unusual motion patterns and large amounts of background texture. Unlike High Definition Television (HDTV), individual frames from an angiogram sequence may be extracted and closely scrutinized. All of these requirements have triggered the development of compression methods, specifically designed for dynamic medical data.

8.3.2.2. *Modified video coders*

In angiogram video sequences, temporal correlation in the high frequency image bands results partially from the high frequency noise and partially from the layered nature of the data in which multiple overlaid, semi-transparent layers all move differently. For this reason, constructive use of temporal redundancy in angiogram sequences is an absolute necessity. One solution has been suggested in [GIB 01b] by integrating a low-pass spatial filtering process to the MC feedback loop in conventional DCT based video coders. The two key stages in the modified feedback loop are the MC block and the low-pass spatial filter. The low pass filter employs a Gaussian operator which removes the majority of the high frequency spatial texture. This is essential for angiogram images, because in its absence the MC proves completely counter productive. The MC block is selected according to one of the following schemes: block matching, global compensation or no compensation.

To reduce blocking artefacts often visible when DCT is used in conjunction with block-based motion estimation methods, some solutions [TSA 94] [HO 96] propose using a block-matching approach, to estimate image motion, followed by a global wavelet compression method, instead of the more conventional block-based DCT arrangement. Nevertheless, wavelet-based coding functionalities enable, not only the avoidance of blocking effects, but also the support of object-based bit allocation [BRE 01] and progressive decoding, at the cost of a fuzzy or smoothing effect (see section 8.3.2.3).

In their comparative study, Gibson *et al.* [GIB 01b] have shown that the use of MC is not beneficial in the case of 2D+t medical image compression. This is principally due to the displacement nature of the considered organs; the amount of motion present in dynamic medical sequences is relatively small compared to the motions present in standard video scenes. Furthermore, block-based MC depends on the assumption that objects only move from frame to frame but do not deform or rotate. This assumption is not always satisfied in video applications since, for example, objects may rotate or shrink as they move away from the camera. In the same study, efficiency of 2D DCT and wavelet approaches has been reviewed. Experimental results for these intra-frame compression techniques have shown that wavelet and DCT based algorithms gave very similar results, both numerically and visually. All of these reasons led researchers to define new compression approaches more adapted to 2D+t medical sequences [GIB 04].

8.3.2.3. *2D+t wavelet-based coding systems limits*

More recent studies were oriented towards scalable 2D+t wavelet based coding techniques, inspired by 3D wavelet based coding approaches [LOW 95]. To reduce the computational complexity and the huge amount of memory space, needed to calculate the 2D+t wavelet coefficients, dynamic image sequences are divided into groups of images (GOP: *Group Of Picture*). These groups are compressed independently.

Although they are well adapted to natural video scenes, wavelet-based coding techniques are not suitable for dynamic medical data sets, such as video angiograms [MEN 03]. These images are highly contrasted: very sharp edges are juxtaposed with a smooth background. The edges spread out in the whole subband structure generating a distribution of non-zero coefficients whose spatial arrangement cannot be profitably used for coding. This is principally due to the fact that wavelets are not suitable descriptors of images with sharp edges.

8.4. Irreversible compression of volumetric medical data sets

8.4.1. *Wavelet-based intra coding*

With the widespread use of the Internet, online medical volume databases have gained popularity. With recent advances in picture archiving communication systems (PACS) and telemedicine, improved techniques for interactive visualization across distributed environments are being explored. Typically, data sets are stored and maintained by a database server, so that one or more remote clients can browse the datasets interactively and render them. In many cases, the client is a low-end workstation with limited memory and processing power. An interactive user may be willing to initially sacrifice some rendering quality or viewing field in exchange for real-time performance. Thus, one of the fundamental needs of a client is breadth in terms of interactivity (such as reduced resolution viewing, ability to view a select subsection of the volume, view select slices, etc.) and a pleasant and real-time viewing experience (immediate and progressive refinement of the view volume, etc.). Such a setup enforces the use of hierarchical coding schemes that allow progressive decoding. Out of all the lossless compression methods presented in section 8.2, only HINT and GRINT approaches allow progressive reconstruction. However, since no binary allocation process is made in these two compression schemes, the compression ratios attained are considerably low.

The wavelet transform has many features that make it suitable for progressive transmission (or reconstruction). The implementation via the lifting steps scheme is particularly advantageous in this framework. It provides a very simple way of constructing non-linear wavelet transforms mapping integer-to-integer values. This is very important for medical applications because it enables a lossy-to-lossless coding functionality; i.e. the capability to start from lossy compression at a very high compression ratio and to progressively refine the data by sending detailed information, up until the point where a lossless decompression is obtained. We will not return to the description of 2D compression algorithms, supporting progressive transmission. At present, we only mention 2D wavelet-based embedded image coding algorithms, which have extended our knowledge of 3D coding of the volumetric medical images:

- EZW and SPIHT coders;
- SPECK algorithm;
- QT-L algorithm;
- EBCOT coder.

It is important to note that these techniques support lossless coding, all the required scalability modes, as well as ROI coding.

8.4.2. *Extension of 2D transform-based coders to 3D data*

8.4.2.1. *3D DCT coding*

To take into account the inter-image correlation in volumetric images, DCT-based coding systems have been extended to 3D coding [VLA 95]. A 3D image can then be seen as a volume of correlated content to which 3D DCT transformation is applied for decorrelation purposes. The 3D JPEG-based coder is composed of a discrete cosine transform, followed by a scalar quantizer and finally a combination of run-length coding and adaptive arithmetic encoding. The basic principle is simple: the volume is divided into cubes of 8x8x8 pixels and each cube is separately 3D DCT-transformed, similar to a classical JPEG-coder. Thereafter, the DCT coefficients are quantized using a quantization matrix. In order to derive this matrix, we have to consider two options. One option is to construct quantization tables that produce an optimized visual quality based on psycho-visual experiments. It is worthwhile mentioning that JPEG uses such quantization tables, but this approach would require elaborate experiments in order to come up with reasonable quantization tables for volumetric data. The simplest solution is to create a uniform quantization matrix. This option is motivated by the fact that uniform quantization is optimal or quasi-optimal for most of the distributions. Currently, the uniform quantizer is optimal for Laplacian and exponential input distributions; otherwise the differences with respect to an optimal quantizer are marginal. A second possibility involving quantizers that are optimal in the rate-distortion sense is discussed elsewhere.

The quantized DCT-coefficients are scanned using a 3D space-filling curve, i.e. a 3D instantiation of the Morton-curve, to allow for the grouping of zero-valued coefficients and thus to improve the performance of the run-length coding. This curve was chosen due to its simplicity compared to that of 3D zigzag curves. The non-zero coefficients are encoded using the same classification system as for JPEG.

The coefficient values are grouped into 16 main magnitude classes (ranges), which are subsequently encoded with an arithmetic encoder. Finally, the remaining bits to refine the coefficients within one range are added without further entropy coding.

The adopted entropy coding system shown in Figure 8.5 is partially based on the JPEG architecture, except that the Huffman coder is replaced by an adaptive arithmetic encoder which tends to have a higher coding efficiency. The DC coefficients are encoded with a predictive scheme: apart from the first DC coefficient the entropy coding system encodes the difference between the current DC coefficient and the previous one. The AC coefficients are encoded in the form of pairs (RUN, BEEN WORTH), where "RUN" specifies the amount of zeros preceding the encoded symbol, designated by the "BEEN WORTH" term. The range of the encountered significant symbol is encoded, using an arithmetic encoder with a similar (AC) model as in the case of the DC coefficients. Nevertheless, similar to their 2D counterparts, 3D DCT-based compression techniques are not able to satisfy the requirements of progressive transmission and perfect reconstruction.

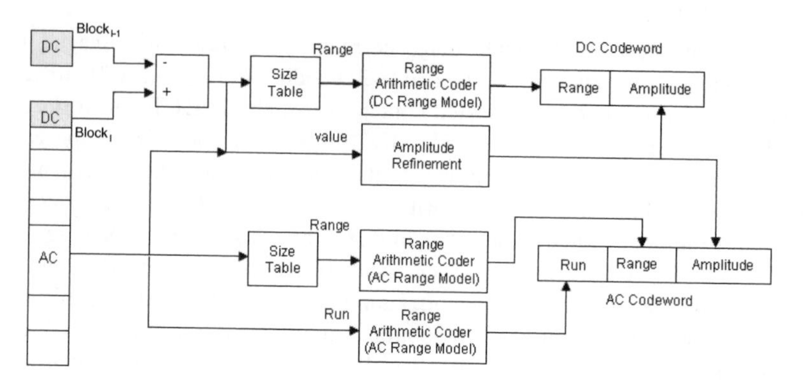

Figure 8.5. *3D DCT-based coder*

8.4.2.2. *3D wavelet-based coding based on scalar or vector quantization*

Before describing the 3D wavelet-based coding techniques, it is important to note that these techniques support lossless coding, all the required scalability modes, as well as ROI coding and this is a significant difference with respect to the 3D DCT technique presented above, which is not able to provide these features.

For all 3D wavelet based coders adapted to medical applications, a common wavelet transform module was designed that supports lossless integer lifting filtering, as well as finite-precision floating-point filtering. Different levels of decomposition for each spatial direction (x-, y-, or z- direction) are supported by the wavelet module. In [LOW 95] [WAN 95], a 3D separable wavelet transform is used to remove inter-slice redundancy, while in [WAN 96], different sets of wavelet filters are used in the (x,y) plane and z direction, respectively, to account for the difference between the intra- and inter-slice resolution. In fact, the distance between adjacent pixels in the same 2D image varies from 0.3 to 1 mm, while the distance between two successive images in IRM or CT sequences varies from 1 to 10 mm. This has led to the common consensus that the use of the full 3D data correlation potentially improves compression. The obtained wavelet sub-volumes are independently quantized, either by uniform scalar quantization or vector quantization, and finally entropy coded. It is worth noting that many works have been investigated in order to define a quantization policy which ensures the highest decoding quality, for a given rate, over the entire image volume.

8.4.2.3. *Embedded 3D wavelet-based coding*

Since contemporary transmission techniques require the use of concepts such as rate scalability, quality and resolution scalability, multiplexing mechanisms need to be introduced to select from each slice the correct layer(s) to support the currently required Quality-of-Service (QoS) level. However, a disadvantage of the slice-by-slice mechanism is that potential 3D correlations are neglected. Evaluation of wavelet volumetric coding systems that meet the above-mentioned requirements has shown that these coder types deliver the best performance for lossy-to-lossless coding.

3D wavelet-based embedded image coding algorithms typically apply successive approximation quantization (SAQ) to provide quality scalability and facilitate the embedded coding. The resolution scalability is a direct consequence of the multi-resolution property of the DWT. Both resolution and quality scalability are provided by the multi-layered organization of the code-stream into packets. In what follows, we will restrict ourselves to the description of 3D wavelet-based embedded coding techniques that employ zero tree- or block-based structures. It is worth mentioning that 3D SPIHT (Set Partitioning into Hierarchical Trees) [KIM 00] and 3D SB-SPECK (SuBband-based Set Partitioned Embedded bloCK coding) [WHE 00] are the well-known representatives of the family of inter-band embedded 3D coding approaches.

In the case of 2D images, these strategies use the inter-band dependencies of the wavelet coefficients to provide a progressive improvement of the reconstructed image quality. More recently, other 3D wavelet-based coding algorithms have been developed, taking into account the intra-band dependencies between wavelet coefficients. Among these algorithms, a new coder called the 3D Quadtree Limited (3D QT-L) combines the basic principles of quadtree and block-based coding.

The investigations performed in the context of the new standard for 2D image coding, JPEG 2000 (based on the wavelet transform), like random access to regions of interest and lossy-to-lossless coding, triggered the ISO/IEC JTC1/SC29/WG1 committee to develop the JP2000 3D coder equipped with a 3D wavelet transform. The latter is one of the functionalities provided by the latest Verification Model software (from V7.0 on), which was added to support multi-spectral image coding. More recently, the ISO/IEC JTC1/SC29/WG1 committee decided to develop JP3D (i.e. Part 10 of JPEG 2000 standard), which gives support to 3D encoding mechanisms. The MLZC (Multi Layered Zero Coding) algorithm was proposed recently in [MEN 03] and provides high coding efficiency, fast access to any 2D image of the dataset and quality scalability. Its structural design is based principally on the layered zero coding (LZC) algorithm coder [TAU 94] designated to video coding. The main differences between LZC and the MLZC algorithm concern the underlying subband structure and the definition of the conditioning terms.

8.4.2.3.1. 3D set partitioning in hierarchical trees

The 3D SPIHT [KIM 00] implementation uses balanced 3D spatial orientation trees (Figure 8.6). Therefore, the same number of recursive wavelet decompositions is required for all spatial orientations. If this is not respected, several tree nodes are not linked within the same spatial location, and consequently the dependencies between different tree-nodes are destroyed and thus the compression performance is reduced. The examined 3D SPIHT algorithm follows the same procedure as its 2D homologous algorithm, with the exception that the states of the tree nodes, each embracing eight wavelet coefficients, are encoded with a context-based arithmetic coding system during the significance pass. The selected context models are based on the significance of the individual node members, as well as on the state of their descendants.

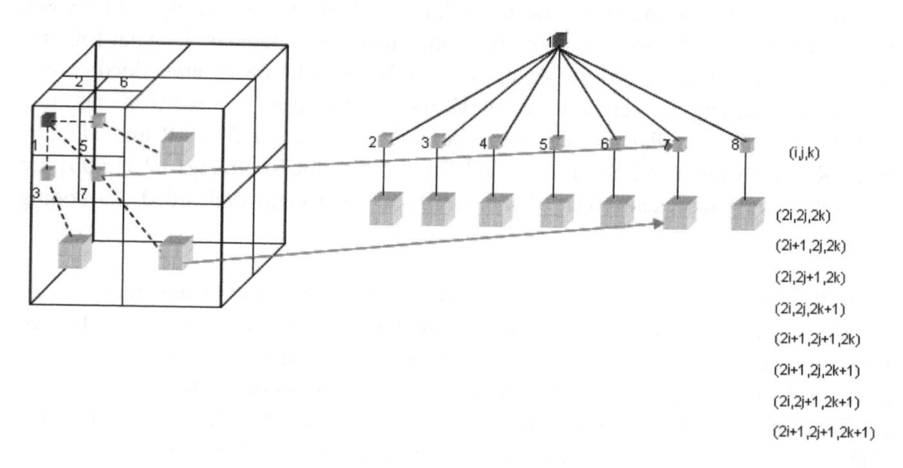

Figure 8.6. *Parent descendant inter-band dependency*

8.4.2.3.2. Cube splitting

The cube splitting technique is derived from the 2D square partitioning coder (SQP) applied to the angiography sequences [MUN 99]. In the context of volumetric encoding, the SQP technique was extended to a third dimension: from square splitting towards cube splitting. Cube splitting is applied to the wavelet image in order to isolate smaller entities, i.e. sub-cubes, possibly containing significant wavelet coefficients. Figure 8.7 illustrates the cube splitting process.

During the first significance pass S, the significance of the wavelet image (volume) is tested for its highest bit-plane. The wavelet image is spliced into eight sub-cubes (or octants). When a significant wavelet coefficient, in a descendent cube, is encountered, the cube (Figure 8.7a) is spliced into eight sub-cubes (Figure 8.7b), and so on (Figure 8.7c) up until the pixel level. The result is an octree structure (Figure 8.8d), with the SGN symbol indicating the significant node and the NS symbol designating the non-significant node. In the next significance pass, the non-significant nodes that contain significant wavelet coefficients are refined further. As we may observe, all branches are accorded equal importance.

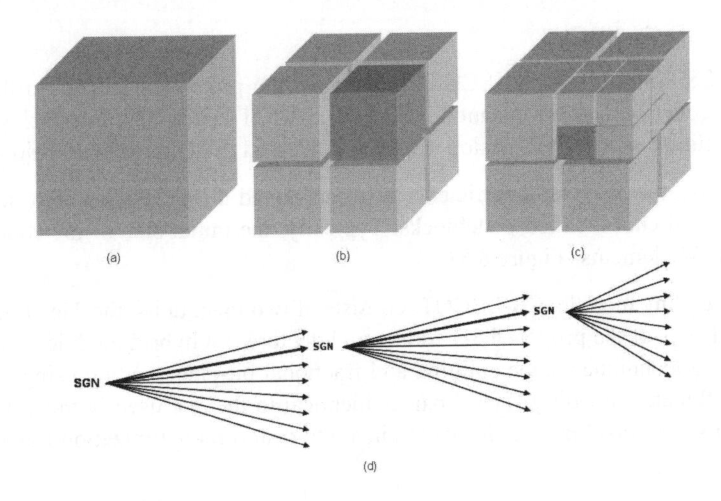

Figure 8.7. *Cube splitting according to the wavelet coefficients significance*

When the complete bit-plane is encoded with the significance pass, the refinement pass R is initiated for this bit-plane, refining all coefficients marked as significant in the octree. Thereafter, the significance pass is restarted in order to update the octree by identifying the new significant wavelet coefficients for the current bit-plane. During this stage, only the previously non-significant nodes are checked for significance, and the significant ones are ignored since the decoder has already received this information. The described procedure is repeated, until the complete wavelet image is encoded or until the desired bitrate is obtained. To encode the generated symbols efficiently, a context-based arithmetic encoder has been integrated.

8.4.2.3.3. 3D QT-L

The QT-L coder has also been extended towards 3D coding. The octrees corresponding to each bit-plane are constructed following a similar strategy as for the cube splitting coder. However, the partitioning process is limited in such a way that once the volume of a node becomes smaller than a predefined threshold, the splitting process is stopped, and the entropy coding of the coefficients within such a significant leaf node is activated. Similar to the 2D version, the octrees are scanned using depth-first scanning. In addition, for any given node, the eight descendant nodes are scanned using a 3D instantiation of the Morton curve. For each bit-plane, the coding process consists of the non-significance, significance and refinement passes. For the highest bit-plane, the coding process consists of the significance pass only. The context-conditioning scheme and the context-based entropy coding are similar to their 2D counterparts.

8.4.2.3.4. 3D CS-EBCOT

The CS-EBCOT coding [SCH 00] combines the principles utilized in the cube splitting coder with a 3D instantiation of the EBCOT coder. The interfacing of the cube splitting coder with a version of EBCOT adapted to 3D is outlined below:

– firstly, the wavelet coefficients are partitioned EBCOT-wise into separate, equally sized cubes, called codeblocks. Typically, the initial size of the codeblocks is 64x64x64 elements (Figure 8.8);

– the coding module, CS-EBCOT, consists of two main units, the Tier 1 and Tier 2 parts: Tier 1 of the proposed 3D coding architecture is a hybrid module combining two coding techniques: cube splitting and fractional bit-plane coding using context-based arithmetic encoding. Tier 2 part is identical to the one used in the 2D coding system. Its concerns the layer formation in a rate-distortion optimization framework.

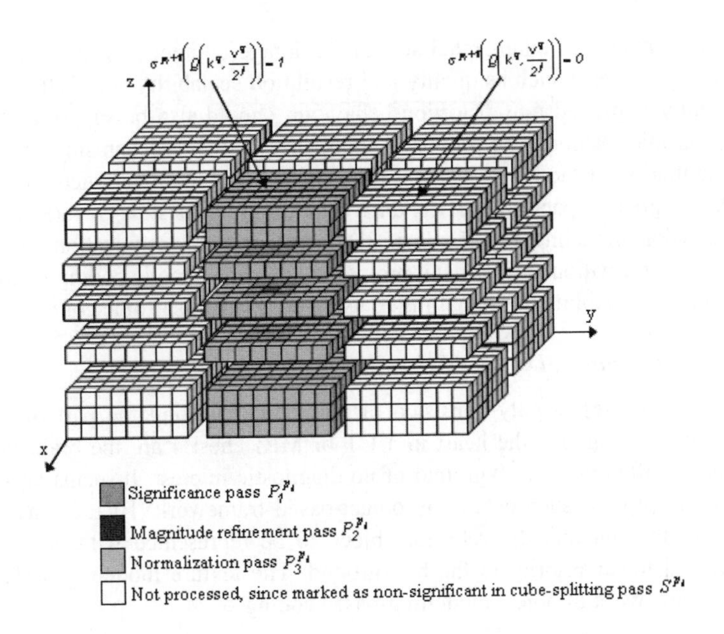

Figure 8.8. *Cube-splitting based on wavelet coefficient significance*

To use pixel correlation with its neighborhood, according to the three spatial directions, CS-EBCOT uses 3D context models for the adaptive arithmetic coding which is identical to those utilized in the previously mentioned encoders.

8.4.2.3.5. JP3D

Part 10 of JPEG 2000 (JP3D) [BIS 03] is a work item that provides extensions of JPEG-2000 for rectangular 3D hyper-spectral and volumetric medical data sets with no time component. It provides a specification for the coding of 3D data sets that are as isotropic as possible; i.e. identical processing capabilities will be provided in all three dimensions. The proposed volumetric coding mechanism is based on a 3D instantiation of the EBCOT coder. To assure the adoptability of the Part 10 extension of the JPEG 2000 standard, two important requirements are imposed; i.e. the *backward compatibility* requirement and the *isotropic functionality* requirement. The *backward compatibility requirement* demands that any compliant Part 1 or Part 2 code-stream needs to be recognized as a compliant JP3D code-stream, respecting the original semantic interpretation of all markers, parameters and bit-streams. In particular, when given a compliant Part 1 or Part 2 code-stream and legal decoding options, a JP3D decoder will produce the same reconstructed image as a Part 1 or Part 2 reference decoder. Secondly, the new or extended technology, included for JP3D, shall provide the same, i.e. *isotropic functionality* in all 3 coordinate directions, while remaining compatible with the syntax used in Part 2 that describes volumetric images as enumerated sets of 2D image components. It is evident that typical requirements such as quality and resolution scalability and ROI, apart from additionally improved rate-distortion behaviour, should also be supported by JP3D in an isotropic fashion. The proposed volumetric coding mechanism is based on a 3D instantiation of the EBCOT coder [SCH 05]. All of the mentioned requirements have had a great impact on JP3D coding technology: consisting of rate control and rate-distortion modeling for 3D data, 3D rate/distortion optimization, coding and code-stream formation for 3D data, provision of 3D arithmetic coding contexts and coding passes, enabling the signalling of user-specified anisotropic context models.

8.4.2.4. *Object-based 3D embedded coding*

Medical images usually consist of a region representing the part of the body under investigation (i.e. the heart in a CT or MRI chest scan, the brain in a head scan) on an often noisy background of no diagnostic interest. It seems very natural therefore to process such data in an object-based framework [BRE 01] assign high priority to the semantically relevant object, to be represented with up to lossless quality, and lower priority to the background. The texture modeling perfectly fits into the framework of object modelling-based coding.

In the context of object-based coding, this means that texture modeling can be applied only to objects for which the constraint of being able to recover the original data without loss can be relaxed. This naturally maps onto the medical imaging field: objects which are diagnostically relevant should be selectively encoded by one of the methods mentioned above with the remaining objects (or even the background) replaced by a suitable synthetic model. Although local structures and regular edges can be effectively coded, it is worth pointing out that similar to other pre-segmentation approaches where ROIs are identified and segmented before encoding, this method does not allow direct access to the regions of interest. To alleviate the pre-segmentation problem, region scalable coding schemes, based on post-segmentation approaches have been developed. The 3D MLZC (Multi Layered Zero Coding) algorithm [MEN 03] proposes a fully 3D wavelet-based coding system allowing random access to any object at the desired bitrate. Continuous wavelet-based segmentation is used to select objects that should be coded independently. Although this approach, also referred to as the *spatial approach,* has the advantage that the ROI information coincides with the physical perception of the ROI, extra complexity is inserted at the level of the wavelet transform (shape-adaptive transform).

Although it is possible to compress an image set in intra mode, which uses the intra image correlation, a more efficient compression can be achieved by exploring the correlation between slices. However, the experiments carried out in [MEN 00] have shown that the benefit of an inter-image correlation depends on the slice distance of the image set. The smaller the slice distance, the better the correlation in the slice direction, and the better the compression performance. In general, effective results are produced for image sets with slice distances smaller than 1.5 mm. The same experimental results have demonstrated that the use of 3D context models is not profitable if it is integrated into a 3D wavelet compression scheme.

8.4.2.5. *Performance assessment of 3D embedded coders*

The lossy and lossless compression performances of the embedded wavelet-based coders, described previously, were evaluated according to a set of volumetric data obtained with different imaging modalities [SCH 03a], including: positron emission tomography, PET (128x128x39x15bits), computed tomography, CT1 (512x512x100x12bits), CT2 (512x512x44x12bits), ultrasound, US (256x256x256x8bpp), and MRI data, MRI (256x256x200x12bpp). Lossless coding results are reported for most of the techniques discussed so far: CS, 3D QT-L, 3D SPIHT, 3D SB-SPECK (only for the CT2 and MRI data-based), CS-EBCOT and JPEG 2000. For obvious reasons the 3D DCT coder was not included in the lossless compression test due to the lossy character of its DCT front-end. Additionally, the coding results obtained with the JPEG 2000 coder equipped with a 3D wavelet transform (JPEG2K-3D) are reported. For all the tests performed in lossless coding (as well as for lossy coding later), typically a 5-level wavelet transform (with a lossless 5x3 lifting kernel) was applied to all spatial dimensions, except for the low-resolution PET image (4 levels). The same number of decompositions in all dimensions was used to allow fair comparison with the 3D SPIHT algorithm. Table 8.1 shows the increase in percentage of the bitrate achieved in lossless compression, with the reference technique taken as the algorithm yielding the best coding results for each test volume.

	US	PET	CT1	CT2	MRI
CS-EBCOT	0.59	0.49	0.0	0.0	0.0
3D QT-L	0.0	0.0	0.82	0.37	0.43
CS	1.7	2.2	1.29	1.34	1.83
JPEG 2000 3D	6.24	2.12	3.11	2.16	5.70
3D SPIHT	1.52	6.87	3.21	4.60	2.48
SB-SPEK				4.15	10.44
JPEG 2000	15.62	18.81	7.05	4.01	19.49
Best coder	3D QT-L	3D QT-L	CS-EBCOT	CS-EBCOT	CS-EBCOT

Table 8.1. *The increase of the lossless bitrate in terms of percentage, with the reference technique taken as the algorithm yielding the best coding results for each test volume*

Based on Table 8.1, we can observe that for the US and PET volumes, the 3D QT-L coder delivers the best coding performance, while for the other three volumetric data, the CS-EBCOT performs best. If we refer to the average increase in percentage taking the CS-EBCOT coder as the reference, then we will observe that the 3D QT-L yields a similar performance, since the average difference between the two is only 0.1%. The CS coder follows it, with a difference of 1.45%. The 3D SPIHT and the JPEG2K-3D coders provide similar results, with an average difference of 3.56% and 3.65% respectively. Finally, the average difference increases up to 7.07% and 12.78% for the 3D SB-SPECK and JPEG 2000 coders. We also observe from Table 8.1 that the relative performance of several techniques is heavily dependant on the data set involved. For example, 3D SPIHT delivers excellent results for the US, CT1 and MRI sets, while for the other ones the performance is relatively poor. JPEG 2000 yields the worst coding results of all, except for the CT2 image, which has a low axial resolution. The results of the 3D SB-SPECK have been reported only for the CT2 and MRI data sets, and the results are situated in between JPEG2K and JPEG2K-3D for the MRI volume.

In summary, these results lead to the following important observations for lossless coding:

– CS-EBCOT and 3D QT-L deliver the best lossless coding results on all images;

– the 3D wavelet transform as such significantly boosts the coding performance;

– as spatial resolution and consequently inter-slice dependency diminishes, the benefit of using a 3D de-correlating transform and implicitly a 3D coding system decreases.

Lossy coding experiments were carried out on the five volumetric data sets for the aforementioned coders, and in addition, the 3D DCT-based coding engine is included [SCH 03b]. The peak signal-to-noise ratio (PSNR) is measured at seven different bitrates: 2, 1, 0.5, 0.25, 0.125, 0.0625 and 0.03125 bits-per-pixel (bpp). Similar to the lossless coding experiments, the performance of the lossy wavelet coders is evaluated. In a first evaluation, we observe that the 3D QT-L coder outperforms all the other wavelet coders in the whole range of bitrates. For example, the 3D QT-L coder yields on the US data set at 1.00 bpp a PSNR of 38.75dB, which is 0.5 dB better than the JPEG 2000 3D, and 1.43 dB better than CS-EBCOT. At higher rates (2bpp) the differences between them increase up to 0.88 dB and 1.45 dB for the JPEG 2000 3D and CS-EBCOT coders respectively. The 3D QT-L and the JPEG 2000 3D algorithms perform equally well at low rates (0.125 bpp) on the US data set, and outperform the CS-EBCOT coder with 0.37 dB and 0.25 dB respectively. A similar ranking according to their performance can be achieved by taking into account the result obtained on the MRI data set. At 0.125bpp the results are in order 52.01 dB, 51.61 dB and 51.17 dB for the 3D QT-L, JPEG 2000 3D, and CS-EBCOT respectively. Note that at lower rates CS-EBCOT gives slightly better results (0.03125bpp – 46.52 dB) than JPEG 2000 3D (0.03125bpp – 46.22 dB) but still less than those provided by 3D QT-L with 46.75 dB at the same rate. Similarly, the 3D QT-L outperforms on the CT1 data set the next rated wavelet coder JPEG 2000 3D, but the differences between them are smaller: from 0.27 dB at 1bpp, to 0.57 dB at 0.03125bpp.

The results obtained on the PET volumetric data indicate that at rates below 0.25 bpp the 3D QT-L coder outperforms all the other coders. However, at higher rates the JPEG 2000 3D outperforms the 3D QT-L coder on this data. If we look at the JPEG 2000 standard, we note that this coder typically delivers poor coding results on the PET, US volumes and MRI data; on CT1 it yields good results at rates higher than 0.25bpp, but the results are modest at lower rates. However, for CT2, JPEG 2000 is the best coder at high bitrates, and is only beaten by JPEG 2000 3D and SB-SPECK at low-bitrates.

8.5. Conclusion

This chapter provides an overview of various state-of-the-art lossy and lossless image compression techniques, which were applied to dynamic and volumetric medical images. Although both of the datasets considered are composed of 2D image sequences, the coding performance of lossy compression has been shown to be heterogenous with the image representation or characteristics. For example, since video angiogram sequences are highly contrasted, very sharp edges are juxtaposed to a smooth background. Wavelet-based coding techniques are not suitable for this kind of data. The edges spread out in the whole subband structure generating a distribution of non-zero coefficients whose spatial arrangement cannot be profitably used for coding.

Conversely, in the case of volumetric medical sequences, the combination of the 3D wavelet transform with an *ad-hoc* coding strategy provides a high coding efficiency. Another important advantage of 3D wavelet transforms is the ability to produce a scalable data stream which embeds subsets progressing in quality all the way up to a lossless representation which is very important for medical applications. The *zero-trees* principle, as well as other set partitioning structures such as *quad-trees*, allow for the grouping of zero-valued wavelet coefficients, taking into account their intra or inter-band dependencies. Therefore, by using such models, we can isolate interesting non-zero details by immediately eliminating large insignificant regions from further consideration. Before passing through the entropy coding process, the significant coefficients are quantized with successive approximation quantization to provide a multiprecision representation of the wavelet coefficients and to facilitate the embedded coding.

It is obvious that diagnostic zone selection according to the "ROI" principle supported by cubic splitting compression systems is not very reliable. This is mainly due to the fact that the particular structures that characterize them are not modeled with sufficient precision. The hybrid compression techniques recently implemented propose hybrid coding systems of embedded object oriented type.

The progressive compression systems highlighted in this chapter can be approved by different strategies developed in [KIM 98], [XIO 03] and [KAS 05]. The latter reference also integrates a novel MC technique that addresses respiratory and cardiac activity movement. Cardiac image compression is a more complex problem than brain image compression, in particular because of the mixed motions of the heart and the thorax structures. Moreover, cardiac images are usually acquired with a lower resolution than brain images. This highlights certain problems induced by the spatio-temporal image acquisition that hamper MC in dynamic medical image compression schemes. The integration of multiple complementary data, generated in "listmode" acquisition mode, into a common reference provides information relevant to compression research. For example, in a cardiac image coding framework, based on MC, additional physiological signals such as ECG-gated film sequences as well as the respiration movement are required to take account of heart motion and deformation; for more details, see [VAN 05].

8.6. Bibliography

[AIA 96] AIAZZI B., ALBA P. S., BARAONTI S., ALPARONE L., "Three dimensional lossless compression based on a separable generalized recursive interpolation", *Proc. ICIP'96,* Lausanne, Switzerland, p. 85-88, 1996.

[ASS 98] VAN ASSCHE S., DENECKER K., PHILIPS W., LEMATHIEU I., "Lossless compression of three-dimensional medical images", *PRORISC'98*, p. 549-53, 1998.

[BAS 95] BASKURT A., BENOIT-CATTIN H., ODET C., "On a 3-D medical image coding method using a separable 3-D wavelet transform", *Proc. SPIE Medical Imaging*, vol. 2431, p. 173-183, 1995.

[BRE 01] BRENNECKE R., BURGEL U., RIPPIN G., POST F., RUPPERCHET HJ., MEYER J., "Comparison of image compression viability for lossy and lossless JPEG and Wavelet data reduction in coronary angiography", *Int J. Cardiovasc. Imaging*, vol. 17, p. 1-12, 2001.

[BRI 03] BRISLAWN C., SCHELKENS P., "Embedded JPEG 2000 Part 10: Verification model (VM10) users' guide", ISO/IEC JTC1/SC29/WG1, 2003.

[CLU 00] CLUNIE D.A., "Lossless compression of grayscale medicale images – effectiveness of traditional and state of the art approaches", *Proceedings of SPIE – The International Society for Optical Engineering*, vol. 3980, p. 74-84, 2000.

[GAL 91] LE GALL D., "MPEG: A video compression standard for multimedia applications", *IEEE PIMRC'95*, vol. 1, p. 120-124, 1995.

[GIB 01a] GIBSON D., TSIBIDIS G., SPANN M., WOOLLEY S., "Angiogram video compression using a wavelet-based texture modelling approach", *Human Vision and Electronic Imaging VI, Proc. of SPIE*, vol. 4299, 2001.

[GIB 01b] GIBSON D., SPANN M., WOOLEY S., "Comparative study of compression methodologies for digital angiogram video", *ISPACS 2000*, Honolulu, Hawaii, 2000.

[GIB 04] GIBSON D., SPANN M., WOOLEY S., "A Wavelet-based region of interest encoder for the compression of angiogram video sequences", *IEEE Trans. on Information Technology in Biomedicine*, vol. 8, no. 2, 2004.

[HO 96] HO B., TSAI M., WEI J., MA J., SAIPETCH P., "Video compression of coronary angiograms based on discrete wavelet transform with block classification", *IEEE Trans. on Medical Imaging*, vol. 15, p. 814-823, 1996.

[KAS 05] KASSIM AA., YAN P., LEE WS., SENGUPTA K. "Motion compensated lossy-to-lossless compression of 4-D medical images using integer wavelet transforms", *IEEE Trans. on Medical Imaging*, vol. 9, p. 132-138, 2005.

[KAT 98] KATSAGGELOS A.K., KONDI L.P., MEIER F.W., OSTERMANN J., SCHUSTER G.M., "MPEG-4 and rate-distortion-based shape-coding techniques", *Proc. of the IEEE*, no. 6, p. 1126-1154, 1998.

[KIM 98] KIM Y-S., AND KIM W-Y., "Reversible decorrelation method for progressive transmission of 3-D medical image", *IEEE Trans. on Medical Imaging*, vol. 17, p. 383-394, 1998.

[KIM 00] KIM. B-J., PEARLMAN W. A., "Stripe-based SPIHT lossy compression of volumetric medical images for low memory usage and uniform reconstruction quality", *Proc. ICASSP*, vol. 4, p. 2031-2034, June 2000.

[MEN 00] MENEGAZ G., "Model-based coding of multi-dimensional data with applications to medical imaging", PhD dissertation, Signal Processing Lab. (LTS), Ecole Polytechnique Fédérale de Lausanne, Switzerland, 2000.

[MEN 03] MENEGAZ G., THIRAN J. P., "Three-dimensional encoding two-dimensional decoding of medical data", *IEEE Trans on Medical Imaging*, vol. 22, no. 3, 2003.

[MUN 99] MUNTENANU A., CORNELIS S.J., CRISTEA P., "Wavelet-based lossless compression of coronary angiographic images", *IEEE Trans. on Medical Imaging*, vol. 18, p. 272-281, 1999.

[RAB 02] RABBANI M., JOSHI. R., "An overview of the JPEG2000 still image compression standard", *Signal Processing Image Communication*, vol. 17, no. 3, 2002.

[RIG 96] RIGOLIN V.H., ROBIOLIO P.A., SPERO L.A., HARRAWOOD B.P., MORRIS K.G., FORTIN D.F., BAKER W.A., BASHORE T.M., CUSMA J.T., "Compression of digital coronary angiograms does not affect visual or quantitative assessment of coronary artery stenosis severity", *Am J Cardiol*, vol. 78, p. 131-136, 1996.

[SCH 00] SCHELKENS P., GIRO X., BARBARIEN J., CORNELIS J., "3-D compression of medical data based on cube-splitting and embedded block coding", *Proc. ProRISC/IEEE Workshop*, p. 495-506, December 2000.

[SCH 03a] SCHELKENS P., MUNTEANU A., BARBARIEN J., MIHNEAI G., GIRO-NIETO X., CORNELIS J., "Wavelet coding of volumetric medical datasets", *IEEE Trans. on Medical Imaging*, vol. 22, no. 3, March 2003.

[SCH 03b] SCHELKENS P., MUNTEANU A., "An overview of volumetric coding technologies", *ISO/IEC JTC1/SC29/WG1, WG1N2613*, 2002.

[SCH 05] SCHELKENS P., MUNTAEANU A., CORNELIS J., "Wavelet coding of volumetric medical data set", *ICIP 2005*, Singapore, 2005.

[TAU 94] TAUBMAN D., ZKHOR A., "Multirate 3-D subband coding of video", *IEEE Trans. Image Processing*, vol. 3, p. 572-588, September 1994.

[TSA 94] TSAI M., VILLASENOR J., HO B., "Coronary angiogram video compression", *Proc. IEEE Conference on Medical Imaging*, 1994.

[VAN 05] VANDERBERGUE S., STAELENS S., VAN DE WALLE R., DIERCKX R., LEMATHIEU I., "Compression and reconstruction of sorted PET listmode data", *Nucl. Med. Commun.*, vol. 26, p. 819-825, 2005.

[VLA 95] VLAICU A., LUNGU S., CRISAN N., PERSA S., "New compression techniques for storage and transmission of 2-D and 3-D medical images", *Proc. SPIE Advanced Image and Video Communications and Storage Technologies*, vol. 2451, p. 370-377, February, 1995.

[WAN 95] WANG J., HUANG H.K., "Medical image compression by using three-dimensional wavelet transformation", *Proc. SPIE Medical Imaging*, vol. 2431, p. 162-172, 1995.

[WAN 96] WANG J., HUANG. H.K., "Three-dimensional medical image compression using a wavelet transform with parallel computing", *IEEE Trans. on Medical Imaging*, vol. 15, no. 4, p. 547-554, August 1996.

[WHE 00] WHEELER F.W., "Trellis source coding and memory constrained image coding", PhD thesis, Dept. Elect., Comput. Syst. Eng., Renselaer Polytech. Inst., Troy, New York, 2000.

[XIO 03] XIONG Z., WU X., CHENG S., HUA J., "Lossy-to-lossless compression of medical volumetric data using three-dimensional integer wavelet transforms", *IEEE Trans. on Medical Imaging*, vol. 22, p. 459-470, 2003.

[YOO 04] YOON S.H., LEE J.H., KIM J.H., ALEXANDER W., "Medical image compression using post-segmentation approach", *ICASSP 2004*, p. 609-612, 2004.

Chapter 9

Compression of Static and Dynamic 3D Surface Meshes

9.1. Introduction

Static and dynamic volume data has been used in fluid mechanics, aeronautics, and geology for a long time now, and nowadays it is becoming more widely used in medical imagery to analyze the complex functions such as those in the lungs [PER 04], [FET 03], [FET 05], [SAR 06] and the heart [ROU 05], [ROU 06], [DIS 05] (Figure 9.1).

3D medical imagery is generally visualized by volume rendering [LEV 88]. An observer point of view is selected, its viewing rays cut through the 3D volume perpendicularly to the visualization plane, which is a 2D projection of the volume. The voxels with the same gray level make an isosurface of the same opacity, chosen according to the tissue we wish to view. The gray level gradient allows us to know the normal to the isosurface. Shading based on the transmission, reflection and diffusion of light, gives the desired volume rendering (Figure 9.2a).

An alternative is surface rendering also known as "geometric". Here, we visualize an isosurface with a predefined gray level. This isosurface is extracted by segmentation of the volume of the data (voxels): a binary volume is constructed. In comparison, volume rendering can display weak surfaces without binary decision. Geometric rendering generally considers the isosurface to be opaque. It reflects the ray according to the normal to the surface, with light diffusion displaying volume

---

Chapter written by Khaled MAMOU, Françoise PRÊTEUX, Rémy PROST and Sébastien VALETTE.

rendering. This technique can be used for non-overlapping embedded surfaces, these surfaces being semi-transparent.

Isosurfaces are generally triangulated (Figures 9.2b and 9.2c). A triangular mesh is generally constructed by scanning the adjacent slices of the binary volume [LOR 87] or by covering the surface by a wave front propagation starting from a seed voxel [WOO 00], or by deforming an initial low-resolution mesh and refining the triangles through subdivisions [LOT 99]. The Marching Cubes (MC) algorithm is the most widely-used in the medical imaging field [LOR 87], (see Appendix A in section 9.6).

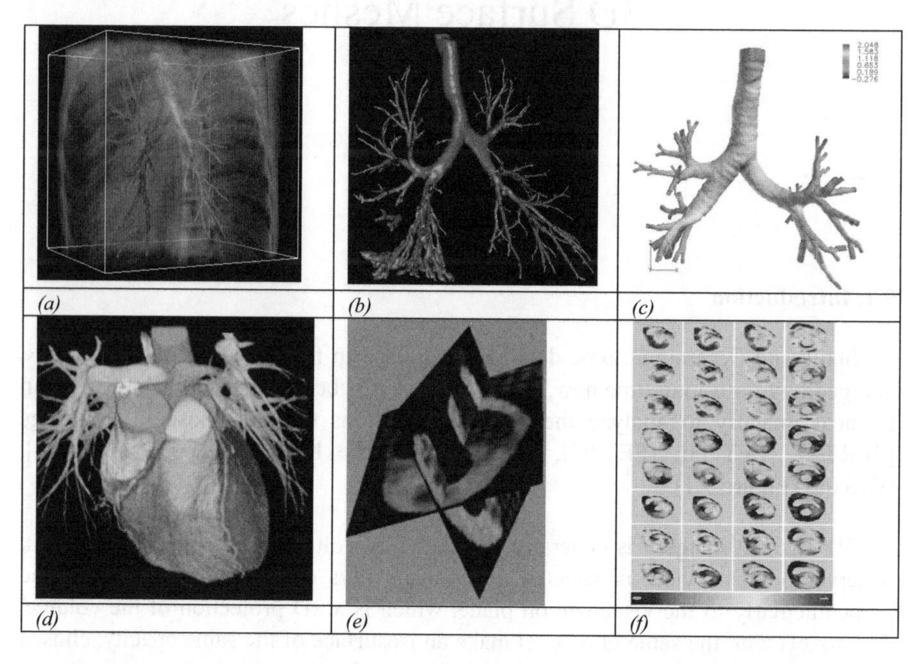

Figure 9.1. *Examples of 3D visualizations in medical imaging: (a) semi-transparent rendering of the lungs and bronchial tree in computed tomography [PER 04], (b) pathological bronchial tree [SAR 06], (c) 3D pressure map of the airway mesh model [FET 05], (d) volume rendering of the heart, (e) visualization of myocardial function [ROU 05], and (f) myocardial radial contraction in tagged MRI [ROU 05]*

Volume rendering requires careful management of the region opacity and relatively complex calculations, but it has the advantage of requiring no segmentation. Surface rendering is quick, particularly when the triangular mesh has been calculated and stored in advance. Computer graphics hardware is developed for fast processing of this type of data. Storing or sending a 3D object is much cheaper, in terms of bits, for a surface than for a volume. Nevertheless, the cost of storing meshed surface data remains prohibitive, especially in the case of animated 3D meshes, where hundreds of frames are required to describe a short sequence lasting only a few seconds. Therefore the development of efficient compression techniques based on compact and hierarchical representations is essential to allow 3D meshes transmission, management and visualization.

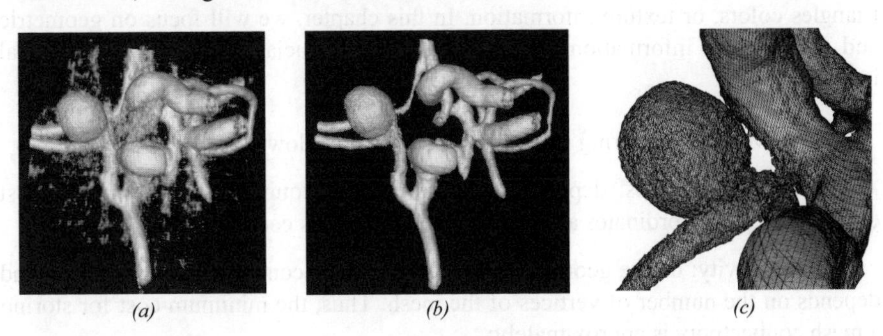

Figure 9.2. *3D visualization of a cerebral vascular artery with an aneurism [BON 03]: a) volume rendering, b) surface rendering of (a), c) wire rendering of (b)*

This chapter is devoted to the compression of static and dynamic 3D meshes. After a summary of the definitions and properties of 3D triangular meshes (section 9.2), the key approaches for static mesh compression are presented and the multiresolution geometric wavelet techniques are examined closely (section 9.3). Section 9.4 presents a critical analysis of dynamic mesh compression techniques (sequences of meshes). After a description of the mathematical methods underlying the four main types of approach, their respective performances are compared and contrasted on the criteria of their computational complexity, genericity and functionalities such as progressive transmission or scalability. Finally, an application to the pulmonary function analysis in high resolution computed tomography illustrates and concludes this method overview.

9.2. Definitions and properties of triangular meshes

The representation of a 3D object using a surface mesh requires storing two types of information:

– geometry: the coordinates of the vertices;

– connectivity: the description of how the vertices are connected, which takes the form of a list of triangles, therefore describing the surface of the mesh.

The connectivity is described by a graph – which may or may not be planar – with corresponding vertices and edges. The connectivity is the topological data which should not be confused with the topology of the surface represented or approximated by the mesh.

These two types of data are sufficient to represent the shape of the underlying object. Additional information can enrich this representation, such as vertices or triangles colors, or texture information. In this chapter, we will focus on geometric and connectivity information, which are the most crucial in the field of medical imaging.

The naive cost for storing mesh data is estimated below:

– geometry: this cost depends on the precision required by the user. In most cases, the vertex coordinates are represented by integers coded with 10 or 16 bits;

– connectivity: unlike geometry, the cost of storing connectivity is not fixed, and depends on the number of vertices of the mesh. Thus, the minimum cost for storing a mesh connectivity is approximately:

$$c = 3|F|\log_2(|V|) \tag{9.1}$$

where |F| is the number of triangles and |V| is the number of vertices.

Here, we will restrict our discussion to 2-manifold triangular meshes, i.e. those where an edge is shared by one (boundary triangle) or two triangles.

EULER-POINCARE FORMULA– This formula links the number of vertices |V|, the number of edges |E|, the number of triangles |F| and the number of boundaries |B| to the Euler number χ of the surface:

$$|V| - |E| + (|F| + |B|) = \chi \tag{9.2}$$

$$\chi = 2(1 - g) \tag{9.3}$$

where g is the genus of the surface.

For example, we have (g = 0, B = 0), (g = 1, B = 0), (g = 0, B = 2) for a surface homeomorphic to a sphere, a torus and an open cylindrical tub, respectively.

THE VALENCY (DEGREE) OF A VERTEX– The number d(v) of neighbors of a given vertex v is called the valency (or degree) of v (Figure 9.3). We should note that d(v) is equal to the number of triangles adjacent to v, if v is an interior vertex, and exceeds this number by one for boundary vertices.

The valency of a vertex is between two (three for a closed mesh) and infinity. However, its value is limited by the number of edges. For a mesh without boundaries, homeomorphic to a sphere, we have a relationship known as "handshaking":

$$\sum_i d(v_i) = 2|E| \quad [9.4]$$

and

$$2|E| = 3|F| \quad [9.5]$$

By expressing the number of edges in the handshaking formula according to the number of vertices, using the latter relationship and the Euler-Poincare formula, then dividing by the number of vertices, the average valency is achieved:

$$\frac{1}{|V|}\sum_i d(v_i) = 6 - \frac{12}{|V|} \quad [9.6]$$

Thus, the average valency of large meshes is asymptotically six. We can note that the valency of vertices of meshes constructed using the MC algorithm cannot exceed 12 [LEE 06].

REGULAR AND IRREGULAR VERTICES– The interior vertices v, with $d(v_I) = 6$ and the outer vertices v_B with $d(v_B) = 4$ will be called "regular". All others will be called "irregular" (Figure 9.3).

REGULAR MESH– A mesh is referred to as regular if all its vertices have a valency of 6 (respectively 4). Only meshes topologically equivalent to a torus or an open cylinder can be regular.

MESH WITH REGULAR SUBDIVISION CONNECTIVITY ONE TO FOUR– A mesh with regular subdivision connectivity 1 to 4, which we will write as 1:4, is a mesh constructed from an initial mesh called the "base mesh", all of whose triangles are subdivided into four triangles. This mesh is often described as "subdivision mesh", without mentioning 1:4.

VALENCY OF VERTICES OF ONE TO FOUR SUBDIVISION MESHES – The valency of the vertices of the base mesh remains unchanged by the subdivisions; for the inner vertices it is 6 (regular vertices) and it is 4 for the boundary vertices (regular vertices).

SEMI-REGULAR MESHES– 1:4 subdivision meshes are known as "semi-regular".

IRREGULAR MESHES– A mesh is described as irregular if it is neither regular nor semi-regular. This is the case in practice for the vast majority of meshes, such as – for example – those constructed using the MC algorithm.

GEOMETRIC QUALITY OF MESHES– Well shaped triangles are important, so we generally desire triangles whose shape is as close as possible to the equilateral triangle. Nevertheless, this depends greatly upon the application in question. The equilateral triangle is desirable for digital calculations using finite element methods, the processing of geometric signals, calculation of discrete curvature, whereas in terms of surface approximation, triangles elongated in the direction of the minimal curvature are, for example, better adapted. There are many objective criteria for measuring the quality of a triangle [FRE 99]. We should note that on a plane, a vertex neighbored only by equilateral triangles has a valency of six.

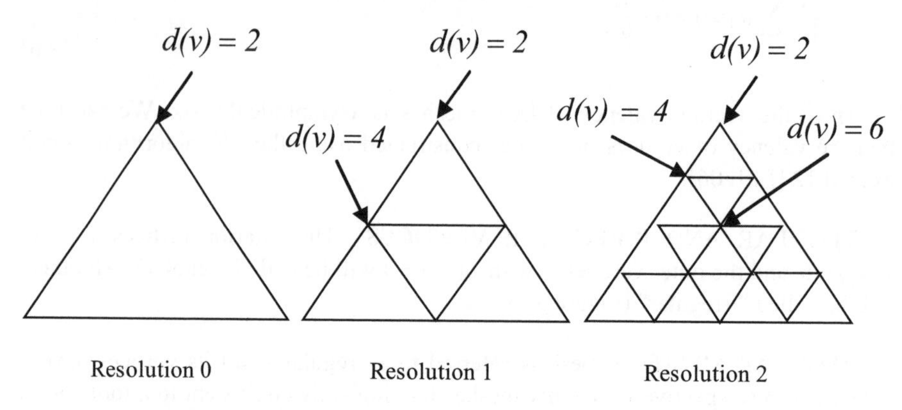

Figure 9.3. *1 to 4 subdivision mesh and valency of the vertices*

9.3. Compression of static meshes

An overview of recent compression algorithms is presented in [ALL 05a]. The different approaches share a common feature to make the coding effective: they consider together the description of the mesh connectivity with the description of its geometry. These two types of information are strongly linked: a vertex v_1, having as its neighbors (in terms of connectivity) a group of vertices $\{v_1, v_2, v_3, \cdots\}$, will be located in the 3D space close to these same neighbors. The geometric description of the mesh (the coordinates of the vertices) can therefore be effectively transmitted according to the connectivity description.

We can classify the compression algorithms for 3D meshes into two categories:

– single resolution approaches: the mesh connectivity is coded only once, following a compact code;

– multiresolution approaches: the original mesh is repeatedly simplified until a minimum resolution mesh is reached. A compact code is used for progressive reconstruction of the original mesh, from the low resolution mesh.

9.3.1. *Single resolution mesh compression*

Single-resolution compression methods are usually based on coding the adjacency graph of the mesh simultaneously with the contextual coding of the vertices coordinates. The coding of the graph at full resolution is generally carried out by a deterministic scanning of the mesh. The encoding therefore consists of generating successive codes, allowing for the reproduction of this scanning at the decoder end.

9.3.1.1. *Connectivity coding*

From a theoretical point of view, an enumeration of planar graphs [TUT 62] has shown that the minimum theoretical limit of the cost of coding the connectivity of planar graphs, with uniform probability distribution of the possible triangulations, is 3.245 bits per vertex.

For a mesh obtained by 1:4 regular subdivisions of all the triangles of a base mesh, the cost of coding the connectivity is that of the base mesh. For example, for a 0-genus surface, the smallest base mesh is a tetrahedron (four vertices and four triangles), the cost of coding its connectivity is: *12log2(4)=24* bits. At each subdivision level, the number of triangles is multiplied by 16. Thus at the subdivision level j, $j > 0$, the resulting number of triangles is: 4^{j+1} and the coding cost per triangle is $6/4^{j}$. Finally, it follows from this that the coding cost per vertex vanishes as the subdivision level grows.

The pioneering works in connectivity coding were carried out by Deering *et al.* [DEE 95], who propose a single scan of the mesh to convert triangle data into a triangle strip ("orange peeling" method). It results in a non-linear coding cost versus the number of vertices in the mesh.

Touma and Gotsman [TOU 98] proposed mesh connectivity encoding by visiting all the vertices. The binary flow generated is therefore made up quite exclusively of the valency of the visited vertices, to which are added some incident codes, generally of very low frequency. Although this approach offers no guarantee about the coding cost, because of the unpredictable incident codes, it gives very good results. It has been shown that the results achieved are very close to optimal [GOT 03].

Rossignac [ROS 99] introduced Edgebreaker, a coder based on a canonical scanning of the edges. The data flow is made up of five symbols, guaranteeing a coding of 2 bits/triangle for a mesh homeomorphic to a sphere.

More recently, Poulalhon and Schaeffer [POU 03] proposed a coding algorithm whose results meet the optimal limit (3.245 bits per vertex). This approach has the advantage of guaranteeing a constant coding cost. However, applied to the meshes generally encountered, the performance of this algorithm remains below the valency-based coders [TOU 98].

9.3.1.2. *Geometry coding*

Compressing the geometry is generally carried out using differential coding, where the canonical path of the graph allows for an effective prediction of the coordinates of the vertices. For example, the prediction rule known as "parallelogram prediction" [TOU 98] predicts the coordinates of the vertex of the triangle adjacent to the triangle already coded (seed triangle at the fist step of the algorithm) (Figure 9.4) by the following formula:

$$\hat{\mathbf{v}}_4 = \mathbf{v}_1 + \mathbf{v}_2 - \mathbf{v}_3 \qquad [9.7]$$

where $\mathbf{v}_i = [x_i \quad y_i \quad z_i]^T$ represents the coordinates of the vertex $\mathbf{v}_i^j$.

The prediction error $\boldsymbol{\varepsilon} = \mathbf{v}_4 - \hat{\mathbf{v}}_4$ is quantized, and then coded with an entropic coder. We should note that this error is zero if triangle two is in the plane of triangle one, i.e., if the two triangles form a parallelogram.

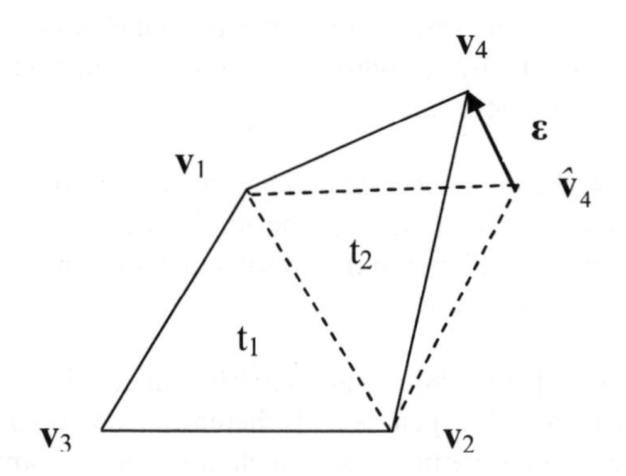

Figure 9.4. *Prediction for the parallelogram method*

9.3.2. *Multi-resolution compression*

9.3.2.1. *Mesh simplification methods*

Multiresolution compression of 3D meshes allows progressive transmission, via repeated refinement of the mesh. The careful choice of the refinement methods allows a low coding cost of the relationship between the obtained resolution levels. Several coarsening/refinement models have been used.

The "progressive meshes" algorithm [HOP 96] is used by [PAR 00]. This compression involves edge collapses merged into batches in order to create several resolution levels.

In [COH 99], some vertices are deleted, and their neighborhood is remeshed. A deterministic remeshing technique enables a reconstruction of the mesh with a low coding cost, at around 3 bits per triangle. This algorithm has recently been improved in [ALL 01], where the authors use two distinct vertex removal stages. The first stage deletes vertices whose valency lies between three and six, and the second deletes only those with a valency equal to three. The combination of these two steps keeps the mesh close to regular during the simplifications, and gives an average connectivity coding cost of the of 3.7 bits/vertex and 11.4 bits/vertex on average for the geometry of meshes natively coded at 10 bits/vertex.

New advances [GAN 02], [PEN 05] suggest the multi-resolution coding of meshes with fewer constraints on the mesh genus. These coders are in fact as capable of processing arbitrary topology meshes as "triangle soups".

9.3.2.2. *Spectral methods*

The spectral approach has similarities with orthogonal transform coding (discrete cosine transform, for example), which is widely used in image and signal coding. Karni and Gotsman [KAR 00] apply a spectral decomposition by projecting the geometry of the mesh on the eigenvectors taken from the diagonalization of the Laplacian operator of the connectivity graph. Although the irregular sampling of meshed surfaces does not enable an exact analogy with discrete cosine transform (DCT) – which is applied to regularly-sampled signals – the authors say that a geometric decorrelation is sufficient. The results show a progressive representation with a satisfactory rate/distortion function for smooth objects. This method does, nevertheless, have some drawbacks: to limit the eigenvectors calculation cost, the mesh must be split into blocks (similar to images split into blocks for DCT in JPEG), which results in a distortion between the different blocks. Finally, only the geometry is progressively reconstructed; the connectivity remains unchanged during data transmission.

9.3.2.3. *Wavelet-based approaches*

9.3.2.3.1. Subdivision surfaces and wavelets

The approaches based on wavelet expansion are fundamentally linked to subdivision surfaces, which are introduced in order to smooth the meshes [CAT 78], [DOO 78], [LOO 87]. An arbitrary mesh M^0 is subdivided, which increases its number of vertices. Then a smoothing follows, depending upon the type of subdivision, as shown in the example of Figure 9.5.

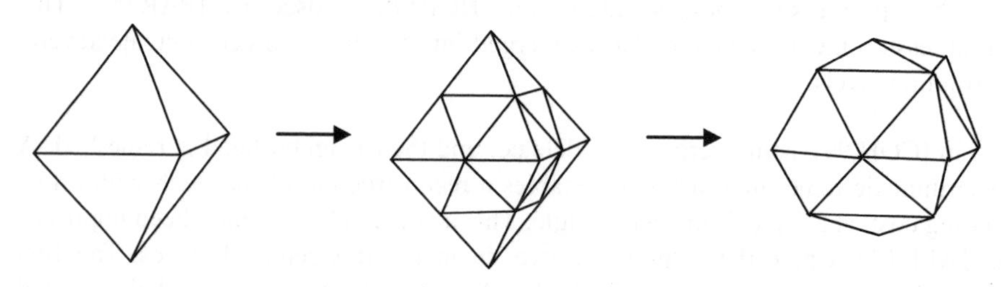

Figure 9.5. *Example of mesh subdivision followed by smoothing*

Moreover, surface subdivisions allow the mesh to be represented as a sum of scaling functions suitable for refinement [STO 96], and well-suited for multi-resolution wavelet analysis (see Chapter 2).

In this way, Lounsbery *et al.* extended the wavelet expansion to triangular meshes, allowing a multi-resolution analysis [LOU 97] (see also [STO 96]). Lounsbery considers meshes with subdivision connectivity (section 9.2). The coarsening of a mesh at the level j spatial resolution is achieved by merging all of its triangles, by groups of four (noted 4:1); the inverse process of subdivision. Thus, a mesh with the resolution j-1 is built, and its number of triangles is a quarter that of the mesh with resolution j. The mesh with the lowest resolution is called the "base mesh" (the coarsest mesh). The coefficients of the wavelets are related to the loss of geometric details due to the 4:1 triangles merging.

9.3.2.3.2. Multiresolution analysis by geometric wavelets

In this section, the background on wavelet representation is summarized (see also Chapter 2), and after we will detail its extension to meshes, using the notation by Lounsbery *et al.*

We will first consider a nested set of vector spaces, called "approximation spaces":

$$V^0 \subset V^1 \subset \cdots V^j \subset \cdots V^J$$

where the geometric resolution of the functions of the vector spaces V^j grows with j. The basis functions $\phi_i^j(x)$ of V^j are called scaling functions.

Next we define the vector space W^j for each vector space V^j, as the complement V^j of in V^{j+1}:

$$V^j \oplus W^j = V^{j+1} \quad [9.8]$$

$$V^j \cap W^j = \phi \quad [9.9]$$

The basis functions $\psi_i^j(x)$ for W^j are the wavelets. In order for a function of the space V^j to be the best approximation of a function of the space V^{j+1} in the least squares sense, the wavelets must all be orthogonal to all the scale functions of V^j. Lounsbery *et al.* decided to represent the surface of a triangular mesh as a sum of "hat" scale functions (B-Spline of order 1). A hat function takes the value 1 on a vertex and vanishes at its neighbors (1-neighborhood). It interpolates linearly the surface between two vertices (Figure 9.6a).

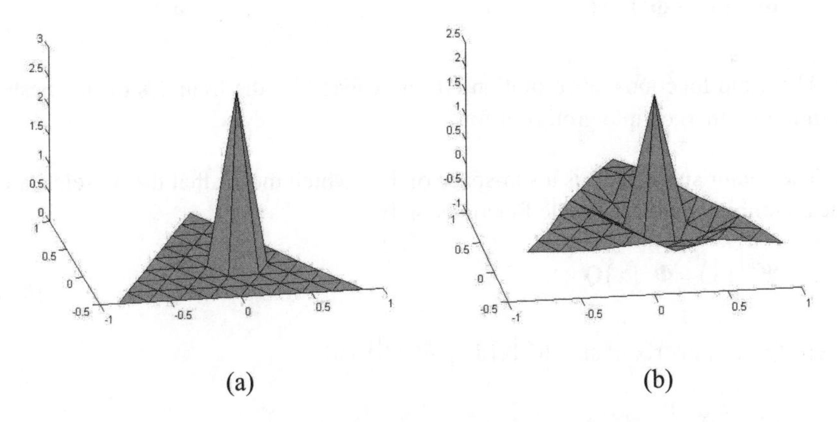

Figure 9.6. *(a) Lazy wavelet and (b) lazy wavelet after 1-ring lifting*

In this way, they are defined on a space defined by the connectivity of the processed mesh. The hat functions fit well with mesh representation, since the surface of triangular meshes is linear between the vertices.

For a mesh M^j made up of $|V^j|$ vertices, represented by a function $S^j(x)$, the corresponding vector space V^j will be of dimension $|V^j|$.

Given $\mathbf{v}_i^j = \begin{bmatrix} x_i^j & y_i^j & z_i^j \end{bmatrix}^T$ as the coordinates of the vertex $\mathbf{v}_i^j$, we have:

$$S^j(x) = \mathbf{\Phi}^j(x)\mathbf{C}^j \quad [9.10]$$

with

$$\mathbf{\Phi}^j(x) = \begin{bmatrix} \phi_0^j(x) & \phi_1^j(x) & \cdots & \phi_{|V^j|-1}^j(x) \end{bmatrix} \quad [9.11]$$

$$\mathbf{C}^j = \begin{bmatrix} \mathbf{v}_0^j & \mathbf{v}_1^j & \cdots & \mathbf{v}_{|V^j|-1}^j \end{bmatrix}^T \quad [9.12]$$

where $\mathbf{C}^j$ is the vector of scaling coefficients.

The vector space V^{j-1} is a sub-space of V^j which implies that the scale functions of V^{j-1} are a linear combination of those of V^j. There is therefore a matrix $\mathbf{P}^j$ of size $|V^j| \text{x} |V^{j-1}|$ satisfying:

$$\mathbf{\Phi}^{j-1}(x) = \mathbf{\Phi}^j(x)\mathbf{P}^j \quad [9.13]$$

The scale functions, at resolution j-1, are carried by the triangles of the mesh at resolution j, merged into groups of four.

The vector space W^{j-1} is a sub-space of V^j, which means that the wavelets are a linear combination of the scale functions of V^j:

$$\mathbf{\Psi}^{j-1}(x) = \mathbf{\Phi}^j(x)\mathbf{Q}^j \quad [9.14]$$

where $\mathbf{Q}^j$ is a matrix of size $|V^j| \text{x} \left(|V^j| - |V^{j-1}|\right)$ and

$$\mathbf{\Psi}^{j-1}(x) = \begin{bmatrix} \psi_0^{j-1}(x) & \psi_1^{j-1}(x) & \cdots & \psi_{|V^j|-|V^{j-1}|-1}^{j-1}(x) \end{bmatrix} \quad [9.15]$$

Lounsbery uses, at first, wavelets which are the basis functions of the odd index of V^j (scale functions). These wavelets are known as “lazy wavelets”. Lazy wavelets are orthogonal between themselves, but not orthogonal to the scale functions, as shown in Figure 9.7 in 1D. Scale functions and lazy wavelets are the primal base functions of a bi-orthogonal base, where the dual base functions are Dirac functions [SCH 95] (Figure 9.7).

The approximation $S^{j-1}(x)$ of $S^j(x)$ on the space V^{j-1} is written as in [9.10]

$$S^{j-1}(x) = \mathbf{\Phi}^{j-1}(x)\mathbf{C}^{j-1} \qquad [9.16]$$

The details lost in this approximation – i.e. the details lost during coarsening from the resolution j of the mesh (M^j) to the resolution j-1 – belong to W^{j-1} and are represented on the wavelet basis of this vector space:

$$S^j(x) - S^{j-1}(x) = \mathbf{\Psi}^{j-1}(x)\mathbf{D}^{j-1} \qquad [9.17]$$

where $\mathbf{D}^{j-1}$ represents the vector of the wavelet coefficients.

From [9.16] and [9.17] it follows:

$$S^j(x) = \mathbf{\Phi}^{j-1}\mathbf{C}^{j-1} + \mathbf{\Psi}^{j-1}\mathbf{D}^{j-1} \qquad [9.18]$$

$$S^j(x) = \mathbf{\Phi}^j\left(\mathbf{P}^j\mathbf{C}^{j-1} + \mathbf{Q}^j\mathbf{D}^{j-1}\right) \qquad [9.19]$$

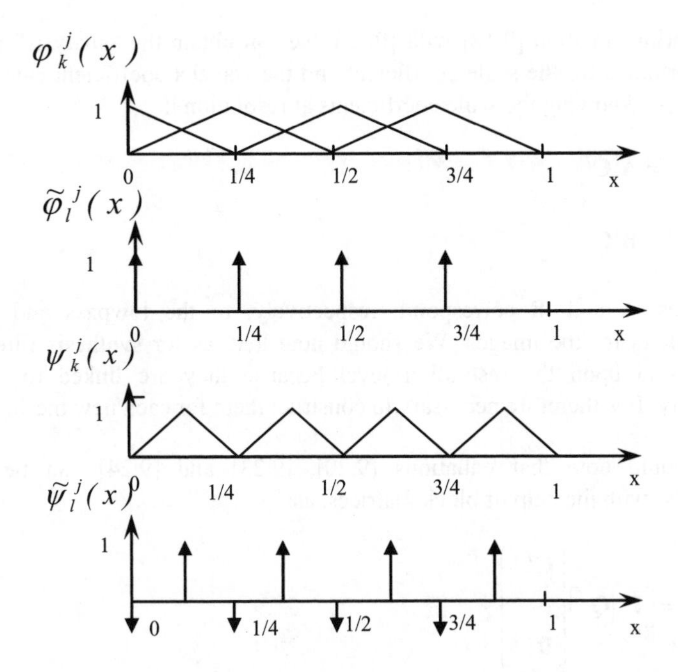

Figure 9.7. *1D illustration of wavelet bases*

By comparing with [9.10], we can obtain a formula for reconstructing the scale coefficients at the resolution j, knowing those of the lower resolution j-1 and the corresponding wavelet coefficients:

$$\mathbf{C}^j = \mathbf{P}^j\mathbf{C}^{j-1} + \mathbf{Q}^j\mathbf{D}^{j-1} \qquad [9.20]$$

Matrices $\mathbf{P}^j$ and $\mathbf{Q}^j$ correspond, respectively, to the lowpass and highpass synthesis filters used in image wavelet-based compression techniques.

The complementary spaces V^{j-1} and W^{j-1} mean that the scale functions at resolution j are a unique linear combination of the scale functions at the resolution j-1 and the wavelets, which we write as follows:

$$\mathbf{\Phi}^j(x) = \mathbf{\Phi}^{j-1}(x)\mathbf{A}^j + \mathbf{\Psi}^{j-1}(x)\mathbf{B}^j \qquad [9.21]$$

Thus, we have, following [9.10]:

$$S^j(x) = \mathbf{\Phi}^j(x)\mathbf{C}^j = \mathbf{\Phi}^{j-1}(x)\mathbf{A}^j\mathbf{C}^j + \mathbf{\Psi}^{j-1}(x)\mathbf{B}^j\mathbf{C}^j \qquad [9.22]$$

Comparing equation [9.18] with [9.22], we can obtain the "analysis" formulae, i.e., the formulae for the scale coefficient and the wavelet coefficient calculation at resolution j-1, knowing the scale coefficients at resolution j:

$$\mathbf{C}^{j-1} = \mathbf{A}^j\mathbf{C}^j \qquad [9.23]$$

$$\mathbf{D}^{j-1} = \mathbf{B}^j\mathbf{C}^j \qquad [9.24]$$

Matrices $\mathbf{A}^j$ and $\mathbf{B}^j$ correspond, respectively, to the lowpass and highpass analysis filters for the images. We should note that, as for synthesis filters, these filters depend upon the resolution level because they are linked to the mesh connectivity. It is therefore necessary to construct them for each new mesh.

We should note that equations [9.20], [9.23] and [9.24] can be written, respectively, with the help of block matrices, as:

$$\mathbf{C}^j = \left[\mathbf{P}^j \middle| \mathbf{Q}^j\right]\begin{bmatrix}\mathbf{C}^{j-1} \\ - \\ \mathbf{D}^{j-1}\end{bmatrix} \qquad [9.25]$$

and

$$\begin{bmatrix} \mathbf{C}^{j-1} \\ - \\ \mathbf{D}^{j-1} \end{bmatrix} = \begin{bmatrix} \mathbf{A}^{j-1} \\ - \\ \mathbf{B}^{j-1} \end{bmatrix} \mathbf{C}^{j} \qquad [9.26]$$

From this, we obtain the relationship linking the synthesis and analysis filters:

$$\left[\mathbf{P}^{j} \middle| \mathbf{Q}^{j}\right]^{-1} = \begin{bmatrix} \mathbf{A}^{j-1} \\ - \\ \mathbf{B}^{j-1} \end{bmatrix} \qquad [9.27]$$

Figure 9.8 illustrates the geometric meaning of the lazy wavelet coefficients.

With the lazy wavelet, coarsening from resolution j to j-1 is simply a sub-sampling of the meshed surface (hence the name "lazy" wavelet). The obtained approximation is of poor quality because the scaling functions and the lazy wavelets are not orthogonal.

A better approximation – in the least squares sense – of M^{j} by M^{j-1} can be achieved if the wavelets are orthogonal to the scaling functions. It is therefore necessary to define an inner product coherent with the formalism of geometric wavelets. Lounsbery *et al.* defined the inner product between two functions f and g, defined on the surface of a mesh M, by the following equation:

$$\langle f,g \rangle = \sum_{\tau \in \Delta(M)} \frac{k^{j}}{area(\tau)} \int f(s) g(s) ds \qquad [9.28]$$

where τ is a triangle, $\Delta(M)$ the set of triangles forming the mesh M and $k^{j} = 4^{-j}$ is a constant which translates the division of the area of a triangle by four when we change the resolution level using a subdivision 1:4 of all the triangles. We should note that this definition assumes that the mesh triangles all have the same area, which is not the case for usual subdivision meshes. Therefore, the use of this definition in the following calculations will be more accurate if the mesh's triangles have similar areas.

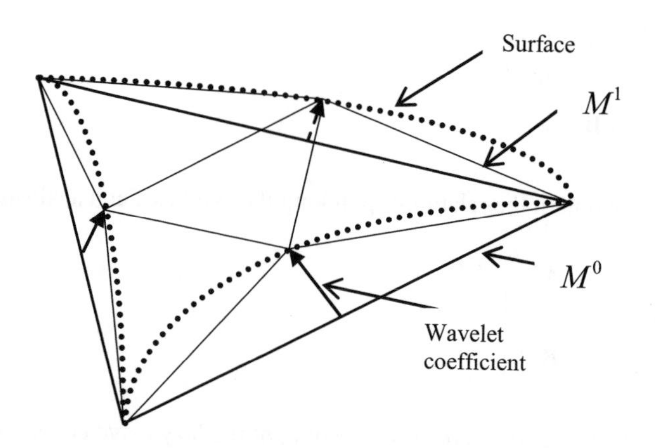

Figure 9.8. *Illustration of the lazy wavelet coefficients*

In order to achieve an optimal approximation, Lounsbery *et al.* used the lifting scheme [SWE 96], so as to make the wavelets orthogonal (as much as possible) to the scaling functions. This is performed by subtracting from each wavelet a weighted sum of all the scaling functions (or a fraction of them), in order to create a new base of wavelets which we will write as $\psi^j_{lift}(x)$:

$$\mathbf{\Psi}^{j-1}_{lift}(x) = \mathbf{\Psi}^{j-1}(x) - \mathbf{\Phi}^{j-1}(x)\boldsymbol{\alpha}^j \quad [9.29]$$

where $\boldsymbol{\alpha}^j$ is the matrix of the weighting coefficients.

The coefficients are calculated in such a way as to cancel out all (or a fraction of) the inner products between the scaling functions and the wavelets at a given resolution level. We therefore obtain matrix $\boldsymbol{\alpha}^j$:

$$\boldsymbol{\alpha}^j = \left(\mathbf{I}^{j-1}\right)^{-1} \mathbf{P}^{jT}\mathbf{I}^j\mathbf{Q}^j \quad [9.30]$$

where $\mathbf{I}^j$ is the following inner product matrix:

$$\left\{\mathbf{I}^j\right\}_{m,n} = \left\langle \varphi^j_m, \varphi^j_n \right\rangle \quad [9.31]$$

The lifting scheme leads to a new analysis and synthesis filterbank:

$$\mathbf{A}^j_{lift} = \mathbf{A}^j + \boldsymbol{\alpha}^j\mathbf{B}^j \quad [9.32]$$

$$\mathbf{B}_{lift}^{j} = \mathbf{B}^{j} \quad [9.33]$$

$$\mathbf{P}_{lift}^{j} = \mathbf{P}^{j} \quad [9.34]$$

$$\mathbf{Q}_{lift}^{j} = \mathbf{Q}^{j} - \mathbf{P}^{j}\boldsymbol{\alpha}^{j} \quad [9.35]$$

In order to reduce the computer load the orthogonalization procedure of wavelets is only calculated in the vicinity of each vertex, generally in the 2-ring. It only results in a small approximation quality loss for the mesh M^j at the resolution j-1. Figure 9.6.b illustrates a lifted lazy wavelet.

Note that the base mesh topology (j=0) is the same as that of the full resolution mesh. For example, if the full resolution mesh topology is equivalent to a sphere (g=0) the base mesh will be a tetrahedron. In addition, it can be noted that the full resolution level j index is only known after coarsening up to the base mesh.

9.3.2.3.3. Compression

The wavelet transform has proved to be an effective tool for signal and image compression. In geometry compression, wavelets are also very effective. To compress a subdivision mesh, the base mesh is coded at a low cost in terms of connectivity and geometry, and only wavelet coefficients for each level j should be coded. Note that the connectivity of the successive meshes M^j is implicit at the decompression step due to the 1:4 subdivision connectivity.

Many coding techniques have been proposed in the literature. Khodakovski *et al.* [KHO 00] proposed to code the wavelet coefficients using bit planes in a similar way to [SAI 96]. Using this approach the progressive transmission of mesh is very effective by sending the highest order bits of the largest magnitude coefficients first. In a single resolution scheme Payan and Antononi [PAY 05] adapt the quantization of the wavelet coefficients to the target bitrate. The set of quantization steps minimizes the distortion defined by the mean square error of the reconstructed mesh geometry, at a given bitrate by a Lagrangian minimization.

9.3.2.3.4. Drawbacks and extension

Lounsberry's wavelet approach only applies on regularly subdivided meshes (mesh with a subdivision connectivity; see section 9.2). This constraint corresponds to the image processing field where wavelet expansion applies on a regular sampling grid (square separable or quincunxes) with power of two size (128, 256, 512, etc.) due to the dyadic decomposition used. Thus, a mandatory requirement to apply Lounsberry's wavelet approach to irregular meshes is remeshing this mesh to a subdivision mesh. Figure 9.9 shows an example: the left mesh represents the left ventricle of a dog's heart. Its connectivity is irregular and cannot be expanded on a wavelet basis. A remeshing (see for example [ALL 05b]) allows its connectivity to be changed into a subdivision connectivity preserving its shape (geometry).

Remeshing an object with complex shape or topology is a difficult problem. The pioneering work for remeshing into a semi-regular mesh is proposed by Lee *et al.* [LEE 98]. The first step in remeshing is generally a parameterization of the input mesh. The parameterization flattens (maps) a 3D mesh having a disk topology on a closed plane surface (2D), a disk or a square, by means of a conformal transform (angles preserving transform) or an harmonic transform (elastic transform). Lee *et al.* [LEE 98] propose applying this parameterization by patches. Note that deleting only one edge of sphere topology input mesh is sufficient to obtain a mesh with disk topology. However, cutting an input mesh with complex topology into a disk is a difficult task. The second step is to match this parameterization to the new subdivision connectivity. Finally, the input mesh geometry is transferred to the new mesh. Unfortunately, this last step is not free of geometric distortion. Gu *et al.* [Gu 02] proposed the "geometry images" that are parameterized meshes on a regular rectangular grid, similarly to a digital image. Then, usual image compression techniques can be used for compression of geometry images.

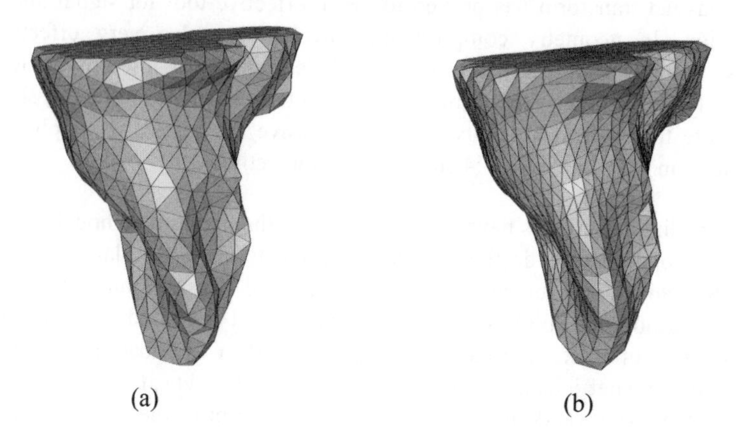

Figure 9.9. *(a) Irregular mesh, (b) mesh (a) remeshed to a semi-regular mesh*

9.3.2.3.5. Irregular subdivision

Another solution, which avoids the need for remeshing, was proposed by Valette and Prost [VAL 99]. They considered an irregular subdivision where a triangle can be subdivided into two, three or four triangles, or remains intact. In this way, a hierarchy of meshes of varying resolutions can be generated from an irregular mesh. This approach requires solving the inverse problem of subdivision, i.e. merging, during the calculation of the connectivity graph of the mesh to be compressed, by the proposed irregular merging of triangles (coarsening step). Edge flips can also occur at this point, in order to ensure that the merging is reversible. By controlling the regularity of the created levels of resolution [VAL 04a], it is possible to store the subdivision information with an effective code, giving very good connectivity compression rates [VAL 04b]. Geometry coding can occur in a similar way as in the regular case: a progressive resolution compression to lossless compression [VAL 04b] or progressive precision compression [VAL 04c]. An essential characteristic of a coder – particularly if its resolution or precision is progressive – is the curve which depicts the tradeoff between the bitrate and the distortion. Figure 9.10 shows the result obtained by the coder described in [VAL 04b], for a simple mesh.

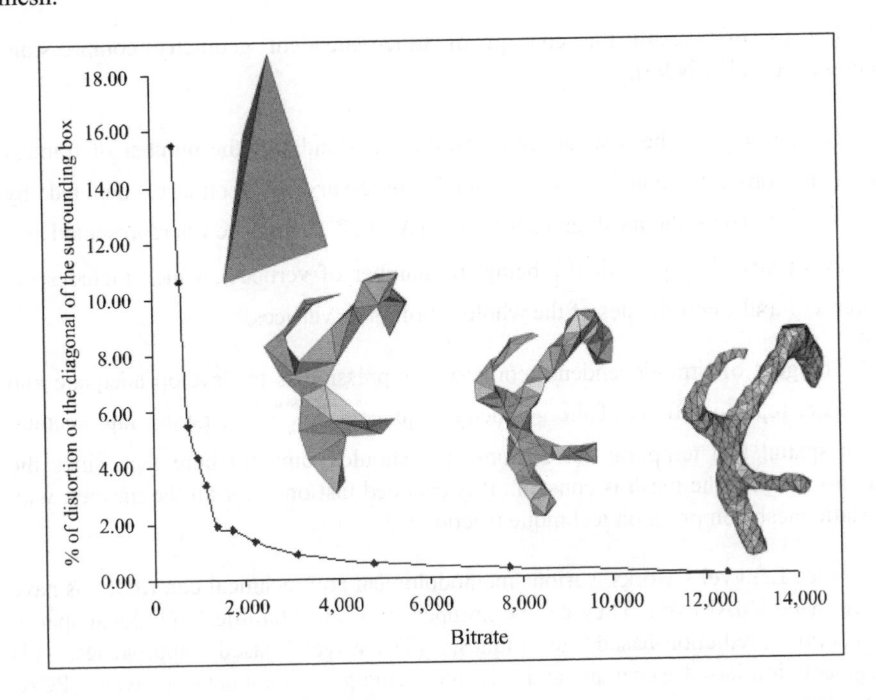

Figure 9.10. *Trade-off between the bitrate and the distortion for a progressive resolution decoding (irregular mesh, algorithm in [VAL 04b])*

9.4. Compression of dynamic meshes

In a wide range of medical applications – such as the numerical simulation of airflows in bronchial trees or the analysis of pulmonary nodulation in computed tomography [PER 04], [FET 04] – only the isosurfaces extracted from volumetric data are taken into account. The generated isosurfaces are usually represented as a set of 3D meshes with an arbitrary time-varying connectivity. The compression of these dynamic surfaces with topological changes remains a widely unexplored subject. We should mention here the pioneering works carried out by Eckstein *et al.* [ECK 06], which extend the static encoder [LEE 03] to compress dynamic isosurfaces. Most of the other compression approaches assume a constant connectivity to be shared by the sequence of meshes. The technique reported in [YAN 04] carries out a semi-regular re-meshing in order to obtain such a dynamic mesh with a constant connectivity.

9.4.1. *State of the art*

Let us first recall the concept of time-dependent geometry compression, introduced in [LEN 99].

Let $(M_t)^{t\in\{0,\dots,F-1\}}$ be a sequence of meshes (*F* stands for the number of frames) with a constant connectivity *T*, and a time-varying geometry, denoted by $(G_t)^{t\in\{0,\dots,F-1\}}$. Here, the mesh geometry $G_t = (X_t, Y_t, Z_t)^t$ at frame t is represented as a vector of size $3\times|V|$ (with $|V|$ being the number of vertices), which includes the three x, y and z coordinates of the whole set of mesh vertices.

The goal of time-dependent geometry compression is to develop adequate and compact representations of the geometry sequence $(G_t)^{t\in\{0,\dots,F-1\}}$, taking into account both spatial and temporal correlations. We should point out here that, since the connectivity of the mesh is constant, it is encoded just once for all the meshes, with a static mesh compression technique (section 9.3).

Since Lengyel's works, various methodological and technical contributions have been made [MAM 05]. They can be grouped into four "families": (1) local spatio-temporal prediction-based techniques, (2) wavelet-based approaches, (3) segmentation-based schemas and (4) the principal component analysis (PCA) encoders.

9.4.1.1. *Prediction-based techniques*

Local spatio-temporal prediction-based compression techniques analyze the animation locally in space and time. The compression of a vertex position at a given temporal instant only involves a local spatio-temporal neighborhood.

As a representative of this family of approaches, let us first introduce the Interpolation Compression schema (IC) [JAN 04], recently adopted by the MPEG-4/AFX standard [BOU 04] [ISO 04]. The IC coder combines a sub-sampling procedure with a local spatio-temporal prediction strategy. The decoder decompresses the key-frames and generates intermediary frames by applying a linear interpolation procedure. Different strategies can be used to select the key-frames, going from a simple uniform sampling process to sophisticated techniques minimizing an error criterion. Such an approach is discussed in [JAN 04]. The authors suggest starting from a minimal sequence made-up of the first and last frame, and iteratively refining it by introducing intermediate frames until a predefined error threshold is reached.

The Dynapack technique [IBA 03] uses a similar spatio-temporal prediction strategy with more elaborate predictors involving a deterministic vertex traversal similar to [TOU 98]. The authors propose two different predictors called the Extended Lorenzo Predictor (ELP) and Replica. The first extends to the dynamic case of the "parallelogram" prediction rule [TOU 98] extensively used for static mesh compression. The ELP perfectly predicts translations (i.e., with zero residual errors). It involves only additions and subtractions. The Replica predictor relies on a more complex local coordinate-based prediction approach in order to perfectly predict rigid movements and uniform scales.

The MPEG-4/AFX-IC and Dynapack techniques offer the advantages of simplicity and low computational cost, which makes them well-suited for real-time decoding applications. However, because of the underlying deterministic traversal of the mesh vertices involved, such approaches do not support more advanced functionalities such as progressive transmission and scalable rendering.

9.4.1.2. *Wavelet-based techniques*

9.4.1.2.1. Regular wavelet-based compression

Wavelet-based representations have been successfully used for image and video compression. Nevertheless, the extension of wavelets, generally defined on regular structures, to 3D meshes with arbitrary connectivities is not completely straightforward (see section 9.3.2.3.4).

A first group of approaches circumvent this problem by applying a re-meshing procedure to the initial irregular mesh connectivity in order to obtain a regular connectivity suited to wavelet compression.

In [BRI 03], the authors introduced the Geometry Video (GV) representation, which converts the dynamic 3D geometry into a sequence of 2D images using a mesh cut and a stretch minimizing parameterization [GU 02] over a 2D square domain. Here, the initial mesh connectivity is completely discarded and replaced by a regular one, obtained by uniformly sampling the parametric domain. The resulting sequence of geometry images is then compressed using traditional video encoding techniques. A global affine motion compensation procedure is first applied. The resulting prediction errors are then compressed using a wavelet-based encoding scheme. The GV encoder offers good compression performances while enabling advanced functionalities such as progressive transmission and scalable rendering. Its main drawbacks are related to the re-meshing procedure involved which may lead to a loss of surface details and tangent plane discontinuities at the level of the cut.

In order to overcome the GV's limitations, Mamou *et al.* have recently introduced the Multi-Chart Geometry Video (MCGV) approach [MAM 06a]. The MCGV technique prevents the re-meshing-related problems by preserving the initial mesh connectivity. The motion compensation stage was improved by applying a more elaborate piecewise affine predictor instead of the global compensation procedure considered in GV. Finally, the MCGV encoder minimizes parameterization distortions by using an atlas parameterization [SAN 03] instead of the restrictive mapping on a 2D square domain.

9.4.1.2.2. Irregular wavelet-based compression

The use of irregular wavelets [DAU 99] for compression purposes requires the storage of the wavelet filters associated with each vertex. This additional parameterization information – costly in terms of bitrate – makes this approach unsuitable for static mesh compression, unlike the approach described in section 9.3.2.3.5. In the case of dynamic meshes with a coherent parameterization throughout the sequence, this information can be deduced from the first frame. In [GUS 04], the authors present the Animation Wavelets Compression (AWC) approach, which makes use of this idea to compress a sequence of meshes with irregular anisotropic wavelets.

Firstly, the AWC method builds a hierarchical structure of progressive meshes [HOP 96] on the frame M_0 of the sequence. This structure is then used to build anisotropic wavelet filters which are exploited to decompose the geometry signal $(G_t)^{t\in\{0,\ldots,F-1\}}$ into a coarse representation and a set of wavelet details for each frame. Finally, these details are quantized, predicted and entropically encoded.

The AWC approach allows low bitrate compression, while offering advanced functionalities such as progressive transmission, scalable rendering and real-time decompression. This approach is, however, ill-suited for meshes with a reduced number of vertices per connected component. In this case, the structure of the progressive meshes does not contain enough hierarchical levels for an efficient wavelet-based decomposition.

9.4.1.3. *Clustering-based techniques*

In [LEN 99], Lengyel introduced the first clustering-based approach for encoding dynamic 3D meshes. The principle consists of splitting the mesh into sub-parts whose motion can be accurately described by rigid transforms. The proposed motion-based segmentation approach selects at random 10% of the mesh's triangles and classifies the vertices with respect to their motion. The animation is finally described by: (1) a partition information, (2) a set of affine motion parameters associated with each cluster and (3) the prediction residuals associated with each vertex.

Lengyel's approach is further improved in [COL 05], where animation is expressed only in terms of rigid transforms (RT). The authors introduce a new weighted least square mesh segmentation algorithm which minimizes the number of clusters under a distortion bound criterion.

In [MAM 06b], Mamou *et al.* propose another extension of Lengyel's coder called the Temporal-DCT-based encoder (TDCT). The TDCT technique is based on a k-means-segmentation-based approach. The resulting partition is used to construct a piecewise affine motion predictor. The residual errors are then compressed by applying a Discrete Cosine Transform (DCT) to the signal's temporal component. The TDCT segmentation procedure requires the user's intervention to specify the number of clusters for the k-means algorithm. To overcome this limitation, the authors introduced in [MAM 06c] a new fully automatic hierarchical segmentation approach. Here, the segmentation is performed using a decimation approach [HOP 96] allowing for the automatic detection of the number of clusters thanks to a global error criterion applied to the sequence. The motion compensation stage is also improved-upon by introducing a skinning model.

A different clustering-based compression scheme is proposed in [ZHA 04]. Here, an octree structure is defined on the mesh bounding box. Next, eight motion vectors are associated with each sub-cube of the octree. The motion of each sub-cube vertex is calculated with a user-defined error threshold by applying trilinear interpolation. The octree structure and the quantized motion vectors are finally arithmetically encoded. An optimized version of this approach, the so-called Dynamic 3D Mesh Coder (D3DMC) [MUL 04], improves this entropic encoding stage by adopting a context-based adaptive approach.

By using semi-global representations in combination with motion models, such clustering-based approaches are able to compactly describe a large category of motions. The compression performances as well as the complexity of calculation are often determined by the segmentation process. The major limitation of these techniques is related to the segmentation procedure which is computationally complex. In addition, such algorithms generally (with the exception of [MAM 06c]) cause disgraceful discontinuities at low bitrates at the level of the cluster boundaries, since different motion models are considered.

9.4.1.4. *PCA-based techniques*

In [ALE 00], the authors introduce a different class of approaches based on a PCA of the mesh deformation field. First of all, a rigid global motion compensation procedure is applied. The elastic movement field is then projected on the basis of the PCA vectors. Finally, the animation is compactly represented with the PCA vectors corresponding to the largest singular values and their associated decomposition coefficients.

In [KAR 04], the authors enhance the PCA approach by introducing a temporal second linear prediction. Combining PCA with linear prediction makes possible a better temporal decorrelation of the animation signal. An additional refinement is introduced in [SAT 05], where the authors propose to partition the mesh vertices into a set of clusters that are optimally adapted to the PCA representation. The introduced clustering-based PCA (CPCA) approach captures more efficiently the local linear behavior of the motion field.

The PCA-based approaches are specifically adapted for long repetitive animation sequences with a small number of vertices compared to the number of frames. However, they suffer from the high computational complexity ($O(T\times|V|^2)$, with T as the number of frames and $|V|$ the number of vertices) of the singular value decomposition algorithm involved.

9.4.1.5. *Discussion*

Table 9.1 summarizes the principles, properties and functionalities of each group of methods.

The computational complexity is evaluated on three levels and categorized as:

– *: low;

– **: medium;

– ***: high.

Since theoretical computational complexity bounds are not available for all the techniques detailed in the previous section, the evaluation given here is based on the calculation times reported in the literature.

Table 9.1 shows that only wavelet-based compression techniques offer both progressive transmission and scalable rendering functionalities. The irregular wavelet compression method seems particularly well-adapted to real-time applications, thanks to its low calculation cost. The PCA-based compression approaches enable the progressive transmission of the compressed flows. However, they have a high calculation cost.

The clustering-based compression techniques (excepting TDCT and skinning) and local prediction approaches support neither progressive transmission nor scalable rendering functionalities.

Approach		**Principle**	**Computational complexity**	**Progressive transmission**	**Scalable rendering**
Local spatio-temporal prediction	**MPEG-4/ AFX-IC**	- Local spatio-temporal prediction	*	No	No
	Dynapack	- Deterministic mesh traversal - Local spatio-temporal prediction	*	No	No
Clustering-based compression		- Motion-based segmentation - Parametric motion model - Temporal prediction	**	Only for the TDCT and skinning techniques	No
Wavelets	**regular**	- Re-meshing - Regular wavelet-based decomposition	***	Yes	Yes
	irregular	- Progressive mesh structure - Anisotropic irregular wavelet decomposition	*	Yes	Yes
PCA compression		- Global affine motion compensation - PCA decomposition - Linear prediction	***	Yes	No

Table 9.1. *Summary of the performances of the main groups of methods*

9.4.2. *Application to dynamic 3D pulmonary data in computed tomography*

Developing accurate and reproducible procedures to assess the individual lung capacity in patients suffering from tumours or chronic obstructive pulmonary diseases is still a challenging issue for clinical diagnosis and therapy follow-up.

The imaging modality making possible such a quantitative investigation is the X-ray computed tomography (CT).

9.4.2.1. *Data*

The protocol for acquiring 3D thoracic data at different pulmonary volumes is as follows:

– 1.25 mm X-ray beam collimation;

– 0.6 mm interval for images axis;

– LUNG reconstruction kernel (filtered back-projection algorithm);

– “Lung” windowing, between -1,000 HU and 200 HU (Hounsfield Units).

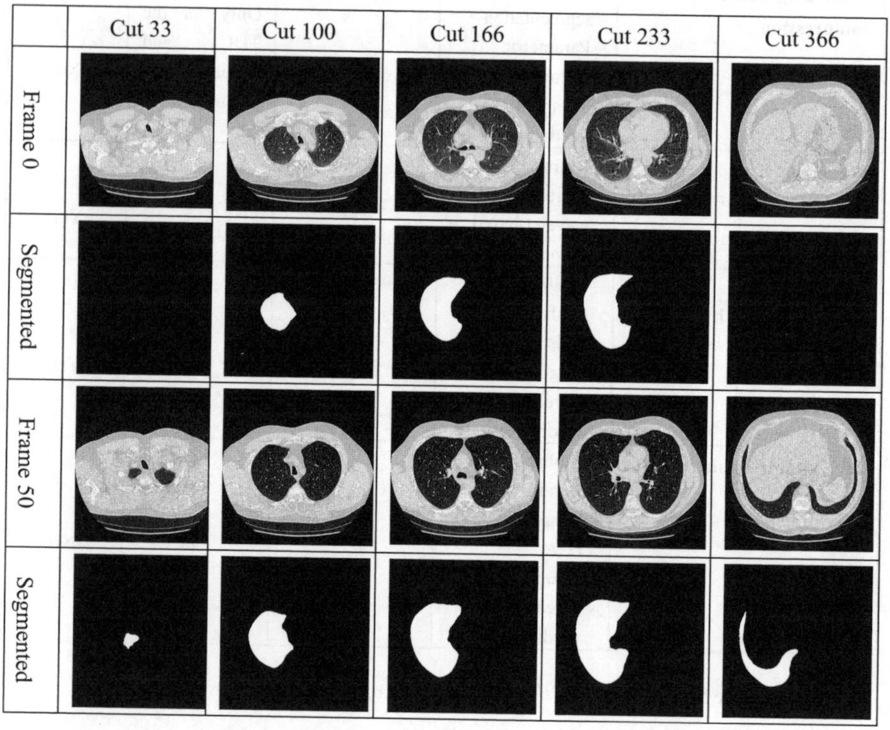

	Cut 33	Cut 100	Cut 166	Cut 233	Cut 366
Frame 0					
Segmented					
Frame 50					
Segmented					

Figure 9.11. *“Lung” sequence: dynamic volume data obtained during respiration. Native and segmented images (right lung)*

An automatic segmentation procedure [PRE 87], [FET 04] of the left and right lungs is applied to each data set. In the case of the "lung" sequence, chosen here for illustration purposes, this results in 50 volumetric frames after a temporal linear interpolation. Each volumetric frame is made up of 490 binary axial images of 512x512 pixels (Figure 9.11).

9.4.2.2. *Proposed approach*

The proposed compression is illustrated Figure 9.12. The principle consists of converting the volume data into a dynamic mesh with a constant connectivity in order to compress it by using the skinning-based encoder. The conversion process is made up of three steps.

Firstly, each volume frame is converted into a 3D mesh (Figure 9.13a) using the MC algorithm introduced in [LOR 87]. Let us note that the connectivities obtained by MC vary significantly from one frame to another, as they are generated independently for each volume data.

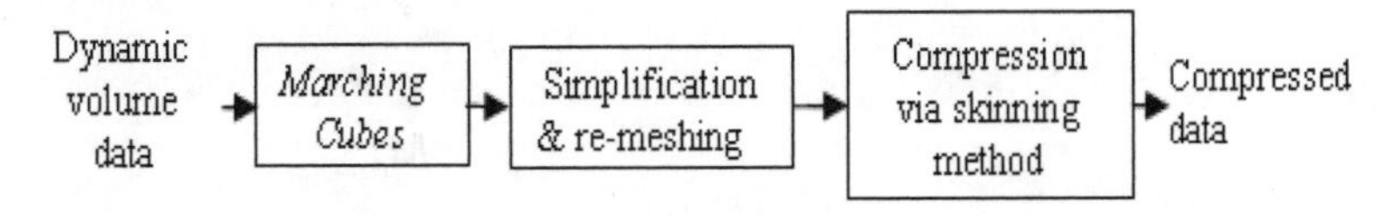

Figure 9.12. *Synoptic schema of the proposed compression approach*

In order to reduce the complexity of the MC generated meshes, we apply the mesh simplification technique [HOP 96] with a modified cost function in order to minimize the volume variations. The obtained meshes are illustrated in Figure 9.13b: the global aspect of the mesh is preserved even though the number of vertices is reduced by 95%.

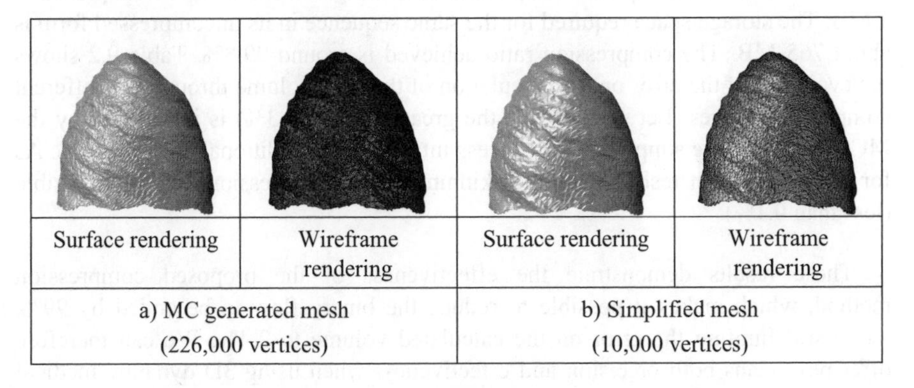

Figure 9.13. *Triangulation and simplification of the volume data*

The last pre-treatment stage before compression is the generation of a constant connectivity for all the frames. We propose an iterative approach which deforms the mesh M_0 generated for the first frame in order to obtain the shapes of the frames $(M_i)_{i \in \{1,\ldots,50\}}$. The approach we propose uses the ray-tracing procedure presented in Figure 9.14. For each step i, the normal vectors associated with the vertices of M_0 are re-calculated. Then, each vertex v of M_0 is projected on M_i by considering its associated normal vector as a direction. In this way, a new re-meshed version of M_i is obtained with the connectivity of M_0. This same process is repeated from one neighbor to another until all the frames are re-meshed (Figure 9.15).

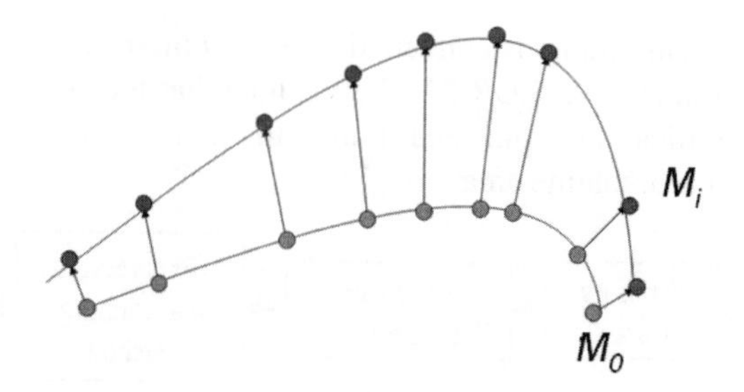

Figure 9.14. *Re-meshing by ray tracing*

9.4.2.3. *Results*

The compression schema described in the previous section has been applied to the "lung" sequence. Figure 9.15 shows the original and compressed versions of some frames in the sequence.

The size of the compressed bitstream describing the whole of the sequence is 1 MB. The storage space required for the same sequence in its uncompressed form is about 765 MB. The compression ratio achieved is around 99.8%. Table 9.2 shows the evolution of the error on the calculation of the lung volume through the different compression stages. Let us note that the greatest error (2.3%) is introduced by the MC algorithm. The simplification process introduces an additional error of 0.1%. As for the error which results from the skinning-based compression, this is negligible (less than 0.1%).

These results demonstrate the effectiveness of the proposed compression method, which makes it possible to reduce the binary flow to be coded by 99%, while still limiting the error on the calculated volume to 2.4%. We can therefore offer physicians both precision and effectiveness when using 3D dynamic medical data within the framework of telemedicine applications.

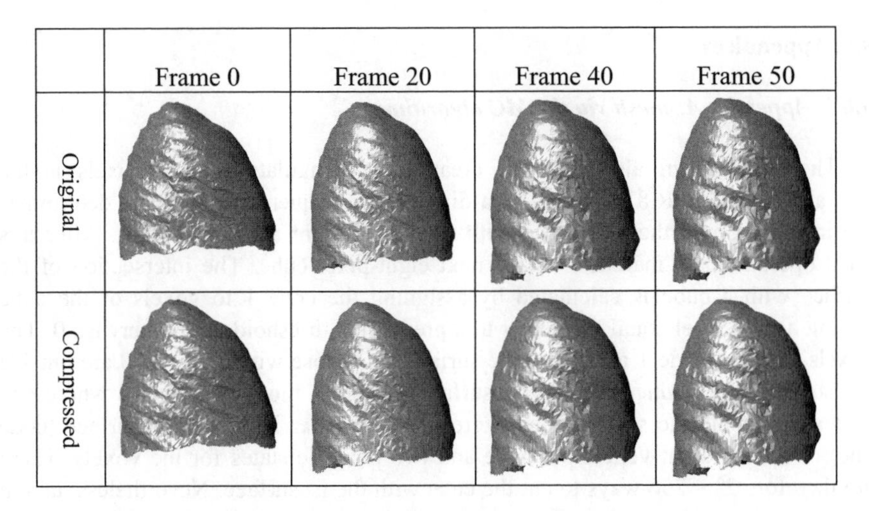

Figure 9.15. *Original and compressed versions of the "lung" sequence*

	Volumetric representation	**MC generated meshes**	**Simplified meshes**	**Compressed meshes**
Volume (liter)	2.15	2.10	2.098	2.11
Error (%)	0	2.3%	2.4%	2.4%

Table 9.2. *The different origins of error arising during the compression process, and their influence on the estimation of the volume of the right lung*

9.5. Conclusion

After the revolution which took place in digital medical imaging during the 1970s, the introduction of data acquisition and processing systems in the 1980s, and the implementation since 1993 of the DICOM (Digital Imaging and Communications in Medicine) standard, medical data has been in continuous evolution: opening up the way for e-health. Based on information and communication technologies, the concept is already being applied in telemedicine, computer-aided surgery, telediagnoses, teleconsultation, shared computerized medical files, medical research networks, etc. Each time, the questions of the representation, manipulation, storage and transmission of 2D, 3D, static or dynamic data are key in the new functionalities available. In this context, the mesh compression techniques presented in this chapter play a key role in tomorrow's medical world.

9.6. Appendices

9.6.1. *Appendix A: mesh via the MC algorithm*

The well-known algorithm for creating a triangulation from voxels is the MC algorithm [LOR 87]. It uses of a divide-and-conquer approach and determines the intersection of the isosurface with a cube of eight adjacent voxels. After this local operation, it "marches" to the next eight-pixel cube. The intersection of the surface with a cube is calculated by assigning the code 1 to voxels of the cube having a gray level equal or greater to a predefined threshold, and otherwise 0. The voxels with the code 1 are inside the surface, and those with the code 0 are outside the surface. We deduce that the isosurface intersects those cube edges where one vertex is outside the surface (code zero) and the other is inside the surface (code one). We have eight voxels per cube and two possible states for the voxels. There are therefore $2^8 = 256$ ways to cut the cube with the isosurface. Nevertheless, taking into account equivalent configurations which are deduced from each other by symmetry, there are only 15 basic configurations (Figure 9.16).

The algorithm acts locally, thus it cannot resolve ambiguities in the data, producing holes in the final mesh. These ambiguities are removed via the continuity of the isosurface [AND 04]. Note that the MC algorithm extracts several meshes in the presence of multiple unconnected isosurfaces. This advantage makes it useful for medical imaging. In addition to the isosurface, this algorithm generally calculates the unit normal vector to each elementary surface in order to facilitate rendering. Nevertheless, it is not necessary to store or transmit these vectors since they can be recalculated from the mesh.

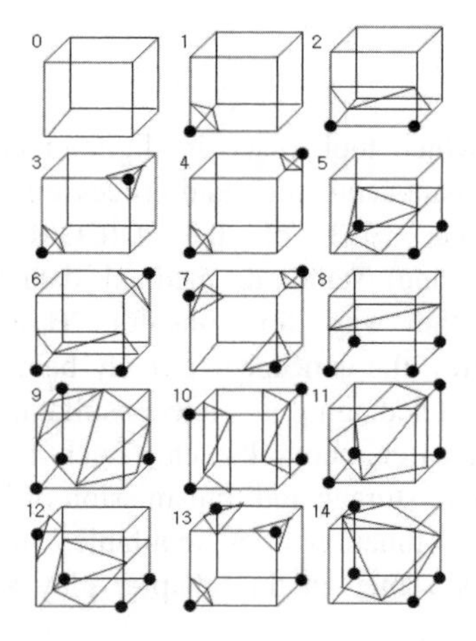

Figure 9.16. *Basic configurations of the Marching Cubes algorithm. The dots represent voxels with a gray level above the predefined threshold*

9.7. Bibliography

[ALE 00] ALEXA M., MÜLLER W., "Representing animation by principal components", *Proceedings of Eurographics 2000*, vol. 19, no. 3, p. 411-418, 2000.

[ALL 01] ALLIEZ P., DESBRUN M., "Progressive Encoding for Lossless Transmission of 3D Meshes", *Proceedings Siggraph '01*, p. 198-205, 2001.

[ALL 05a] ALLIEZ P., GOTSMAN C., "Recent Advances in Compression of 3D Meshes", *Advances in Multiresolution for Geometric Modelling*, Springer-Verlag ISBN 3-540-21462-3, N.A. Dodgson, M.S. Floater and M.A. Sabin (eds.), p. 3-26, 2005.

[ALL 05b] ALLIEZ P.,UCELLI G., CRAIG GOTSMAN C., ATTENE M., "Recent Advances in remeshing of surfaces", in *AIM@SHAPE EU network report*, 2005.

[AND 04] ANDUJAR C., P. BRUNET P., CHICA A., NAVAZO I., ROSSIGNAC J., VINACUA A., "Optimal iso-surfaces", *Proceedings CAD Conference*, p. 503-511, 2004.

[BON 03] BONNET S., PEYRIN F., TURJMAN F., PROST R., "Nonseparable wavelet-based cone-beam reconstruction in 3D rotational angiography", *IEEE Trans. on Medical Imaging, Special Issue on Wavelets in Medical Imaging,* vol. 22, no. 3, 2003, p. 360-367, 2003.

[BOU 04] BOURGES-SEVENIER M., JANG E. S., "An introduction to the MPEG-4 Animation Framework extension", *IEEE Transactions on Circuits and Systems for Video Technology*, vol. 14, p. 928-936, 2004.

[BRI 03] BRICEÑO H. M., SANDER P. V., MCMILLAN L., GORTLER S., HOPPE H., "Geometry videos: a new representation for 3D animations", *ACM Siggraph Symposium on Computer Animation*, p. 136-146, 2003.

[CAT 78] CATMULL E., CLARK J., "Recursively Generated B-spline Surfaces on Arbitrary Topological Meshes", *Computer Aided Design*, vol. 10, no. 7, pp. 350-355, 1978.

[COH 99] COHEN-OR D., LEVIN D., REMEZ O., "Progressive Compression of Arbitrary Triangular Meshes", *Proceedings IEEE Visualisation '99*, p. 67-72, 1999.

[COL 05] COLLINS G., HILTON A., "A rigid transform basis for animation compression and level of detail", *ACM Symposium on Interactive 3D Graphics*, p. 21–29, 2005.

[DAU 99] DAUBECHIES I., GUSKOV I., SCHRÖDER P., SWELDENS W., "Wavelets on irregular point sets", *Royal Society Typescript*, vol. 357, p. 2397-2413, 1999.

[DEE 95] DEERING M., "Geometry Compression", *Proceedings ACM Siggraph '95*, p. 13-20, 1995.

[DIS 05] DISCHER A., ROUGON N., PRÊTEUX F., "An unsupervised approach for measuring myocardial perfusion in MR image sequences", *SPIE Conference on Mathematical Methods in Pattern and Image Analysis*, vol. 5916, p. 126-137, 2005.

[DOO 78] DOO D., SABIN M., "Behaviour of Recursive Division Surfaces Near Extraordinary Points", *Computer Aided Design*, vol. 10, no. 6, p. 356-360, 1978.

[ECK 06] ECKSTEIN I., DESBRUN M., KUO C. C. J., "Compression of time varying isosurface", *Conference on Graphics Interface*, p. 99-105, 2006.

[FET 04] FETITA C., PRETEUX F., "Modélisation et segmentation 3D des nodules pulmonaires en imagerie TDM volumique", *Actes 14ème Congrès Francophone AFRIF-AFIA Reconnaissance des Formes et Intelligence Artificielle (RFIA'2004)*, p. 28-30, 2004.

[FET 05] FETITA C., MANCINI S., PERCHET D., PRÊTEUX F., THIRIET M., VIAL L., "An image-based computational model of oscillatory flow in the proximal part of the tracheobronchial trees", *Computer Methods in Biomechanics and Biomedical Engineering*, vol. 8(4), p. 279-293, 2005.

[FRE 99] FREY P. J., GEORGES P. L., *Maillages, Applications aux éléments finis*, Paris, Hermes Science Publications, 1999.

[GAN 02] GANDOIN P. M., DEVILLERS O., "Progressive Lossless Compression of Arbitrary Simplicial Complexes", *Proceedings ACM Siggraph'02, ACM Transactions on Graphics*, vol. 21, p. 372-379, 2002.

[GOT 03] GOTSMAN C., "On the optimality of valency-based connectivity coding", *Computer Graphics Forum*, vol. 22, p. 99-102, 2003.

[GU 02] GU X., GORLTER S., HOPPE H., "Geometry images", *Proceedings ACM Siggraph'02*, p. 355-361, 2002.

[GUS 04] GUSKOV I., KHODAKOVSKY A., "Wavelet compression of parametrically coherent mesh sequences", *ACM Siggraph Symposium on Computer Animation*, p. 183-192, 2004.

[HOP 96] HOPPE H., "Progressive meshes", *International Conference on Computer Graphics and Interactive Techniques, Siggraph '96*, p. 99-108, 1996.

[IBA 03] IBARRIA L., ROSSIGNAC J., "Dynapack: space-time compression of the 3D animations of triangle meshes with fixed connectivity", *Proceeding Eurogaphics '03*, p. 126-135, 2003.

[ISO 04] ISO/IEC JTC1/SC29/WG11: Standard 14496-16, also known as MPEG-4 Part 16: Animation Framework eXtension (AFX), ISO, 2004.

[JAN 04] JANG E. S., KIM J. D. K., JUNG S. Y., HAN M. J., WOO S. O., LEE S. J., "Interpolator data compression for MPEG-4 animation", *IEEE Transactions on Circuits and Systems for Video Technology*, vol. 14, p. 989-1008, 2004.

[KAR 00] KARNI Z., GOTSMAN C., "Spectral compression of mesh geometry", *Proceedings ACM Siggraph '2000*, p. 279-286, 2000.

[KAR 04] KARNI Z., GOTSMAN C., "Compression of soft-body animation sequences", *Computer & Graphics*, vol. 28, no. 1, p. 25-34, 2004.

[KHO 00] KHODAKOVSKY A., SCHRÖDER P., SWELDENS W., "Progressive Geometry compression", *International Conference on Computer Graphics and Interactive Techniques, Siggraph '2000*, p. 271-278, 2000.

[LEE 98] LEE A. W. F., SWELDENS W., SCHRÖDER P., COWSAR L., DOBKIN D., "MAPS: Multiresolution Adaptive parameterization of Surfaces", *Proceedings Siggraph '98*, p. 95-104, 1998.

[LEE 03] LEE H., DESBRUN M., SCHRODER P., "Progressive encoding of complex isosurfaces", *ACM Siggraph*, p. 471-476, 2003.

[LEE 06] LEE S-W, PROST R., "Enhancement of the connectivity regularity of irregular meshes", *ICSP'06, International Conference on Signal Processing*, Guilin, China, IEEE digital library, vol. 1, p. 85-88, 2006.

[LEN 99] LENGYEL J., "Compression of time-dependent geometry", *ACM Symposium on Interactive 3D Graphics*, p. 89-96, 1999.

[LEV 88] LEVOY M., "Display of surfaces from volume data", *Computer Graphics Hardware Applications*, vol. 8, no. 3, p. 29-37, 1988.

[LOO 87] LOOP C., Smooth Subdivision Surfaces Based on Triangles, Master's Thesis, University of Utah, p. 74, 1987.

[LOR 87] LORENSEN W., CLINE H., "Marching cubes: a high resolution 3D surface reconstruction algorithm", *Computer Graphics*, vol. 21, p. 163–169, 1987.

[LOT 99] LÖTJÖNEN J., P.J. REISSMAN P. J., I.E. MAGNIN I. E., KATILA T., "Model extraction from magnetic resonance volume data using the deformable pyramid", *Medical Image Analysis*, vol. 3, no. 4, p. 387-406, 1999.

[LOU 97] LOUNSBERY J. M., DEROSE T., WARREN J., "Multiresolution Analysis for Surfaces of Arbitrary Topological Type", *ACM Transations on Graphics*, vol. 16, no. 1, p. 34-73, 1997.

[MAM 05] MAMOU K., ZAHARIA T., PRÊTEUX F., "A preliminary evaluation of 3D mesh animation coding techniques", *SPIE Conference on Mathematical Methods in Pattern and Image Analysis*, vol. 5916, p. 44–55, 2005.

[MAM 06a] MAMOU K., ZAHARIA T., PRÊTEUX F., "Multi-chart geometry video: a compact representation for 3D animations", *Symposium on 3D Data Processing, Visualization and Transmission, CD-ROM*, 2006.

[MAM 06b] MAMOU K., ZAHARIA T., PRÊTEUX F., "A DCT-based approach for dynamic 3D mesh compression", *WSEAS Transactions on Information Science and Applications*, vol. 3, p. 1947-1954, 2006.

[MAM 06c] MAMOU K., ZAHARIA T., PRÊTEUX F., "A skinning prediction scheme for dynamic 3D mesh compression", *SPIE Conference on Mathematical Methods in Pattern and Image Analysis*, vol. 6315(02), p. 1:12 2006.

[MUL 04] MÜLLER K., SMOLIC A., KAUTZNER M., EISERT P., WIEGAND T., "Predictive compression of dynamic 3D meshes", IEEE International Conference on Image Processing, vol. 1, p. 621-624, 2005.

[PAR 00] RENATO PAJAROLA, JAREK ROSSIGNAC, "Compressed Progressive Meshes", *IEEE Transactions on Visualization and Computer Graphics*, 6(1) pp. 79-93, January-March 2000.

[PAY 05] PAYAN F., ANTONINI M., "An efficient bit allocation for compressing normal meshes with an error-driven quantization", *Computer Aided Geometric Design, Special Issue on Geometric Mesh Processing*, vol. 22, no. 5, p. 466-486, 2005.

[PEN 05] PENG J., KIM C. S., KUO C. C. J., "Technologies for 3D mesh compression: A survey", *Journal of Visual Communication and Image Representation*, vol. 16, p. 688-733, 2005.

[PER 04] PERCHET D., FETITA C., VIAL L., PRÊTEUX F., CAILLIBOTTE G., SBIRLEA-APIOU G., THIRIET M., "Virtual investigation of pulmonary airways in volumetric computed tomography", *Computer Animation & Virtual Worlds*, vol. 15, p. 361-376, 2004.

[POU 03] POULALHON D., SCHAEFFER G., "Optimal coding and sampling of triangulations", *Proceedings of the 30th International Colloquium ICALP'03, LNCS 2719,* Springer, p. 1080-1094, 2003

[PRE 87] F. PRETEUX, Description et interprétation des images par la morphologie mathématique. Application à l'imagerie biomédicale, PhD thesis, Pierre and Marie Curie University – Paris VI, Paris, France, 1987.

[ROS 99] ROSSIGNAC J., "EdgeBreaker: connectivity compression for triangle meshes", *IEEE Transations on Visualization and Computer Graphics*, vol. 5, no. 1, p. 47-61, 1999.

[ROU 05] ROUGON N., PETITJEAN C., PRÊTEUX F., CLUZEL P., GRENIER P., "A non-rigid registration approach for quantifying myocardial contraction in tagged MRI using generalized information measures", *Medical Image Analysis*, vol. 9(4), p. 353-375, 2005.

[ROU 06] ROUGON N., DISCHER A., PRÊTEUX F., "Region-based statistical segmentation using informational active contours", *SPIE Conference on Mathematics of Data/Image Pattern Recognition, Compression, and Encryption with Applications*, vol. 6315, 2006.

[SAI 96] AMIR SAID A., PEARLMAN W. A., "A New Fast and Efficient Image Codec Based on Set Partitioning in Hierarchical Trees", *IEEE Transactions on Circuits and Systems for Video Technology*, vol. 6, pp. 243-250, 1996.

[SAN 03] SANDER P., WOOD Z., GORTLER S., SNYDER J., HOPPE H., "Multi-chart geometry images", *Eurographics Symposium on Geometry Processing* , p. 146–155, 2003.

[SAR 06] SARAGAGLIA A., FETITA C., PRÊTEUX F., "Assessment of Airway Remodeling in Asthma: Volumetric Versus Surface Quantification Approaches", *MICCAI*, vol. 4191, p. 413-420, 2006.

[SAT 05] SATTLER M., SARLETTE R., KLEIN R., "Simple and efficient compression of animation sequences", *Siggraph Symposium on Computer Animation*, p. 209-217, 2005.

[SCH 95] SCHRÖDER P., SWELDEN W., "Spherical Wavelets: Efficiently Representing Functions on the Sphere", *Proceedings Siggraph '95*, p. 161-172, 1995.

[STO 96] STOLLNITZ E. J., DEROSE T. D., SALESIN D. H., *Wavelets for computer graphics: Theory and Application*, Morgan Kaufmann Publishers, San Francisco, 1996.

[SWE 96] SWELDENS W., *The Lifting Scheme: A Custom-Design Construction of Biorthogonal Wavelets, Applied and Computational Harmonic Analysis*, vol. 3, no. 2, April 1996, p. 186-200, 1996.

[TOU 98] TOUMA C., GOTSMAN C., "Triangle Mesh Compression", *Proceedings Graphics Interface '98*, Vancouver, Canada, p. 26-34, 1998.

[TUT 62] TUTTE W. T., "A census of planar triangulations", *Canadian J. Math.*, vol. 14, p. 21-38, 1962.

[VAL 99] VALETTE S., KIM Y.-K., JUNG H.-Y., MAGNIN I., PROST R., "A multiresolution wavelet scheme for irregularly subdivided 3D triangular mesh", *Proceedings IEEE ICIP '99*, vol. 1, p. 171-174, 1999.

[VAL 04a] VALETTE S., PROST R., "Wavelet Based Multiresolution Analysis Of Irregular Surface Meshes", *IEEE Transations on Visualization and Computer Graphics*, vol. 10, no. 2, p. 113-122, 2004.

[VAL 04b] VALETTE S., PROST R., "Wavelet-based progressive compression scheme for triangle meshes: Wavemesh", *IEEE Transations on Visualization and Computer Graphics*, vol. 10, no. 2, p. 123-129, 2004.

[VAL 04c] VALETTE S., GOUAILLARD A, PROST R., "Compression of 3D triangular meshes with progressive precision", *Computers & Graphics*, vol. 28, no. 1, p. 35-42, 2004.

[WOO 00] WOOD Z. J., "Semi-Regular Mesh Extraction from Volumes", *Proceedings Visualization 2000*, p. 275-282, 2000.

[YAN 04] YANG J. H., KIM C. S., LEE S. U., "Progressive compression of 3D dynamic sequences", *International Conference on Image Processing*, p. 1975-1978, 2004.

[ZHA 04] ZHANG J., OWEN C. B., "Octree-based animated geometry compression", *IEEE Data Compression Conference*, p. 508-517, 2004.

Chapter 10

Hybrid Coding: Encryption-Watermarking-Compression for Medical Information Security

10.1. Introduction

Nowadays, more and more digital images are being sent over computer networks. The works presented in this chapter show how encryption and watermarking algorithms provide security to medical imagery. In order to do this, the images can be encrypted in their source codes in order to apply this functionality at application level. In this way, the encryption and watermarking of images occurs at software level. We can therefore guarantee the protection of a medical image during transmission, and also once this digital data is archived. The subsequent challenge is to ensure that such coding withstands severe treatment such as compression. The quantity of information (entropy) to be sent greatly increases from the original image to the encrypted image. In the case of certain types of medical imagery, large homogenous zones appear. These zones affect the effectiveness of the coding algorithms. Nevertheless, these homogenous zones, useless for any diagnosis, can be safely used for the watermarking of medical images.

When a physician receives a visit from a patient, he often requires a specialist opinion before giving a diagnosis. One possible solution is to send images of the patient, along with a specialist report, over a computer network. Nevertheless, computer networks are complex and espionage is a potential risk. We are therefore faced with a real security problem when sending data. For ethical reasons, medical

---

Chapter written by William PUECH and Gouenou COATRIEUX.

imagery cannot be sent when such a risk is present, and has to be better protected. Encryption is the best form of protection in cases such as this. Many different techniques for the encryption of text already exist. Since ancient times, humanity has attempted to encode secret messages in order to elude wandering, indiscreet eyes and ears. The most basic forays into this field relied upon algorithms which allowed coding and decoding. Over time, the notion of a key arose. Today, encryption systems rely upon algorithms which are available to the world at large, and it is the key, a code which remains confidential, which allows for the encryption and decryption of the message [KER 83].

In section 10.2 we will show how essential it is to ensure the security of medical imagery and data. Then in section 10.3 we will present the standard encryption algorithms and will show, in section 10.4, how these can be suited to medical imagery. Finally, in section 10.5, we will show how it is possible to hide data in these images, while retaining a high level of image quality.

10.2. Protection of medical imagery and data

Developments in techniques for the treatment, sharing and communication of medical imagery, and medical information in general, go hand in hand with an increased risk for information in a digital format. Medical information in general is chiefly made up of the results of analyses, clinical and para-clinical examinations, and personal information [DUS 97]. Possibilities for distant access and the sending of information have increased the chances of leaks, losses and alterations of the information which are also greater due to, or even assisted by, the availability of network surveillance tools and advanced editing tools such as imagery software.

However, it is the consequences brought about by the occurrence of these risks which create the need for the protection of medical information. These consequences, which are not negligible, concern an individual and his health, and the privacy of these. This is why many countries attribute legal and ethical weight to this question; acknowledging patient rights and thereby obliging medical professionals and health centers to ensure the protection of the data in their possession.

10.2.1. *Legislation and patient rights*

Legislation and the medical ethics code accompany the technical evolution and, through a number of important legal texts, recognize patient rights. The first, and best-known, refers specifically to the patient-doctor relationship and concerns medical confidentiality. The guaranteed confidentiality of any information which a patient may exchange with anyone in the healthcare system allows a relationship based on trust to be established. This relationship also enables the healthcare professional to judge the patient's situation as effectively as possible.

The computerization of the health system and the possibilities this offers both in terms of the mechanization of treatments and the sharing of information, has resulted in the widening of the legal coverage of the field, and new laws must be taken into account by healthcare professionals; in particular in France law no. 78-17 of 6th January 1978, known as the "information technology and freedom law", complemented by the law of 1st January 1994, the "law pertaining to the treatment of data, with regard to health sector research". In France, the CNIL (National Commission on IT and Liberty) has the task of ensuring that these laws are respected (articles 6 to 13). These laws, aside from the collection of information, give every citizen – and therefore every patient – the right to control the use of information which concerns them personally [DUC 96]. In particular, the patient has a right to security, and article 29 states that it is the responsibility of the healthcare professional to take "every possible measure to ensure the security of the information, and particularly to ensure that it is not altered, damaged or allowed to reach unauthorized third parties". If this law is not respected, legal measures can be taken (article 226-17 of the penal code). From a practical and technical point of view, working groups such as that put in place by the European Standards Committee TC/251 Medical Information (Working Group III), show that in order for this criteria to be met, the following must be achieved [ALL 94]:

– the *confidentiality* of the data, by restricting access to the rightful owners (the patient and the healthcare professionals dealing with his case, considering the collaborative nature of medical practice and derogations allowed by the law);

– the *integrity* of the information, ensuring that the information has not been modified by anyone but a qualified person in agreed conditions;

– the *availability*, which guarantees access to the data within standard procedure.

10.2.2. *A wide range of protection measures*

Whatever the nature of the information, we can distinguish three types of protection measure for data stored, treated and sent using an information system: the legislation, security policies and protection mechanisms. These measures should be considered together in order to meet the AIC (Availability, Integrity and Confidentiality) requirements for the data. Nevertheless, these measures vary amongst themselves according to the target information system and its context. We can draw a difference between the systems installed at a single practitioner's practice from those of a health center – whether fitted or not with communication applications such as telemedicine applications.

The first type of legal measure aims to discourage those who would infringe either deliberately or accidentally the confidentiality, integrity and availability of data and information systems (e.g. in France, the Godfrain Law no. 88-19 of 05/01/1998). However, such measures are only effective if it is possible to detect the intrusion of a third party into an information system. A healthcare establishment's security policy aims to set the strategy for the implementation and upkeep of the highest security level. This policy decides upon, among other things, various protocols for the usage of information and systems, taking into consideration the risks to the AIC of the data and the specific roles of the various parties present in the hospital framework. Most notably, it is up to the security policy to decide and regulate the use of protection tools and mechanisms. These physical or logical mechanisms are numerous and entail more or less complex procedures to be carried out. The first group deals with the physical protection of the material (restricted access to the rooms concerned, steps to avoid damage from the elements, to prevent theft, etc.), and the second group is integrated into the information systems. The tools which we will study in this chapter fall into this second category; cryptographic tools (encoding and digital signatures) and the watermarking of images. Among the other logic protection mechanisms we can include [COA 03]:

– access control which includes a policy determining those with a right to the information or access to the workings of a system, and technical solutions for the identification of users, such as chip-card systems or biometric screening;

– firewalls whose primary task is to control access to the system both at the entry and exit stages, as soon as the system is connected to a network;

– antivirus systems;

– auditing which allows us to keep a record of the access made to the information by users or computer programs.

It is important to highlight the fact that these mechanisms are complementary, and are therefore to be used alongside one another. Additionally, some of these mechanisms, such as access restriction, rely upon cryptographic systems. The watermarking of images has come into use more recently, and has found its place among the range of tools on offer. Before discussing these techniques, let us return to medical imagery. These images are often produced, stored and communicated with the DICOM standard described in Chapter 4. This standard is more than a simple storage format, and includes very specific "profiles" or procedures, with the aim of guaranteeing the AIC requirements for storage and exchange between DICOM-compatible systems. These profiles are based on cryptographic mechanisms.

10.3. Basics of encryption algorithms

10.3.1. *Encryption algorithm classification*

There are four key objectives for the encryption of digital data:

– *confidentiality* or masking of the data – the most widely-used characteristic – which aims to render the cryptogram unintelligible to anyone without the key;

– *authentication* allows the sender to sign his message, thereby leaving the recipient in no doubt as to who sent the message;

– *integrity* serves to assure the recipient that the message content has not been altered or manipulated since its creation;

– *non-repudiation* is the guarantee that neither of the parties involved will be able to deny having sent or received the message.

The most important objective for medical imagery is, naturally, the first: confidentiality. However, the notion of integrity described in section 10.2., as well as the two others, is also important in the protection of medical imagery.

Encryption algorithms can be separated out according to various characteristics: the systems with a secret key (symmetric systems), illustrated in Figure 10.1, and those with public and private keys (asymmetric systems), shown in Figure 10.2 [DIF 76], [STI 96]. The secret key systems are those which allow encryption and decryption with the same key. It goes without saying that the sender and the recipient must beforehand have exchanged the secret of this key, via a secure method of communication. The systems using a public or asymmetric key can overcome this step by using one key to encrypt the data, and another to decrypt it. Each person should possess a pair of keys, one of which is confidential (the private key) and the other known by the world at large (the public key). In order to write to *B*, all that needs to happen is for the message to be encoded with the public key of *B*, which is known. Upon reception, only *B* will be able to decrypt the message with his private key. In this section, we present several data encryption systems; symmetric block systems, with a secret key (DES and AES), an asymmetric block system, with a public key (RSA); and a stream cipher system.

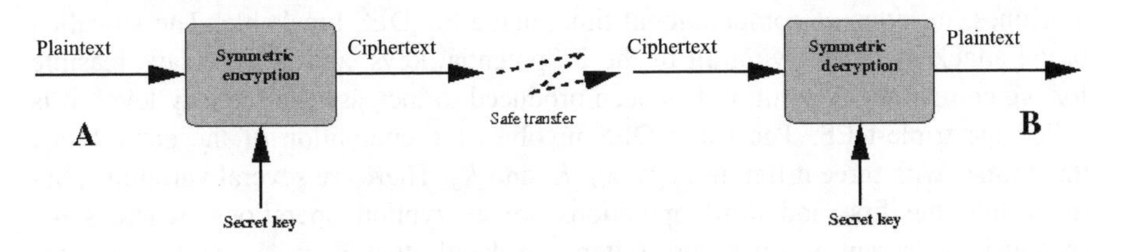

Figure 10.1. *The basis of symmetric encryption*

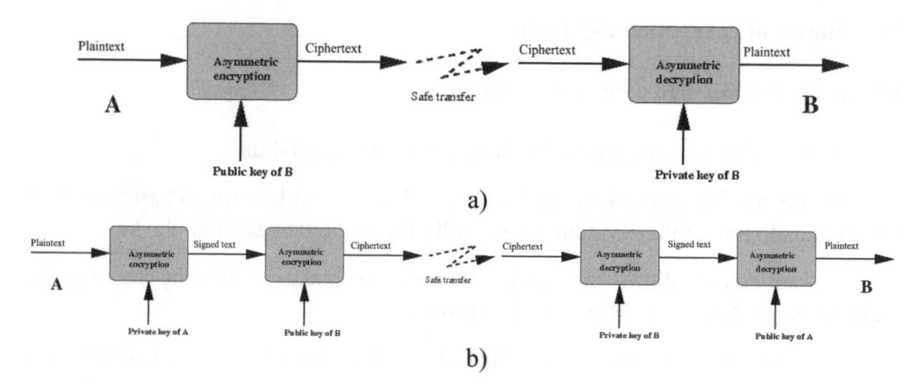

Figure 10.2. *a) Basis of asymmetric encryption; b) double asymmetric encryption, guaranteeing confidentiality and authenticity*

10.3.2. *The DES encryption algorithm*

The DES (Data Encryption Standard) algorithm is one of the standard systems for block encryption (Figure 10.1). Its security relies entirely upon the secrecy of the key, as the algorithm is public. In 1974, the DES algorithm became the first standard of modern cryptography [SCH 95]. The DES algorithm is based on 16 rounds (a collection of stages repeated 16 times) during which a data block of 64 bits is mixed with the key K, which is also encoded on 64 bits. At each of these rounds, a sub-key k_i is calculated from the initial key K (this sub-key will serve to mix up the block's bits). Once the 16 sub-keys have been generated from the secret key, it is possible to cipher (or decipher) a 64-bit block of data. The process begins with an initial permutation (IP) which changes the order of the bits in the initial block, before splitting the result into two blocks of 32 bits, L_0 and R_0. Once the 16 rounds have been passed, and before giving the result, a final permutation must be applied to the block. This permutation is no more than the inversion of the IP. For the decryption, the process is the same, apart from the fact that the sub-keys are used in the opposite order.

Today, even if the algorithm is still respected, it suffers somewhat from the fact that the length of its key is limited to 64 bits. The current performance-levels of machines, in terms of computational time, make the DES breakable. The so-called brutal attack involves trying all of the 2^{64} potential keys, and is nowadays feasible for big computers. A solution has been produced to increase the security level: it is called the triple-DES. The triple DES involves the encryption of the entry block three times with three different keys: K_1, K_2 and K_3. There are several variations, but in general the first and third operations are encryption operations, whereas the second is a decryption operation. Often, we decide that $K_1 = K_3$, which does not allow the whole key to go beyond 128 bits.

10.3.3. *The AES encryption algorithm*

The AES (Advanced Encryption Standard) algorithm is the standard system for block encryption and aims to replace the DES which has become vulnerable. The number of rounds in the AES algorithm depends upon the size of the key and the size of the data blocks. For example, the number of rounds is 9 if the blocks and the key have a length of 128 bits. To encrypt a block of data with AES (Figure 10.3), we must first of all complete a stage called "AddRoundKey", which involves applying an "exclusive OR" (XOR) between a sub-key and the block. After this, we enter into the operation of a round. Each regular round operation involves four steps. The first step is called "SubByte", where each byte of the block is replaced by another value created by an S-box. The second step is called "ShiftRow", where the rows are cyclically moved with different offsets. In the third step, called "MixColumn", each column is treated as a polynomial which is multiplied with a matrix in the GF(2^8) (Galois Field). The final step of a round is again called "AddRoundKey", which is a simple XOR between the given data and the sub-key of the given round. The AES algorithm carries out a final additional stage made up of the "SubByte", "ShiftRow", and "AddRoundKey" stages before producing the final encryption. The process applied to the "plaintext" (original data) is independent of that applied to the secret key, with the latter being called "KeySchedule". This is made-up of two components: the "KeyExpansion" and the "RoundKeySelection" [DAE 02], [AES 01].

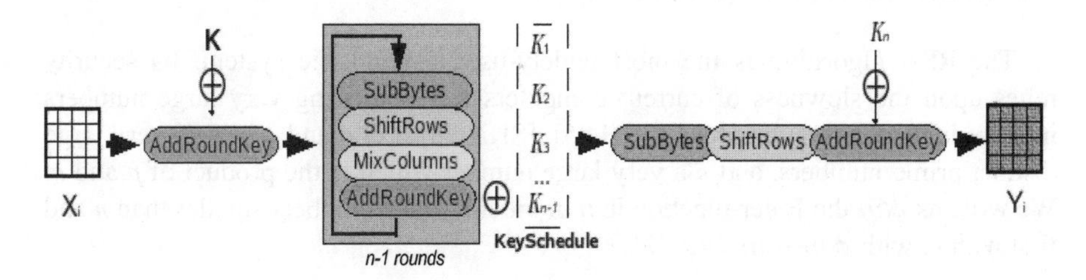

Figure 10.3. *General AES scheme*

The AES algorithm can support the following encryption modes: ECB, CBC, OFB, CFB, CTR, etc. The ECB (Electronic CodeBook) mode is that of the standard AES algorithm as described in document 197 of the FIPS (Federal Information Processing Standards). From a binary sequence $X_1, X_2,..., X_n$ of plaintext blocks, each X_i is encrypted with the same secret key k in order to produce the coded blocks Y_1, Y_2,.., Y_n. The CBC (Cipher Block Chaining) mode adds a step before the encryption. Each encrypted block Y_i is added by an XOR to the new current block X_{i+1} before being encrypted with the key k. An initialization vector (IV) is used for the first iteration. In fact, all the modes apart from ECB need an IV. In the CFB (Cipher

FeedBack) mode, $IV = Y_0$. The dynamic key (or key stream) Z_i is generated by $Z_i = E_k(Y_{i-1})$, $i > 1$, and the encrypted block is produced by $Y_i = X_i \oplus Z_i$. In the OFB (Output FeedBack) mode, as in the CFB, $Y_i = X_i \oplus Z_i$ but $VI = Z_0$ and $Z_i = E_k(Z_{i-1})$, $i > 1$. The input data is encrypted after an XOR with the output Z_i. The CTR (counter) mode has characteristics very similar to those of OFB, but it also allows for a random access for the decryption. It generates the following dynamic key by encryption of the successive value provided by a counter. This counter can be a simple function which produces a pseudo-random sequence. In this mode, the output of the counter is the entry of the AES algorithm.

Even if AES is a block encryption algorithm, the OFB, CFB and CTR modes operate like stream ciphers. These modes require no particular measure concerning the length of messages. Each mode has its own advantages and disadvantages. In the ECB and OFB modes, for example, any change in the plaintext X_i results in a modification in the corresponding encrypted block Y_i, but the other encrypted blocks are not affected. On the other hand, if a plaintext X_i is changed in the CBC and CFB modes, then Y_i and the new encrypted blocks will be affected. These properties mean that the CBC and CFB modes are useful for authentication, and the ECB and OFB modes treat each block separately. As a result, we can note that the OFB mode does not propagate noise, whereas the CFB mode does.

10.3.4. *Asymmetric block system: RSA*

The RSA algorithm is the most widely-used asymmetric system. Its security relies upon the slowness of current computers for factorizing very large numbers into products of prime numbers [SCH 95], [SHA 78]. Let p and q be two very large distinct prime numbers, and n a very large number which is the product of p and q. We write as $\phi(n)$ the Euler function in n in order to have numbers smaller than n and first with n, with $\phi(n) = (p-1)(q-1)$.

The public key/private key pair will reside in two numbers, d and e associated with n. e is first calculated randomly between 2 and $\phi(n)$ and must be prime with $\phi(n)$. The (n,e) pair is the public key. Then d is calculated such as $d = e^{-1} \text{mod}(n)$. The extended Euclidian algorithm allows the calculation of this inversion, even in the case of very large numbers. The (n,d) pair is the private key. The use of keys for the encryption and the decryption is as follows. If m is the original message (lower than n, otherwise it is cut), we encrypt it with the public key (n,e) raising it to the power e, modulo n. We obtain the encrypted message $m' = m^e \text{mod}(n)$. For the decryption, we need the private key second key (n,d). By raising the encrypted message to the power d modulo n and as d and e are inverted modulo n, we obtain:

$$(m')^d \bmod(n) = (m^e \bmod(n))^d \bmod(n) = m^{ed} \bmod(n) = m \qquad [10.1]$$

For example, if Bob wishes to send a message to Alice, he converts his message into numbers, and cuts the message into blocks of a size smaller than n. For each block m_i, using Alice's public key, Bob calculates and codes the block as follows:

$$c_i = m_i^{\ e} \bmod(n), \qquad [10.2]$$

with i, the position of the block in the text, $i \in [1,N]$, if N is the number of blocks.

Alice, with her private key, can then decrypt the message by performing:

$$m_i = c_i^{\ d} \bmod(n) \qquad [10.3]$$

Thus, the RSA method differs from the symmetric encryption systems in that it uses two different keys for encryption and decryption (Figure 10.2). One of these two keys, the public key, is meant to be known to everyone, and the other, the private key, is known to only one individual. The RSA algorithm can allow either encryption with a public key, in which case only the recipient will be able to decrypt the message with his private key, or encryption with one's own private key (signature). In this case, everyone can read the message thanks to the public key, but the sender was able to sign the message, since he is potentially the only person who could have encrypted it with his private key. A double encryption, using a public key/private key therefore makes it possible to combine a signature with confidentiality (Figure 10.2b).

Unfortunately, RSA is a very slow algorithm; much slower than any symmetric system, and even more so because the numbers used are very large. Moreover, it is easily breakable today, even for 512 bits numbers. It is currently advisable to use keys 1,024 bits long. It is therefore preferable to use it to send a secret key in a secure way, which will allow the message to be decrypted, with AES faster than RSA.

10.3.5. *Algorithms for stream ciphering*

Algorithms for stream ciphering can be defined as algorithms for encryption by block, where each block has a unitary dimension (1 bit or 1 byte) or is relatively small. Their main advantages are their very high speed and their ability to change each symbol of the plaintext. With a stream cipher algorithm, it is possible to encrypt each character of the plaintext separately, using an encryption function which varies each time (these algorithms therefore need memories). In general, algorithms for stream cipher are made up of two stages: the generation of a dynamic key (key stream) and the encryption output function using the dynamic key.

When the dynamic key is created independently of the plaintext and the ciphertext, the stream cipher algorithm is synchronous. With a stream cipher algorithm, the sender and receiver have to synchronize using the same key at the same position. Synchronous stream ciphers are used in environments where error is common, because they have the advantage of not propagating errors [GUI 02]. Concerning active attacks, such as the insertion, deletion or copying of digits of the ciphertext by an active adversary, these attacks immediately result in a loss of synchronization. The encryption process of a synchronous stream cipher is described in Figure 10.4a, where *f()* is the function which determines the following state, *g()* is the function generating the dynamic key, and *h()* is the encryption output function:

$$\begin{cases} s_{i+1} = f(K, s_i) \\ z_i = g(K, s_i) \\ c_i = h(z_i, m_i) \end{cases}, \qquad [10.4]$$

where K is the key, s_i, m_i, c_i and z_i are respectively the ith state, plaintext, ciphertext and dynamic key. The decryption process is shown in Figure 10.4b.

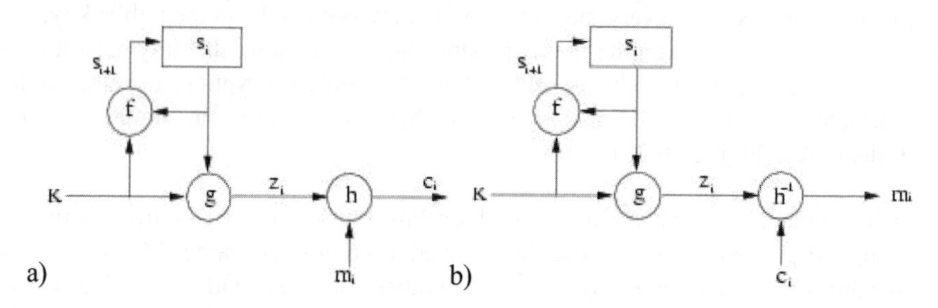

Figure 10.4. *Synchronous stream cipher: a) encryption, b) decryption on the right*

When the dynamic key is generated from the key and a certain number of previous ciphertext, the stream cipher algorithm is called asynchronous, also known as a self-synchronous stream cipher. The propagation of errors is limited to the size of the memory. If digits of the ciphertext are erased or added to, the receiver is able to resynchronize himself with the sender, by using the memory. As for active attacks, if an active adversary modifies any part of the digits of the ciphertext, the receiver will be able to detect this attack. The encryption process of an asynchronous stream cipher is described in Figure 10.5, where *g()* is the function which generates the dynamic key, and *h()* the encryption output function:

$$\begin{cases} z_i = g(K, c_{i-t}, c_{i-t+1}, \ldots, c_{i-2}, c_{i-1}) \\ c_i = h(z_i, m_i) \end{cases}, \qquad [10.5]$$

where K is the key, m_i, c_i and z_i are respectively the ith plaintext, the ciphertext, and the dynamic key. We can observe in equations [10.5] that the dynamic key depends upon the previous t digits of the ciphertext. In order to resist statistical attacks, function *g()*, which generates the dynamic key, must produce a wide period sequence, with good statistical properties which can be called pseudo-random binary sequences. The decryption process is illustrated in Figure 10.5.

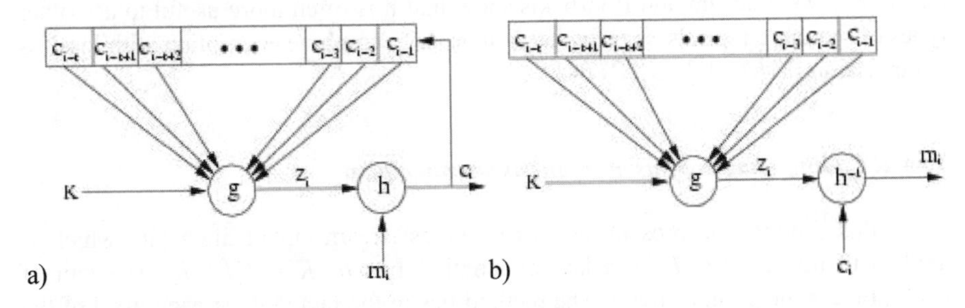

Figure 10.5. *Asynchronous stream cipher: a) encryption, b) decryption on the right*

10.4. Medical image encryption

In this section, we will demonstrate how it is possible to apply the above algorithms to medical images in the gray level. In the case of a 1D medical signal, the standard coding algorithms can be applied directly. However, because of the bidimensional characteristic of images, and their size, these standard algorithms must be modified in order to be used effectively on medical images. The aim of image encryption is to obtain an image in the same format and without increasing the size above that of the original image. The encryption of images is considered as a source coding to process this functionality at the application level. Due to this, if a user does not possess the key, he does at least have access to an image in a known format. By carrying the encryption step up to the application level, it is possible to proceed, for example, towards a region of interest of the image. In the case of large images, it therefore becomes unnecessary to decrypt the whole image if we only want to view one particular area. The compression stage should also be taken into account during the image encryption stage.

10.4.1. *Image block encryption*

In the case of block encryption, the length of the blocks is fixed, and varies from 64 bits (8 pixels) to 192 bits (24 pixels). From the 2D information of an image, several pixel grouping solutions are possible. With the aim of withstanding a downstream compression as well as possible, or compressing at the same time as coding, it is useful to group the pixels with their nearest neighbors (in rows, columns, or blocks). Each block of pixels is encrypted separately. The encrypted block obtained will then come to replace the original block in the image. In this chapter, the route taken for scanning the blocks is carried out only in a linear manner (scan line). Manniccam and Bourbakis show that it is often more useful to use other types of scanning (spirals, zigzags, etc.) in order to combine encryption with lossless compression [MAN 01], [MAN 04].

10.4.2. *Coding images by asynchronous stream cipher*

In this section, we present an asynchronous stream cipher algorithm which is applied to images. Let K be a key of length k bits b_i, $K = b_1 b_2 ... b_k$. The unit of encryption is the pixel (1 byte). The method lies in the fact that for each pixel of the image, the encryption depends upon the original pixel, the value of the key K, and the $k/2$pixels previously encrypted. In order to use equations [10.5], we have $t = k/2$. For each pixel p_i of the original image, we calculate the value of the pixel p'_i of the encrypted image using the following equation:

$$\begin{cases} z_i = \left(\sum_{j=1}^{k/2} \alpha_j p'_{i-j} \right) \bmod (256) \\ p'_i = (z_i + p_i) \bmod (256) \end{cases} \qquad [10.6]$$

with $i \in [0,\ldots,N-1]$ where N is the number of pixels in the image, k is the length of the key with $k \in [1,N]$, and α_j is a sequence of $k/2$ coefficients generated from the secret key K [PUE 01a].

The encryption principle is the same as that shown in Figure 10.5. Equations [10.6] have a recurrence of the order $k/2$, corresponding to half of the length of the key [PUE 01b]. Coefficients α_j are integer values included between -2 and +2 such as:

$$\begin{cases} \alpha_j = \beta_j - 1 & \text{si } \beta_j \in \{0,1,2\} \\ \alpha_j = \pm 2 & \text{si } \beta_j = 3 \end{cases} \qquad [10.7]$$

with $\beta_j = 2b_{2j-1}+b_{2j}$, where b_{2j-1} and b_{2j} are two consecutive bits of the secret key K. In addition, the probability density of the α_j must be uniform in order to reduce the transmission errors during the decryption stage. The sign in front of the coefficients equal to 2 depends on coefficients α_j in order to obtain:

$$\frac{1}{k/2}\sum_{j=1}^{k/2}\alpha_j \approx 0 \qquad [10.8]$$

Considering that the encryption of a pixel is based on the $k/2$ pixels previously encrypted, we cannot encrypt the $k/2$ first pixels of the image. It is necessary to associate the α_i coefficients with a sequence of $k/2$ virtual encrypted pixels p'_{-i}, for $i \in [1,\ldots,k/2]$. This pixel sequence corresponds to an initialization vector (IV). In consequence, an IV is coded in the key: $k/2$ values of virtual pixels which allow us to encrypt the $k/2$ first pixels of the image as though they had predecessors. The length k of the key K must be big enough to guarantee maximum security. Equation [10.9] presents the decryption procedure. In the decryption procedure, we must apply the process in reverse. We can note that the function which generates the dynamic key is the same as equation [10.6]:

$$\begin{cases} z_i = \left(\sum_{j=1}^{k/2}\alpha_j p'_{i-j}\right)\bmod(256) \\ p_i = (p'_i - z_i)\bmod(256) \end{cases} \qquad [10.9]$$

10.4.3. *Applying encryption to medical images*

Starting out with the image in Figure 10.6a, we have applied the DES algorithm by blocks of 8 pixels in a row, with a 64-bit key to obtain the image in Figure 10.6c. We can observe the appearance of textures (Figures 10.6c-e). The reason for this phenomenon lies in the appearance of large homogenous zones (black in this case) on the medical images. At the level of the histograms (Figures 10.6d-f), we observe the strong presence of gray levels corresponding to the encryption of the gray levels of the homogenous zones. The encryption is therefore very poor for two reasons: firstly because it is easy to guess the nature of the medical image (an ultrasound), but mainly because the availability of the value of the plaintext block (the pixels were all black), and after encryption (the gray levels dominating in the encrypted image) is a precious clue for cryptanalysts. Block encryption algorithms therefore present us with serious problems when images contain homogenous zones.

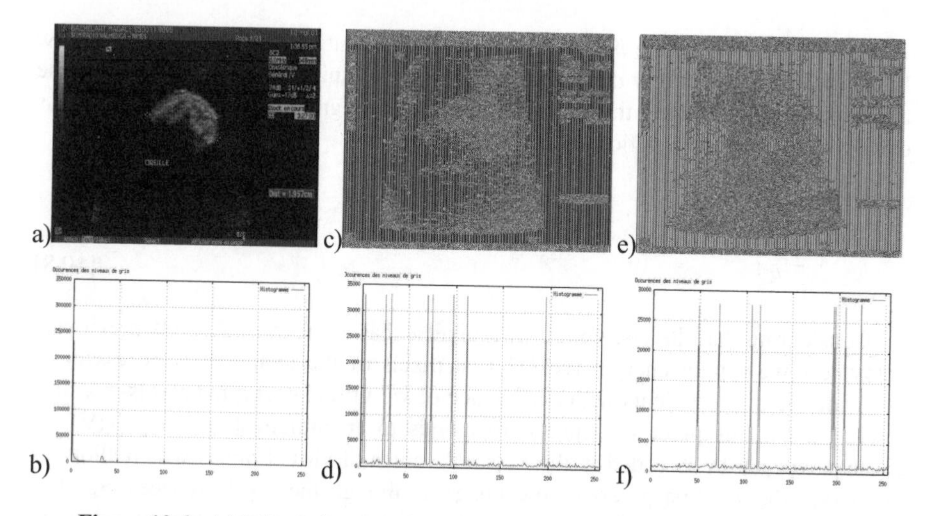

Figure 10.6. *a) Medical ultrasound image (442KB), with large homogenous zones, encrypted image; c) encrypted by DES algorithm (block of 8 pixels with a 64-bit key); e) by AES algorithm (block of 8 pixels with a 128-bit key); b), d) and f) histograms*

From the original image, Figure 10.7a (396x400 pixels), we have applied a stream cipher algorithm with a 128-bit key. Figure 10.7c illustrates the values obtained for the dynamic key z_i generated by equation [10.6]. We can note that (Figure 10.7d) the probability of the appearance of each value is practically equal. Consequently the function generating the dynamic key *g()* produces a sequence with a large period and good statistical properties. From equations [10.6], we obtain an encrypted image (Figure 10.7e), and we can see that the initial image is no longer visible at all. By comparing the histogram of the initial image (Figure 10.7b), with the histogram of the encrypted image (Figure 10.7f), we can see that the density of probability of the gray levels is more or less identical. As a result, the entropies of the encrypted images are very high (around 8 bits/pixel).

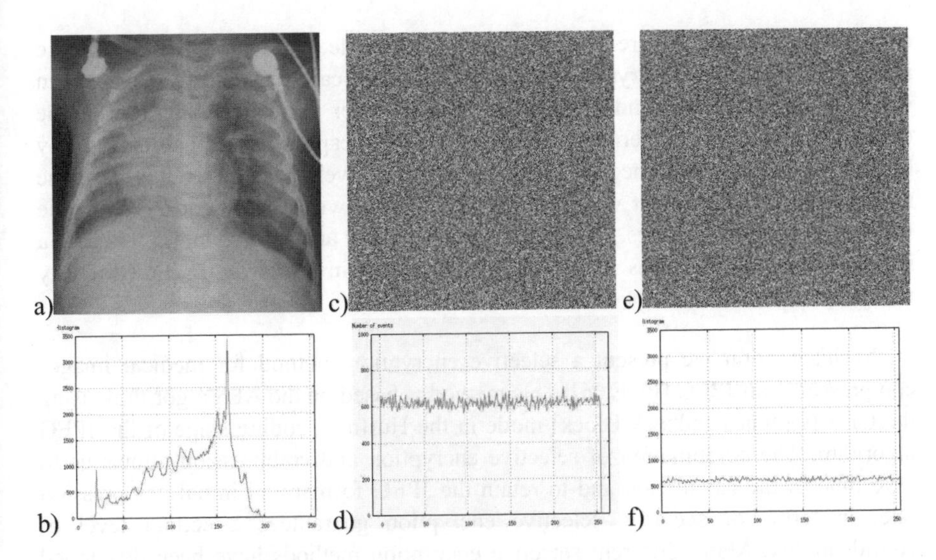

Figure 10.7. *a) Original image, b) histogram of original image, c) image of the dynamic key z_i, d) histogram of the values of the dynamic key z_i, e) final encrypted image with the coding algorithm by asynchronous stream cipher, with a 128-bit key, f) histogram of the encrypted image*

The stream cipher method has one major advantage over other encryption systems used in medical imagery. As the result of the encryption of the previous pixels is taken into account for each pixel to be encrypted, the problem of homogenous zones is solved. We are no longer dealing with block encryption systems, where two identical original blocks give the same encrypted block. We can observe that whatever the type of image with or without homogenous zones, no texture appears in the encrypted images. In conclusion, in the case of stream cipher algorithms, the homogenous zones are no longer visible either in the image or the histogram. The stream cipher method also carries another advantage: as the calculations which make it up are small in number, it proves to be very quick; even more so than AES. For example, a 7 MB image is encrypted (or decrypted) in 5 s with a standard PC, rather than the 15 s required for algorithms using a block encryption.

10.4.4. *Selective encryption of medical images*

Another way to ensure confidentiality is to adapt the protection level according to the application and the time available. It is in this second approach that we find selective encryption where users can apply a security level which can vary according to requirements [NOR 03]. Many applications can be protected with only selective encryption; the images are therefore partially visible, without revealing all the information. Selective encryption can be useful in the case of medical images taken with a medical device and needing to be sent over a network in order to be diagnosed remotely. Furthermore, the device used for capturing medical images may be located in an ambulance or some other mobile vehicle, and in this case the transmission is carried out via the intermediary of a wireless network. Due to the vital nature of these images, they must be sent quickly and safely, and in this case a selective encryption seems to be the best solution (in terms of the time/security ratio).

In this section we present a selective encryption method for medical images compressed with JPEG [PUE 06]. This method is based on the AES algorithm, using the OFB (Output Feedback Block) mode in the Huffman coding stage of the JPEG algorithm. The combination of selective encryption and compression allows us to save time in the calculation and to retain the JPEG format and initial compression rate. In terms of security, selective encryption guarantees a certain level of confidentiality. Many different selective encryption methods have been developed for images coded by DCT. Tang [TAN 96] proposes a technique called zigzag permutation, which can be applied to videos or images. Although his method offers a good level of confidentiality, it does decrease the compression rate. [DRO 02] describes a technique which encrypts a selected number of AC coefficients; the DC coefficients are not encrypted as they carry important visible information and are highly predictable. For this method, the compression rate is constant (compared to compression only) and retains the binary flow format. However, the compression and the encryption are carried out separately, so the method is slower than a simple compression. [FIS 04] presents a method where the data is organized in a binary flow form which can be regulated. Recently, Said has shown the strength of partial encryption methods by testing attacks which use the non-encrypted information of an image alongside a small image [SAI 05].

Let $E_K(X_i)$ be the encryption of a block X_i of n bits using the secret key K with the AES algorithm in OFB mode. In the description of the method, we will suppose that $n = 128$ and X_i is a non-empty plaintext. Let us suppose that $D_K(Y_i)$ is the decryption of a ciphertext Y_i using the secret key K. The encryption is applied at the same time as the entropic coding procedure during the creation of the Huffman vector. The method works in three steps illustrated in Figure 10.8: the construction of the plaintext X_i, the encryption of X_i to create Y_i and the substitution of the original Huffman vector with the encrypted information [ROD 06]. It should be mentioned that these operations are carried out separately for each DCT block quantified.

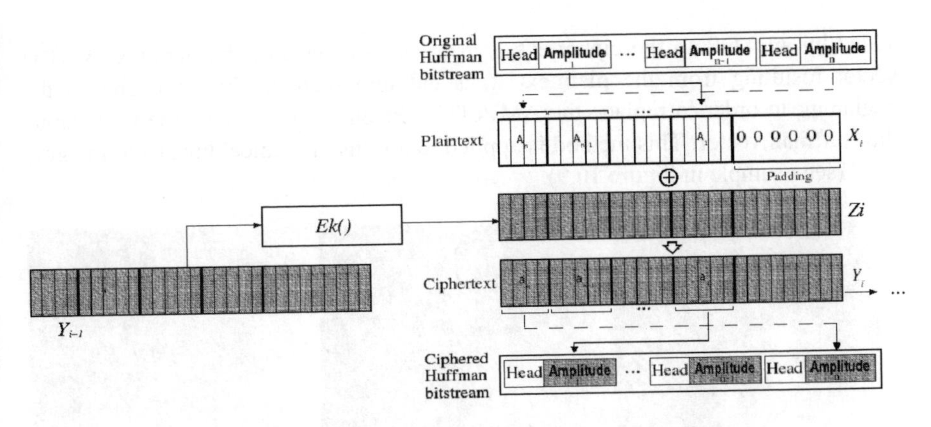

Figure 10.8. *Global overview of the proposed method*

To construct the plaintext X_i, we take the non-zero AC coefficients of the current block i accessing the Huffman vector from its end towards its beginning in order to create {HEAD, AMPLITUDE} pairs. From each HEAD we obtain the size of AMPLITUDE in bits. Only the AMPLITUDES (A_n, A_{n-1},..., A_1) are taken into account to build vector X_i. The final length of the plaintext L_{Xi} depends both on the homogenity ρ of the block and a given constraint C, with $C \in \{128, 64, 32, 16, 8\}$ bits. This means that a block with a large ρ will produce a small L_{Xi}. The Huffman vector is processed as long as $L_{Xi} \leq C$ and the DC coefficient is not reached. Next, the padding function is applied, $p(j) = 0$, where $j \in \{ L_{Xi},..., 128\}$, in order to fill in the X_i vector with zeros, if necessary.

At the stage where X_i is encoded with AES in OFB mode, the dynamic key Z_{i-1} is used as a parameter by the AES encryption in order to obtain a new dynamic key Z_i. For the first iteration, the IV is created from the secret key K with the following strategy: the secret key K is used as a seed for a Pseudo Random Number Generator (PRNG). This K is divided into 16 sections of 8 bits each. The PRNG produces 16 random numbers which define the formation order of the IV. Next, each Z_i is added by an XOR with the plaintext X_i to generate the encrypted block Y_i.

The final step is the substitution of the initial information with the encrypted information in the Huffman vector. As in the first step (the construction of the plaintext X_i), the Huffman vector is read backwards, but the coded vector Y_i is read starting from the beginning and moving to the end. Knowing the length, in bits, of each AMPLITUDE (A_n, A_{n-1},..., A_1), these sections are cut in Y_i to replace the AMPLITUDE in the Huffman vector. The total quantity of bits must be L_{Xi}. This procedure is carried out for each block. Any homogenous blocks are only slightly coded, or not at all. The use of the OFB method for coding allows for the generation of the independent Z_i. During the OFB-mode decryption stage, the dynamic key Z_i is added by an XOR to the ciphertext Y_i in order to regenerate the plaintext X_i. The vector resulting from the plaintext X_i is cut into sections from the end to the beginning in order to replace the AMPLITUDES in the Huffman code to generate the Huffman vector. This method is applied to dozens of medical images in the gray level (see example in Figure 10.9).

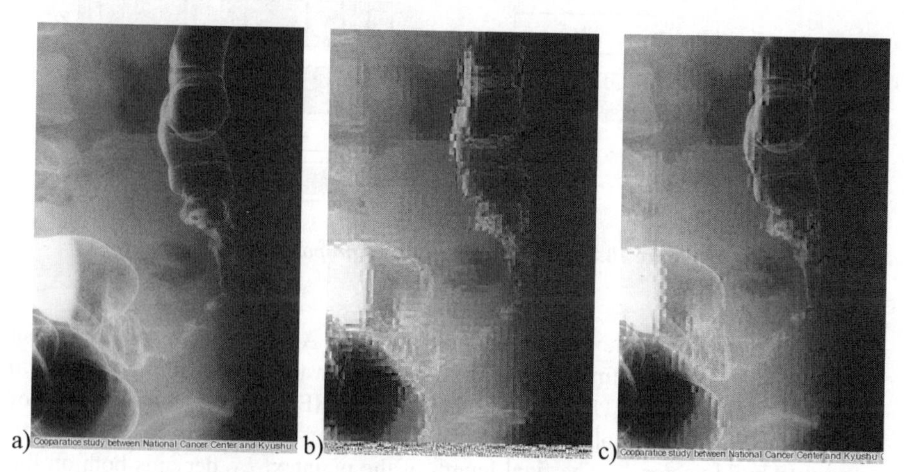

Figure 10.9. *From left to right: original medical image of a colon cancer, 320x496 pixels; encrypted image for C = 128; encrypted image for C=8*

The JPEG algorithm has been used with the online sequential coding system for a quality factor (QF) of 100%. Five values were applied for constraint C (128, 64, 32, 16, and 8). For the encryption, the AES algorithm was used with the stream cipher mode OFB with a key of 128 bits in length. The original medical image of 320x496 pixels (Figure 10.9) is compressed so that all the encrypted images are of the same size: 43.4 KB. The encrypted coefficients are distributed in the 2480 8x8 blocks in the image. This means that there are no totally homogenous blocks. For $C = 128$, a maximum of 128 bits encrypted per block, 26,289 AC coefficients have been encrypted, which is an average of 33 encrypted bits per block. The percentage of encrypted bits in the image as a whole is 22.99%. That means, in the spatial domain, 136,038 modified pixels, which means 85.71% of the coded pixels. The PSNR is of 23.39 dB for $C = 128$. For $C = 8$, the quantities of AC coefficients and bits encrypted are respectively 6,111 and 16,765. The percentage of encrypted bits in terms of the whole image is 4.7%. This constraint gives a number of modified pixels rising to 76.1% of all the pixels in the image. The PSNR is then 30.90 dB. As the images show, selective encryption of the JPEG image produces block artefacts. These artefacts are at the borders between blocks, which often interfere with the HVS. Since the frequential transformation and the quantification of the pixel-blocks are processed separately, any continuity between the values of pixels in neighboring blocks is broken during the coding. One of the advantages of this method is that it is possible to decrypt the 8x8 pixel blocks of the image individually (using the OFB mode for AES encryption). In order to form a remote diagnosis, the doctor needs to view regions of interest at a high resolution, where the background can be partially encrypted. We should note that confidentiality is linked with the ability to guess the values of the encrypted data (cryptanalysis). In terms of security, it is therefore preferable to encrypt the bits which seem the most random [PUE 05].

10.5. Medical image watermarking and encryption

10.5.1. *Image watermarking and health uses*

The watermarking of images comes under the more general heading of information hiding: a message is embedded into a document, a host which may be text, sound, video or images. For images, the difference signal between the original image and its watermarked version corresponds to the watermark signal associated with the embedded message. Care must be taken to ensure that the watermarked host document has the same value as the original host document. Figure 10.10 gives an example of a watermarking chain.

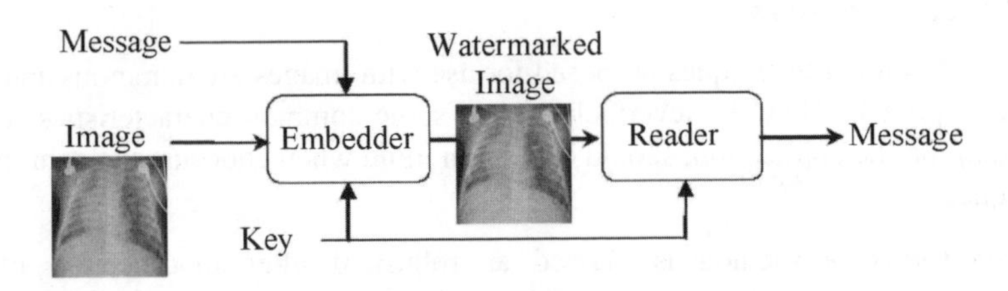

Figure 10.10. *A watermarking chain*

With the link established between the message and its host, we see steganography, watermarking and fingerprinting. Steganography is a form of secret communication where the host, who serves as the hidden communication channel, is of little interest to the message's recipient. Watermarking and fingerprinting have come to the fore since 1995, to meet the needs of managing DRM (Digital Rights Management). Watermarking involves the insertion of a code identifying the owner, and fingerprinting involves the insertion of a trace linked to the buyer. Since then, other applications have been proposed for information security, such as copy control, restriction of access and integrity control. For more details on watermarking applications and methods, see [COX 02]. Henceforth, the terms "watermarking" and "data hiding" will be used interchangeably.

For medical imagery, several cases making use of watermarking can be identified [COA 00]:

– the authenticity of images with the insertion of data confirming the origin and the fact that a certain image refers to a particular patient;

– controlling the integrity of images, by putting control information, such as a digital signature, within the image (section 10.3);

– the addition of meta-data (data hiding), allowing the content of images to be enriched by attaching a semantic description of the content [COA 05].

Another, more detailed, scenario combines authenticity with the integrity control of the images, and aims to establish a solid link between these images and the corresponding test results [COA 06]. In information protection [COA 03], watermarking is complementary to the mechanisms discussed above, as it merges the protection information and the image to be protected into one entity: a watermarked image.

10.5.2. *Watermarking techniques and medical imagery*

10.5.2.1. *Characteristics*

The data-hiding techniques proposed for use with images are numerous and vary in their approach. They do nevertheless have some common characteristics which, depending on the application, should be kept in mind when choosing the appropriate technique:

– *robustness*: a method is classed as robust if after modification of the watermarked image (a "washing" attack or simple image processing) the hidden information can still be accessed and understood. This property is essential for identifying images which undergo treatment or lossy compression;

– *capacity*: this measurement expresses the embedding rate as the number of bits buried per pixel of the image (bpp, for "bit per pixel") and therefore gives an indication of the message size which can be embedded in an image. Data-hiding techniques aim to optimize this parameter and then add useful information to the images;

– *security*: in some cases access to the watermark and its contents must be restricted; as for cryptography, there are symmetric and asymmetric watermarking methods;

– *complexity*: this is an indication of the calculation time needed for embedding and extraction; the complexity plays an important role when treating large image volumes;

– *invisibility*: this is important for medical images, since a watermark must not interfere with the interpretation of an image, in order to avoid affecting the diagnosis. With this in mind, certain watermarking methods have been proposed specifically for medical imagery;

– the need for the original image for decoding is also one of these characteristics; we say that a method is "blind" if the original image is not required in order to extract the watermark; integrity control applications are not possible with non-blind watermarking methods, the question of knowing whether the original image has been modified has no meaning in such cases.

One method can meet certain applications independently, but not simultaneously. A compromise must be reached between capacity, robustness and invisibility. A stronger watermark will better resist alterations to the signal introduced by the compression, or hacking attempts; but its presence will be more obvious to the user and its capacity will be reduced.

10.5.2.2. *The methods*

In their principles, watermarking methods proposed for medical imagery are only very slightly different from other methods which we can refer to as "classic" methods. They make use of particular adaptation strategies specifically for medical imagery.

In the habitual schemes, two types of algorithm can generally be identified. The first involves additive methods. Starting from a message (a sequence of bits), they generate a signal which is added to the image or a transformation of the image (DCT, DWT, etc.). A technique involving spread spectrum links each bit b_j of the message with the value d_j = 1-2b_j thus multiplies this quantity by a pseudo-random sequence W_j low in energy which is then added to the image I to produce the watermarked image: $I_w = I + \alpha d_j W_j$. α is a parameter of insertion or incrustation strength (robustness parameter). The embedding of a message of N bits adds to the image the watermark $W = \sum_{j=1}^{N} \alpha d_j W_j$. The presence of this watermark is checked by correlation techniques, which implies the orthogonal nature of the pseudo-random sequences W_j. The sign of each correlation product gives the value of the embedded bit. The embedding of a large message can lead to a partially visible watermark W. In order to ensure the invisibility of the watermark, psychovisual criteria are used to adapt the insertion strength to the image locally.

The second group covers substitution methods which, in order to embed a bit of the message, replace a piece of information from the image (its gray levels or a transformation thereof) with a symbol from a dictionary. Detection therefore takes place with a simple re-reading. The method of Least Significant Bit (LSB) substitution is the simplest. This method simply replaces the LSB of an image's gray levels with those of the message to be watermarked. For a gray level pixel *p(n,m)*, this means associating the binary value 1 with the odd values of *p(n,m)*, and the value 0 with the even values. This method is far from robust (it is therefore fragile) but does offer a capacity of 1 bpp. More elaborate versions of this approach have since appeared, following the Costa schema [COS 03]. These methods known as informed [COX 02] are based upon structured dictionaries which contain the values that the blocks of pixels will take to carry information.

For medical imagery, three strategies have been established with the key aim of preserving the image interpretation. These are the methods above, methods involving region of non-interest watermarking and reversible watermarking methods.

The first methods produce watermarks which replace part of the information in an image. Using them requires careful attention to ensure that the watermark does not interfere with the diagnostic information. The first solutions proposed were techniques secretly modifying the LSB of certain pixels or coefficients of the transform of an image. These are methods with a large capacity, introducing only a slight damage to the original signal but very fragile nevertheless. More recently, robust techniques have been tested with, during the experiments, the involvement of a practitioner giving a threshold of insertion force which should not be crossed [PIV 05]. More generally speaking, the problem with the automatic evaluation of the maximum authorized distortion level is a pertinent question. This problem is far from helped by the wide diversity of signals in the healthcare sector (see Chapter 3) and the availability for practitioners of tools which allow, for example, the isolation of a certain part of the dynamic of an image. These ranges of gray levels vary according to the user, and some may find that the watermark is visible.

One strategy, suggested to optimize the performance of the above methods in terms of robustness and capacity without further damage to the image, is based on the existence in the image of regions with little or no interest for in the image interpretation. These techniques, known as region of non-interest watermarking, more often than not place the watermark in the black background of the image [COA 01]. Robustness can be achieved, with the watermark not masking any important information, although a strong watermark can be a hindrance for the physician during his image analysis. The embedding strength must be regulated.

The final approach concerns reversible watermarking methods. The idea is to be able to remove the watermark from the image, thereby restoring the exact same gray levels as in the original image. These techniques make it possible to update the contents of a watermark. This is not the case for the previous methods, where one watermark would have to be added to another. The drawback is that the watermarked image is no longer protected once the watermark has been removed.

Such techniques have benefited from progress in recent years; [COA 05], [COA 06], [CAV 04], [COL 07]. The techniques developed have variable performances depending on the host type to be watermarked and with performances lower than any non-reversible methods. Otherwise, these techniques are very rarely robust, and the desire to maximize the capacity often leads to highly visible watermarks: the watermark must be removed before the image can be used.

Watermarking medical imagery is in its early stages at the moment, with the key difficulty encountered being the level of distortion which it introduces. We can however remain hopeful that the work carried out on improving the quality of image compression (Chapter 5) will lead to solutions allowing the full benefits of watermarking to be appreciated.

10.5.3. *Confidentiality and integrity of medical images by data encryption and data hiding*

The applications of watermarking medical imaging are numerous. In this section, we aim to illustrate the combination of cryptography and watermarking in secure image exchange. We saw in section 10.3 that the encryption process could be either symmetric or asymmetric, by block or by stream. Whereas asymmetric algorithms are not appropriate for image encryption due to their calculation time, block algorithms present security problems (due to homogenous zones) and problems with the data integrity. Figures 10.11 demonstrate this problem. The AES block algorithm [AES 01] with a 128-bit key has been applied to the original image (Figure 10.11a) in order to obtain the encrypted image (Figure 10.11b). If the encrypted image is modified during the transfer, it is not necessarily possible to detect this alteration. For example, in Figure 10.11c a small region of the encrypted image has been copied and pasted onto another zone of the image. After decryption, it is possible to view the images, but their integrity cannot be guaranteed, as shown in Figure 10.11d.

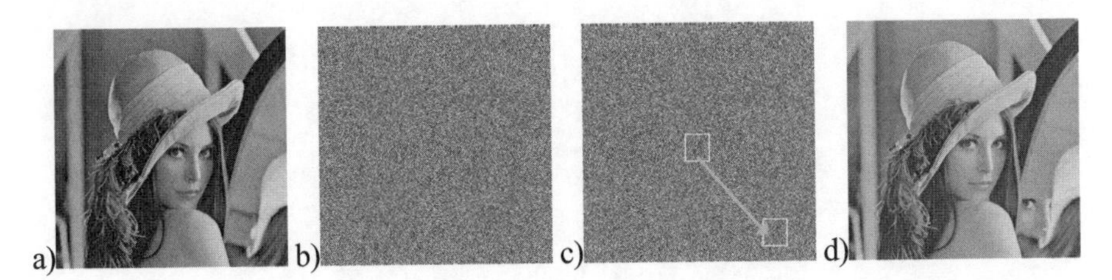

Figure 10.11. *a) Original Lena image, b) image encrypted by AES by 128-bit block, c) copy of a region of the encrypted image, pasted onto another zone, d) decryption of (c)*

In order to solve the integrity problem, it is possible to combine a stream cipher algorithm with a secret key for the image and an asymmetric algorithm to encrypt the secret key. A substitutive watermarking method (section 10.5) then allows for the embedding of the encrypted key into the encrypted image [PUE 04], [PUE 07]. If person A sends an image over a network to person B, sender A will use a stream cipher algorithm with the secret key K to encrypt the image. To send key K, A can encrypt it using an algorithm with a public key such as RSA. Let $pub(e,n)$ be the public key and $priv(d,n)$ the private key for RSA with $e = d^{-1} mod(n)$, so A has his public and private keys $pub_a(e_a,n_a)$ and $priv_a(d_a,n_a)$, and B has his public and private keys $pub_b(e_b,n_b)$ and $priv_b(d_b,n_b)$. As a result, A generates a secret key K for this session and encrypts the image with the stream cipher algorithm. Next, A ciphers the key with the RSA algorithm using his private key $priv_a$ in order to achieve a key K':

$$K' = K^{d_a} \bmod(n_a) \qquad [10.10]$$

This key K' is encrypted a second time with RSA using the public key pub_b of the recipient B to generate K'':

$$K'' = K'^{e_b} \bmod(n_b) \qquad [10.11]$$

The size of the message to be embedded into the image depends upon the size of the recipient's public key and is known to sender A and recipient B. We can therefore calculate the embedding factor and calculate the number of blocks required for the embedding. This key K'' is embedded into the ciphered image. Finally, A sends the image to B as shown in Figure 10.12. This procedure of K encryption with $priv_a$ and pub_b ensures the authenticity, and only B can decrypt the image. The embedding of the key into the image makes the method autonomous and guarantees its integrity. If, during transfer, the image is attacked, then it is no longer possible to extract the right key on reception, and so the image cannot be decrypted.

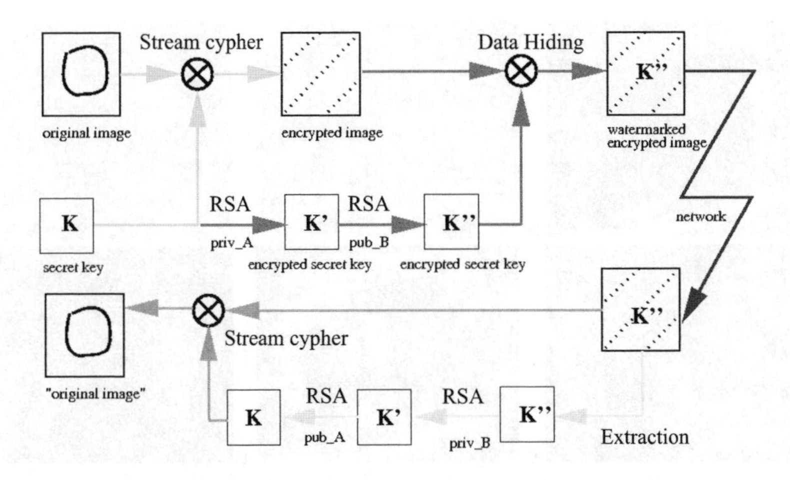

Figure 10.12. *Combination of secret key encryption, public key encryption and a watermarking method*

Person B receives the encrypted and watermarked image, and can then extract the encrypted key K''. He can then identify the sender, A, and decrypt the key K'' using the private key $priv_b$ and the public key pub_a belonging to A:

$$K = (K''^{d_b} \bmod(n_b))^{e_a} \bmod(n_a) \qquad [10.12]$$

With the acquired key K, B can decipher the image and thus view it. Starting from the original ultrasound image (512x512 pixels), Figure 10.13a, we have applied a stream cipher algorithm with a key K of 128 bits, in order to obtain the encrypted image Figure 10.13b. If this image is decrypted, we can note that there is no difference between it and the original image. The 128-bit key K was encrypted twice with the RSA algorithm in order to obtain K''. Due to the length of B's public key, the length of K'' is in the region of 1,024 bits. Next, using a watermarking technique in the spatial domain based on the LSB substitution, key K'' is embedded into the encrypted image (Figure 10.13c). The embedding capacity is of 1 bit for every 256 pixels. The difference between the watermarked, encrypted image and the original is shown in Figure 10.13d. The pixels used for the embedding are visible, the PSNR=75.14 dB. After the decryption of the watermarked, encrypted image, in Figure 10.13c, we reach the final image shown in Figure 10.13e. The difference between the original image and the final one is shown in Figure 10.13f. This figure shows that the differences between the two images (PSNR = 55.28 dB) are spread throughout the image. Nevertheless, because the average value of the $\alpha(i)$ coefficients is equal to zero, the error due to the watermarking is not increased during the decryption stage.

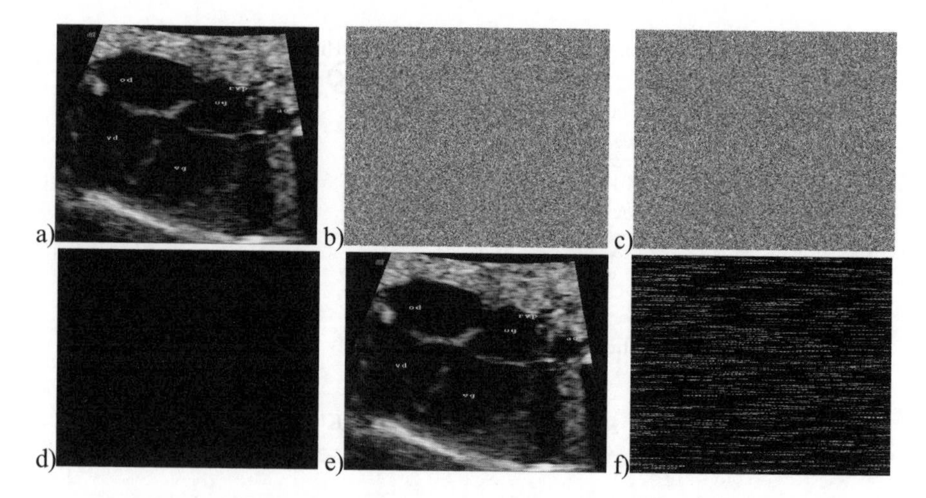

Figure 10.13. *a) Original image, b) encrypted image with a stream cipher algorithm with 128-bit key, c) image (b) watermarked with the secret encrypted key, d) the difference between images (b) and (c), e) decryption of image (c), f) the difference between original image (a) and (e)*

In order to compare the results of this hybrid method, the watermarking method was applied to the encrypted medical image using the AES algorithm with the ECB and OFB modes (stream cipher mode). After decryption, the image watermarked and encrypted by AES in ECB mode shows a great deal of variation compared to the original image (PSNR=14.81 dB). After decryption, the image watermarked and encrypted by AES in OFB mode presents variations which were not diffused by this mode. The final image quality is good (PSNR=52.81 dB) but an overflow problem remains with the OFB AES mode. The black pixels become white, and vice versa. In conclusion, the combination of encryption and watermarking allows for an autonomous transmission system, and guarantees the integrity of the data transmission.

10.6. Conclusion

In this chapter, we have shown that there are many solutions for ensuring security when sending and storing medical images. In current practice, those solutions offered to secure medical data are based on very traditional protection techniques. These old approaches require either the introduction of certain specific mechanisms, or a longer execution time. These traditional approaches are not suitable for real-time applications or for access from a doctor's surgery. Some of the solutions proposed in this chapter can be integrated into systems for sending medical images, if they can be proven robust. The main advantage of all these hybrid approaches is the ability to link several types of coding in one algorithm. In years to come, the appearance of standards in the encryption and watermarking of images will be of great benefit to the safe transmission of medical data.

10.7. Bibliography

[AES 01] AES, Announcing the Advanced Encryption Standard, Federal Information Processing Standards Publication, 2001.

[ALL 94] ALLAËERT F.A, DUSSERRE L., "Security of health system in France. What we do will no longer be different from what we tell", *International Journal of BioMedical Computing*, vol. 1, p. 201-204, 1994.

[CAV 04] CAVARO-MÉNARD C., AMIARD S., "Reversible data embedding for integrity control and authentication of medical images", in *ISIVC'04, Proceedings of 2nd International Symposium on Image/Video Communications*, Brest, July 2004.

[COA 00] COATRIEUX G., MAITRE H., SANKUR B., ROLLAND Y., COLLOREC. R. "Relevance of Watermarking in medical imaging", in *ITAB'00, Proceedings of ITAB*, Washington, USA, November 2000.

[COA 01] COATRIEUX G., SANKUR B., MAÎTRE H., "Strict integrity control of biomedical images", in *SPIE, Proc. Electronic Imaging, Security and Watermarking of Multimedia Contents*, p. 229-240, San Jose, USA, November 2001.

[COA 03] COATRIEUX G., H. MAITRE, "Images médicales, sécurité et tatouage", *Annales des Télécommunications, Numéro Spécial Santé*, vol. 58, p. 782-800, 2003.

[COA 05] COATRIEUX G., LAMARD M., DACCACHE, PUENTES W.J., ROUX. C., "A low distortion and reversible watermark: Application to angiographic images of the retina", in *EMBC'05, Proceedings of Int. Conf. of the IEEE-EMBS*, p. 2224-2227, Shanghai, China, November 2005.

[COA 06] COATRIEUX G., PUENTES J., LECORNU L., CHEZE LE REST C., ROUX. C., "Compliant secured specialized electronic patient record platform", in *D2H2'00, Proceedings of D2H2*, Washington, USA, November 2006.

[COL 07] COLTUC D., "Improved Capacity Reversible Watermarking", *IEEE International Conference on Image Processing, ICIP'2007*, San Antonio, Texas, USA, September 2007

[COS 03] COSTA M.H.M., "Writing on dirty paper", *IEEE Trans. on Information Theory*, vol. 58, p. 782-800, 2003.

[COX 02] COX I.J., MILLER M.L., BLOOM J.A., *Digital Watermarking*, Morgan Kauffman Publishers, San Francisco, CA, 2002.

[DAE 02] DAEMEN J., RIJMEN. V., AES, Proposal: The Rijndael Block Cipher, Technical report, Proton World Int.l, Katholieke Universiteit Leuven, ESAT-COSIC, Belgium, 2002.

[DIF 76] DIFFIE W., HELLMAN M.E., "New directions in cryptography", *IEEE Transactions on Information Theory*, vol. 26, no. 6, p. 644-654, 1976.

[DRO 02] VAN DROOGENBROECK M., BENEDETT R., "Techniques for a selective encryption of uncompressed and compressed images", in *Proceedings of Advanced Concepts for Intelligent Vision Systems (ACIVS) 2002*, Ghent, Belgium, September 2002.

[DUC 96] DUCROT H., "Le dossier médical informatisé face à la Loi Française", *Informatique et Santé : Aspects Déontologiques, Juridiques et de Santé Publique*, vol. 8, p. 87-96, 1996.

[DUS 97] DUSSERE L., Recommandations déontologiques pour le choix de logiciels destinés aux cabinets médicaux. Ordre national des médecins, Conseil National de l'Ordre, Ethique et Déontologie, 1997.

[FIS 04] FISCH M.M., STGNER H., UHL A., "Layered encryption techniques for DCT-coded visual data", in *European Signal Processing Conference (EUSIPCO) 2004*, Vienna, Austria, September 2004.

[GUI 02] GUILLEM-LESSARD S., http://www.uqtr.ca/~delisle/Crypto, visited in 2002.

[KER 83] KERCKHOFFS A., "La cryptographie militaire", *Journal des sciences militaires*, vol. 9, p. 5-38, 1883.

[MAN 01] MANICCAM S.S., BOURBAKIS N.G., "Lossless image compression and encryption using SCAN", *Pattern Recognition*, vol. 34, p. 1229-1245, 2001.

[MAN 04] MANICCAM S.S., BOURBAKIS N.G., "Lossless compression and information hiding in images", *Pattern Recognition*, vol. 37, p. 475-486, 2004.

[NOR 03] NORCEN R., PODESSER M., POMMER A., SCHMIDT H.P., UHL A., "Confidential storage and transmission of medical image data", *Computers in Biology and Medicine*, vol. 33, p. 277-292, 2003.

[PIV 05] PIVA A., BARNI M., BARTOLINI F., DE ROSA A., "Data Hiding Technologies for Digital Radiography", *IEEE Vision, Image and Signal Processing*, vol. 152, no. 5, p. 604-610, 2005.

[PUE 01a] PUECH W., DUMAS M., BORIE J.C., PUECH M., "Tatouage d'images cryptées pour l'aide au Télédiagnostic", in *Proc. 18th. Colloque Traitement du Signal et des Images, GRETSI'01*, Toulouse, France, September 2001.

[PUE 01b] PUECH W., PUECH M., DUMAS M., "Accés sécurisé distance d'images médicales haute résolution", in *Proc. 11th. Forum des Jeunes Chercheurs en Génie Biologique et Médical*, Compiègne, France, p. 72-73, June 2001.

[PUE 04] PUECH W., RODRIGUES J.M., "A new crypto-watermarking method for medical images safe transfer", in *EUSIPCO'04*, Vienna, Austria, 2004.

[PUE 05] PUECH W.,. RODRIGUES J.M., "Crypto-Compression of medical images by selective encryption of DCT", in *EUSIPCO'05*, Antalya, Turkey, September 2005.

[PUE 06] PUECH W., RODRIGUES J.M., DEVELAY-MORICE J.E., "Transfert sécurisé d'images médicales par codage conjoint : cryptage sélectif par AES en mode par flot et compression JPEG", *Traitement du signal (TS), numéro spécial Traitement du signal appliqué à la cancérologie*, vol. 23, no. 5, 2006.

[PUE 07] PUECH W., RODRIGUES J.M., "Method for Secure Transmission of Data", Licence WO 2007/045746, April 2007.

[ROD 06] RODRIGUES J.M., PUECH W., BORS A.G., "A selective encryption for heterogenous color JPEG images based on VLC and AES stream cipher", in *CGIV'06*, Leeds, UK, 2006.

[SAI 05] SAID A., "Measuring the strength of partial encryption scheme", in *ICIP 2005, IEEE International Conference in Image Processing*, Genoa, Italy, vol. 2, p. 1126-1129, 2005.

[SCH 97] SCHNEIER B., *Applied Cryptography*, Wiley, New York, USA, 1995.

[SHA 78] SHAMIR A., RIVEST R.L., ADLEMAN L., "A method for obtaining digital signatures and public-key cryptosystems", *Communications of the ACM*, vol. 21, no. 2, p. 120-126, 1978.

[STI 96] STINSON D., *Cryptographie – Théorie et pratique*, Thompson Publishing, 1996.

[TAN 96] TANG L., "Methods for encrypting and decrypting MPEG video data efficiently", in *ACM Multimedia*, p. 219-229, 1996.

Chapter 11

Transmission of Compressed Medical Data on Fixed and Mobile Networks

11.1. Introduction

An objective of the present chapter will have to be a detailed discussion about the aspect of medical data diffusion (broadcasting or transmission). This encompasses access to medical information, by fixed networks (wired Internet, for example) or mobile networks (wireless communication) or a combination of both fixed and mobile networks (called hybrid networks).

There are numerous situations where this type of information exchange can be encountered, either for remote consultations inside buildings, or outside, or combining the two environments, or even for ambulatory assistance, under the same conditions of geographical variability.

The main difficulty in transmitting such compressed data results from the highly fragile nature of the data. This fragility can arise due to perturbations, which can happen during the transmission, because of errors, packet losses, etc.). At the same time, there can be several other causes of uncertainty in the data. These may arise, for example, as a result of compression techniques, which can also integrate the watermarking phase, or due to the choice of modulation technique, once the bit stream is obtained, from the random access procedure (degradation, unguaranteed rate, etc.), routing, etc. At the end of the chain, a process of quality assessment (as described in Chapter 5) will be carried out, as a function of the usage, to discern whether the received signal is acceptable.

Chapter written by Christian OLIVIER, Benoît PARREIN and Rodolphe VAUZELLE.

The chapter is organized as follows. A brief overview of the existing applications based on the transmission of medical data is presented in section 11.2. This section also highlights the difficulties associated with these methods and the difficulties that may arise in immediate future. Section 11.3 will describe the various networks employed to tackle these situations, with their various specificities and sensitivity to the perturbations. Section 11.4.1 will present a set of existing or possible applications, using the fixed and mobile networks and in section 11.4.2 the effects of the errors or losses on medical images, for various standards of compression techniques, will be demonstrated. Section 11.4.3 will present the effect of the use of usual error correcting codes, at the cost of increased redundancy for the information transmitted. We shall also introduce the *Mojette* transform in this section, along with its application within the domain of medical data n networks. The chapter will be concluded with a discussion of the problem of joint source-channel coding in the context of medical data.

11.2. Brief overview of the existing applications

If the first objective in the communication of medical data is transmission, then the second one is access to the archive or files. It is a well known fact that, with the ever increasing quantity of data, compression becomes necessary.

Sometimes at high rates, the coding procedure may make the information fragile, particularly for an information system environment which is not that robust, either inside the same hospital service or between several services. The "direct access" aspect of the information, from the compressed data, has been mentioned in the preceding chapters and is mentioned in the following discussions. There exist many applications, which are based on medical data transmission. The domain of telemedicine is often mentioned in the same category, if it is a question of exchanging medical data using a simple e-mail.

Generally speaking, the optimal functions of the exchange of messages, files and the sharing of peripherals or even the access to a sophisticated information system, are yet to come into prominence.

Table 11.1 summarizes the existing applications and specifies, for each one, the temporal constraints in terms of transfer of information delay.

Applications	*Constraints*	*Explanations*
Telesurgery control	Critical delay: < 50 ms	The action must be synchronized with the feedback of image
Audio frequency	Delay: 150 ms	With several participants, the interactivity must be perfect
Telephony	Delay: < 200 ms	Interactive duplex is a condition of communication convenience
Image reading	Delay: < 1 s	Over a 5 year period, the consultation time has been reduced from 2 to 0.45 sec/image
Diffusion of medical signals (remote control, e-learning)	Delay of about 1 s	Buffer memory compensates the delay and the loss
Web, transfer of images, multimedia reports	Elastic delay	Few seconds are acceptable

Table 11.1. *Existing medical applications and associated temporal constraints*

11.3. The fixed and mobile networks

In this section, we briefly present the wired and wireless networks, as well as their operating modes, with descriptions of the various stages of the transmission chain for the two cases. These two types of communications can also be mixed together (in hybrid networks) and they are discussed in section 11.3.2.

11.3.1. *The network principles*

11.3.1.1. *Presentation, definitions and characteristics*

In the early 1980s, the beginning of the interconnection required a consensus among manufacturers to enable the interoperability of equipment. In this context, the International Organization for Standardization (ISO) proposed a layer structure of any exchange of data within a computer network using its OSI (Open System Interconnection). Today, this model, which was imposed as a reference in the world of computer networks, also unifies the telecommunication community and the information processing community. Rather than subdividing the activities into different subcategories, it offers a practical conceptual field because of its simplicity in designing systems, involving the relations across different layers. Figure 11.1

summarizes the design from the interface user (level 7) up to the binary transmission (level 1).

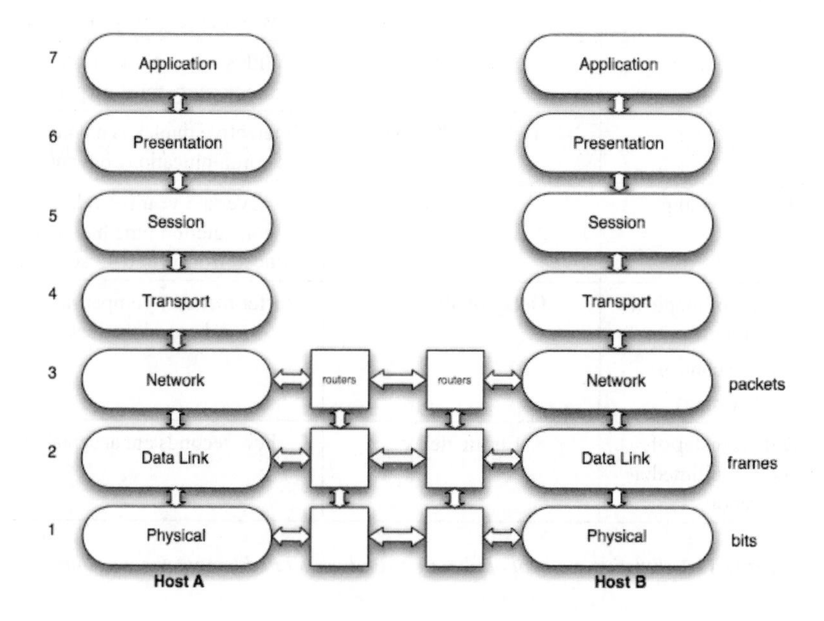

Figure 11.1. *OSI model in 7 layers of two interconnected hosts A and B. The transmission units of layers 1, 2 and 3 are specified on the right of the figure*

In a simple manner, this model can be divided into two categories respectively grouping the applicative layers (levels 5 to 7) and the transport layers (levels 1 to 4). Independent of the temporal constraints mentioned in Table 11.1, the applications are classified according to their mode of information exchange. The types which have become popular in recent times are "client/server", "push", "flow" and "peer-to-peer".

The transportation of information (as shown in Figure 11.1, from levels 1 to 4) is characterized by a fragmentation of the initial volume of information into a series of packets. This mode of transport is different from the continuous flow of data in the commutated circuits, for example, the commutated telephone network. The autonomy of each packet containing a source address and a recipient address authorizes the multiplexing of several users on the same connection and therefore, increases the transmission rates. This same autonomy brings flexibility in the routes through which the packets transit. The routers, as points of interconnection, determine the optimal routes, depending on the state of the network and recopy the packet of a connection to another one. As an after-effect of this process, the multiplication of the paths generates delays, which in turn causes jitters and even the loss of packets with quite significant effects on an initial volume of information weakened by fragmentation. This chapter will make a detailed analysis of the practical impacts of these parameters of quality of service.

11.3.1.2. *The different structures and protocols*

In practice, the Internet, functioning with two protocols of level 3 and 4, is itself imposed. At level 3, the Internet Protocol (*IP*) manages the interconnection networks. Devices called routers interconnect the networks. They analyze the recipient's address in the header of each packet in order to perform the delivery. Two cases usually occur in packet processing, either the router knows the recipient address for a direct delivery or the router does not know the recipient and leaves the task to its default router. At level 4, the TCP (Transmission Control Protocol) involves a connection between the transmitter and the receiver of the packet. It exercises the control and maintains the reliability of the transmission by adjusting its emission rate to the network capacity, by detecting and performing the retransmission of lost packets.

This mode of operation is prohibited for applications sensitive to delays such as telesurgery. In this context, the transport protocol UDP (User Datagram Protocol) is used by acting without connection and without retransmission. Additional mechanisms are needed in this case to increase reliability.

The lower layers are responsible for the transmission of data by adapting themselves to the specifications of the physical medium. Each of these protocols leads to the definition of a format of a particular frame and to a specific address (in the network the IP protocol's role is to erase these specifications). In fixed networks, Ethernet connections are widely deployed. With recent developments, there have been widespread modifications of its use. Although the initial range of a local area network was about 2.5 km, nowadays its use is often limited to a floor of a building (a range of about 25 to 100 m) with a star topology, involving an active repeater type central element (called a "Hub") or switch. Thus, it has been possible to increase the rates to about 100 Mbits/s for fast Ethernet and 1 or 10 Gbits/s for gigabit Ethernet networks. The use of optical fibers facilitates the construction of intra-hospital networks.

11.3.1.3. *Improving the Quality of Service*

The interconnectivity offered by the IP comes at a price. The optimality in the routing algorithms, the relative reliability of transport protocols or even the limited capacities of processing the interconnected nodes lead to delays of transmission, weakening the Quality of Service. To prevent this, two actions are possible, either taken in the heart of the network, or at the terminals. The construction of a network that is capable of applying an intelligent management distinguishing stream or the classes of priority services, is the goal of the “IntServ” and “DiffServ” working groups of the IETF (Internet Engineering Task Force) [BRA 94], [BLA 98]. These approaches assume an end-to-end implementation, which becomes particularly difficult for a wide area network.

The action at the source or destination requires simply a network that is doing its best, e.g. “Best Effort”. In this context, coding with unequal error protection (UEP) is tested today. Contrary to the separation principle (source coding then channel coding), this protection mode considers a coding of a distinct channel for each priority of the source as a function of the channel status. This approach offers good reactivity for no stationary channels in mobile networks. It is facilitated by the developments of the scalable representations of multimedia information (JPEG 2000 for still images or MPEG4/H.264 SVC for video images). Overviews of advanced protection are available in [GOY 01] and [HAM 05]. At the end of the chapter, an example of UEP implementing a function of discrete tomography is detailed.

11.3.2. *Wireless communication systems*

11.3.2.1. *Presentation of these systems*

Wireless communication systems can be characterized according to the context of their use (inside/outside of buildings, range, etc.) and their rates, i.e. implicitly by the proposed services. A considerable evolution of the wireless system has been observed during the last 20 years. Some key elements of this development are mentioned now. At the beginning of the 1990s, standard GSM (Global System for Mobile) systems, functioning digitally at frequencies of 900 and 1,800 MHz, were deployed in Europe. They allowed the transmission of vocal communications and reached rates close to 9.6 Kbits/s.

At that time, they provided a national coverage initially centered on large cities and highways. Gradually, this coverage extended to almost the entire territory and the network became denser in order to increase its capacity in zones with strong traffic. Since 2000, we have experienced an explosion in the standards and proposed services. Indeed, it is no longer only a question of transmitting voice by radio channel, but also of implementing wireless computer peripherals, exchanging computer files, images and, more recently, video.

We can globally classify these wireless systems in three large families: personal networks (WPAN: Wireless Personal Area Network), local networks (WLAN: Wireless Local Area Network) and metropolitan networks (WMAN: Wireless Metropolitan Area Network). For the WPAN, the principal standard is “Bluetooth” to ensure short range connections between peripherals and computers, for example.

We should also mention "Zigbee" which is a recent technology, characterized by data exchanges of low rates and small consumption for networks of sensors for instance. Concerning the WLAN, the norm IEEE *802.11* called Wi-Fi and all its developments in both present and future use (IEEE *802.11a, b, g, n*) represent the main system. They allow information exchange with high rates inside a building within an infrastructure or not, i.e. with or without a fixed access point to a core network.

As an example, the characteristics at the level of the physical layer are 54 Mbits/s for the standards *a* and *g* and 100 Mbits/s at the level of the data link layer for the standard *n* which is based on MIMO (Multiple Input Multiple Output) technology.

In 2003, within the WMAN family, two major evolutions to the GSM standard took place: the GPRS (General Packet Radio Service) qualified for 2.5 G development and the EDGE (Enhanced Data rate for GSM Evolution) associated with a 2.75 G development. Both transmit with a packet mode, like the fixed networks, contrary to the GSM which used a circuit mode. With this main change, GPRS is able to reach net rates ranging around 20 to 30 Kbits/s and the EDGE, net rates around 150 to 200 Kbits/s. These developments, however, do not replace the whole generation.

The third generation (3G) is currently making its appearance. The UMTS (Universal Mobile Transmission System) standard represents an important technological advance because it can transmit much more data simultaneously and should offer a significantly higher rate than those of the previous generations.

This standard is based on the spread spectrum technique called the W-CDMA (Wireless Coded Division Multiple Access) [PRO 00]. It uses the frequency bands located between 1,900 and 2,200 MHz. In theory, it can reach 2 Mbits/s in a fixed environment and 384 Kbits/s in a moving environment. These increased speeds of data transmission enable the provision of new services: visiophony, mobile television, etc. The UMTS is also sometimes called 3GSM, thus underlining the interoperability which was ensured between the UMTS and the standard GSM, which it succeeds.

Finally, we should point out that a specific concept related to non-fixed infrastructure is currently being developed in Europe. It mainly concerns *ad hoc* networks, also known as auto-configurable networks.

In practice, various applications are possible, such as intra-vehicle communications or sensor networks (see section 11.4.1). However, whatever the considered wireless system of communication, the multi-path phenomenon of the transmission channel introduces significant constraints. In addition, we must note that this aspect does not concern fixed networks.

11.3.2.2. *Wireless specificities*

The main difference between fixed and mobile networks lies in the physical layer (see Figure 11.2) and more precisely in the transmission channel.

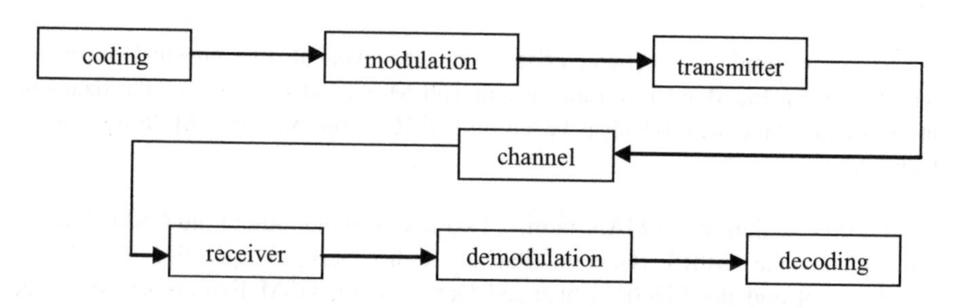

Figure 11.2. *Scheme of the physical layer of a wireless system*

Let us briefly recall the objectives of the various elements of this chain of transmission:

– the coding is decomposed into two operations, the source coding where the objective is the compression of the information, and the channel coding where the goal is to improve the robustness of the compressed signal of the information with respect to the perturbations which it can undergo during its transmission. The principle of the channel coding consists of introducing redundancy, allowing detection and errors correction of bits (described in detail in section 11.4.3.1);

– the modulation is designed for two purposes. Firstly, it ensures the transposition in frequency of the information, and, secondly, it transforms the digital information in an analog signal, which can be sent to the antenna for emission. The principle consists of assigning an amplitude and a phase of an analog signal, to a symbol, as illustrated in Figure 11.3. This figure presents a digital constellation (i.e. representation of a signal in a complex plan) for a Quadrature Amplitude Modulation with 16 states: 16 QAM;

– the radioelectric channel is the medium of transmission which allows the transfer of information between a transmitter and a receiver. Its principal property is the mechanism of multipath trajectories which governs the propagation of the radio waves;

– the digital reception ensures the synchronization and the recovery of rhythm, which facilitates the operations of demodulation and decoding;

– the demodulation and decoding perform the dual operations of modulation and coding.

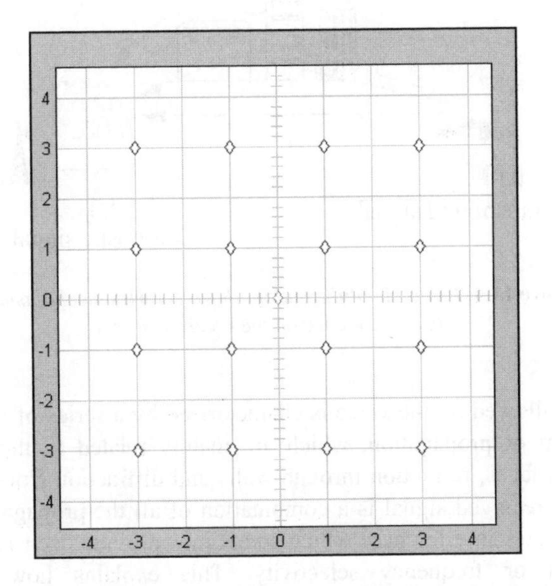

Figure 11.3. *Digital constellation of a 16 QAM modulated signal under ideal conditions*

In this context, special emphasis should be placed on the fact that the channel coding and the modulation are two key elements for the performances of the physical layer. These performances can be defined in different ways, where each of these constitutes a trade-off between the rate and the robustness of a transmission. Transmission performances can also be evaluated using the concept of Quality of Service, which relates to the upper layers of the OSI model. As mentioned above, the main specification of wireless systems compared to wired networks is located at the channel level. This presents spatio-temporal variability which does not exist for wired networks. There are two types of variations: slow variations due to the mask effects, such as an obstacle between the transmitter and the receiver and fast variations related to the multipath trajectories which characterize the channel. This phenomenon corresponds to the fact that a wave generated from a source generally follows different paths before reaching the receiver (Figure 11.4).

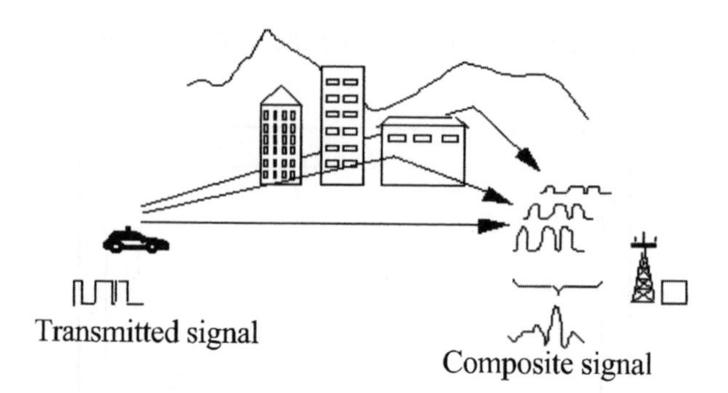

Figure 11.4. *Principle of the multipath trajectories mechanism and its consequence on the received signal*

Each path followed by the wave is characterized by a series of interactions with the environment of propagation, which are mainly related to the phenomena of reflection on surfaces, refraction through walls and diffraction from obstacle edges [VAU 04]. The received signal is a combination of all the propagated waves. This combination creates interferences which sometimes provide deep fading out of the reception level or frequency selectivity. This explains how the multipath phenomenon of the radio-electric channel generates perturbations during the signal transmission at a given time, a position and a frequency (section 11.4.2.2.1).

Various approaches exist to model the behavior of such a channel. They can be stochastic or deterministic [ISK 02] [VAU 04] [COM 06] but they are all based on the experimentation and/or the simulation related to the electromagnetic waves propagation. The statistical channel models generally used are those by "Gauss", "Rice" or "Rayleigh". These models aim to define the variations of the magnitude of the received signal and sometimes the phase.

The Gaussian channel model consists of adding a Gaussian white noise to the received signal; this noise is assumed to represent all the perturbations that influence the signal during its transmission.

The Rice model is used when one of the received paths is dominant. This is generally the case when the transmitter and the receiver are directly visible. The magnitude S of the received signal can be obtained from the following probability law:

$$P(s) = \frac{s}{\sigma^2} e^{-\left(\frac{s^2 + A^2}{2\sigma^2}\right)} I_0\left(\frac{A s}{\sigma^2}\right) \qquad [11.1]$$

where I_0 is the first type Bessel function of zero order, A is the magnitude of the dominant wave and σ^2 is the total power of the signal. The phase can be considered as constant or variable according to a uniform probability law.

In the case where none of the paths is dominant, the Rice model can be replaced by the Rayleigh model, defined by the following probability law:

$$P(s) = \frac{s}{\sigma^2} e^{-\left(\frac{s^2 + A^2}{2\sigma^2}\right)} \quad [11.2]$$

Moreover, it is also necessary to note the possible occurrence of the Doppler phenomenon when the transmitter and/or the receiver are moving. This phenomenon induces frequency shifts in the transmitted signals, directly related to the speed of movement, the carrier frequency and the directions of arrival of the waves, with respect to the direction of moving. As an example, taken from the acoustic field, let us consider an ambulance siren, which is perceived differently as it passes by an observer.

In general, in digital transmission, the channel study leads to characterizing it, in order to facilitate the setting up of appropriate techniques for formatting and restoration of the information. As an example, let us consider the dispersion of the propagated path delays whose value gives an indication of the risk involved in the interferences inter symbols.

11.4. Transmission of medical images

11.4.1. *Contexts*

Let us consider at least three different geographical contexts or situations.

11.4.1.1. *Transmission inside a hospital*

This considers transmission inside a hospital on fixed networks or hybrid networks with a low mobility. The methodologies of remote image reading using fixed networks inside the same service are already well developed. We can consider as an example of hybrid networks employed in this geographical context, the access to the medical data at the patient's bedside in his hospital room, at the time of the physician's visit; this using a simple notepad.

11.4.1.2. *Transmission outside hospital on fixed networks*

Traditionally, this context relates a hospital service to the outside world (to a medical doctor or specialist, other hospitals, etc.). For the purposes of *telemedicine*, specific architectures are proposed for an optimal exchange of all the information which can be useful for the diagnosis, including the hierarchical coding of the image, the inclusion of meta data suitable for the acquisition or the history of the image and the modes of interaction with these data. This is why part 9 of the JPEG 2000 standard extends the concept of the image compression system to a real communicating multimedia system. JPEG 2000 Interactive Protocol (JPIP) specifies the interactions with the elements constituting the JPEG 2000 stream (component, quality layer, levels of resolutions, regions of interest, *metadata*, etc.) (Figure 11.5). JPIP is based on a client/server type of architecture. It is really the stream structure, which ensures the availability of the medical data for the client. By authorizing a remote consultation of a very large image – a size of 64Kx64K pixels is allowed – even on channels with a very small capacity, this protocol is well suited for transmissions outside the hospital on fixed networks. With its recent integration within the DICOM standard, it can be safely assumed that JPIP is going to be implemented rapidly in the near future, in hospital centers and for remote access.

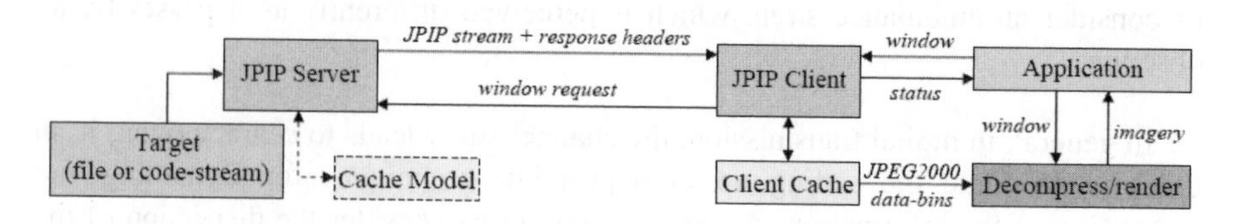

Figure 11.5. *Client/server in JPIP (JPEG 2000 Interactive Protocol) [TAU 03]*

11.4.1.3. *Transmission outside hospital on mobile networks*

These situations will arise where emergency intervention is required to diagnose an injured patient by a medical team, where medical assistance is available at some distance and the environment is quite complex (e.g. involves a complex road network, a mountainous zone or sea navigation, etc.).

11.4.2. *Encountered problems*

Under these transmission modes, a very important factor to be considered is the loss of information. In these situations, the network layer of a fixed infrastructure or the physical layer of a mobile network can be considered as the origin. There can be two main types of impact on the data stream.

11.4.2.1. *Inside fixed networks*

In fixed networks, the loss of packet takes place quite frequently. Currently the packet losses reach more than 5%. These losses can be primarily attributed to the saturation of intermediate equipment as routers and secondarily to parasitic errors in the transmission line (due to interferences, breakdowns, etc.). Inside routers, queue management is generally beyond repair: in fact, the impact of overload can fatally lead to the suppression of packets.

The simplest of these mechanisms is called Random Early Detection (RED). Its name is due to its prevention inside routers. The principle is to put a probability of packet dropping as soon as a minimal threshold of filling is reached. This probability varies linearly according to the filling until a maximum threshold is reached, that occurs under the condition when the queue reaches saturation. If the maximum threshold is reached, the suppression is systematic. Figure 11.6 shows this scheme for preventive management.

These bottlenecks also have an impact on the delay (although their effects are not that severe and they are not beyond repair) and this gives rise to a phenomenon called jitter, which results from the variations between transmission delays.

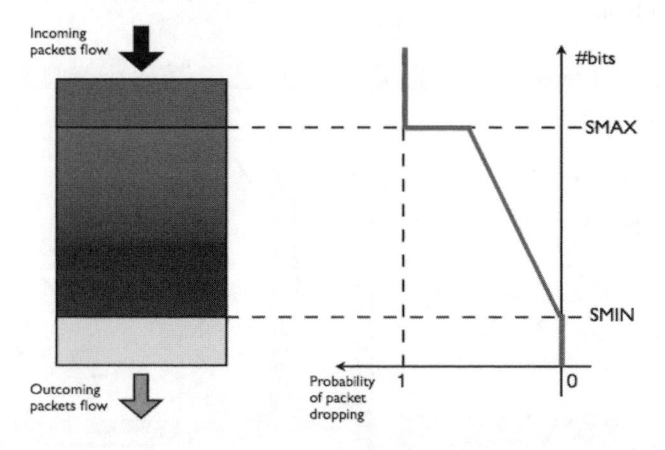

Figure 11.6. *RED queuing algorithm. The probability of packet dropping is a function of filling. Below SMIN, no packet is dropped. Between SMIN and SMAX, the probability increases linearly (random early discards). Above SMAX, all incoming packets are dropped*

11.4.2.2. *Inside mobile networks*

11.4.2.2.1. Difficulties induced by mobility: some errors

As was indicated in section 11.3.2.2, the transmission channel experiences several effects: slow and fast variations from one side and the Doppler shift from the other side. These effects have different consequences on the transmission quality.

The slow variations generated by the masking effects leads to a received power which can get considerably weakened during a relatively long duration. This generates losses by burst during a transmission.

The fast variations due to the multiple path phenomenon induces relatively isolated errors. In fact, it is shown that the deeper the fading the rarer they are [LEE 93]. For example, a fading of 10 dB has a probability of appearance of 10% in time or space, whereas a fading of 30 dB has a probability of only 1%. However, whatever the nature of variations, the degradation of the received power involves a contraction of the digital constellation associated with the received signal (Figure 11.7a).

As far as the Doppler effect is concerned, the Doppler shift leads to a rotation of constellation at a frequency corresponding to this shift (Figure 11.7b). In order to determine the value of the symbols, i.e. to demodulate, it is thus necessary to remove this rotation first.

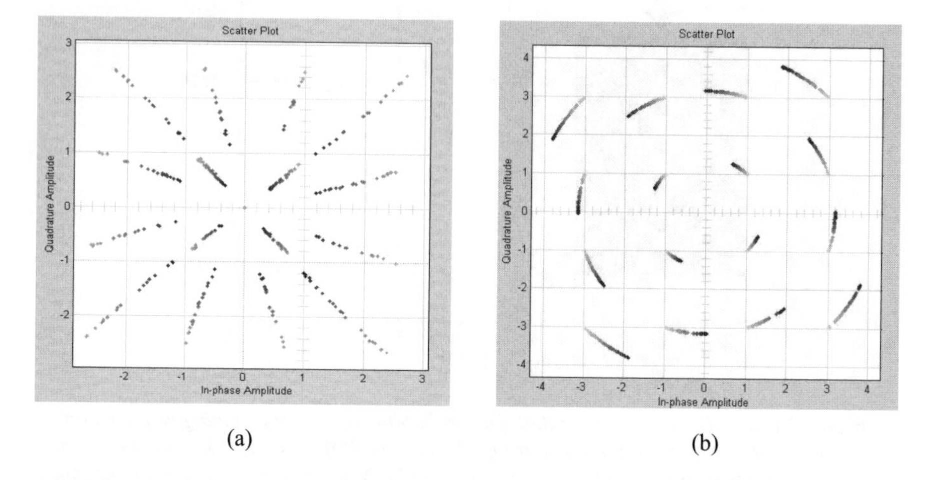

Figure 11.7. *Examples of perturbations of the digital constellation of a 16 QAM modulation (Figure 11.3) induced by the channel: (a) contraction due to a fall of the received power; b) rotation due to the Doppler effect*

In recent systems, the techniques implemented against all these channel effects in time and in space, are based on the estimation of the channel.

There are several techniques. At the emission, for example, a pre-coding of the information can be carried out and at the reception, an equalization [PRO 00] approach is often applied.

11.4.2.2.2. First tests and report

Let us now consider two types of channel models, presented above: the Gaussian channel model and the Rayleigh channel model. The perturbations in the channel can be measured in terms of Binary Error Rates (BER) and the quality of image can be measured in terms of PSNR. The compression rate is noted as $CR = n{:}1$ for a rate of n.

Let us first consider an MRI 256x256 image, compressed by the JPEG and JPEG 2000 standards, which shows the considerable fragility of these two standards, even under the situations of small perturbations in the transmission channel (Figure 11.8).

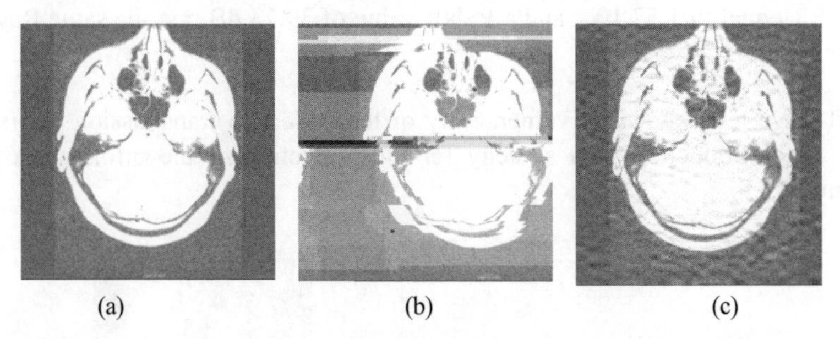

Figure 11.8. *Example MRI images reconstructed after transmission via a Gaussian channel with BER = $2.49.10^{-4}$. (a) Original image; then compressed with CR = 5.6:1, by a JPEG coder (b) and a JPEG 2000 coder (c)*

In the case of images using DICOM, Figure 11.9a shows the average values of the PSNR of the images which can be reconstructed via a Gaussian channel, where the BER is varied. For a BER higher than approximately $1.5.10^{-4}$, the decoding of the files achieved, with DICOM format, appears to be impossible.

The BER limit value of about 10^{-6} was found as being an acceptable binary error rate in order to be able to receive the image without error on this type of channel. Thus, beyond this limit, and under a BER value of about 10^{-4} (Figure 11.9a), and if the files can be opened at the receiving end, the received image inevitably contains some errors.

As an example, in Figure 11.9b, we show an image received with a BER equal to $5.38.10^{-5}$, i.e. with a PSNR = 47.79 dB. We can actually identify the appearance of artefacts in the image, as shown in Figures 11.9b and 11.9d (which shows the zoomed area of a corrupted zone). This may actually lead the doctor to make the wrong diagnosis or conclusion.

In Figure 11.10, we consider several cases of transmission in the Rayleigh channel for the same radiographic image (of size 440x440 pixels) for various BER values and a compression ratio CR = 10:1, with JPEG and JPEG 2000 coding.

We have observed in these tests (with CR = 10:1) that, for the JPEG standard, it is not possible to perform a diagnosis of the reconstructed image for a PSNR below 26 dB. Beyond a negligible BER value of 3.10^{-7}, the JPEG will not allow a diagnosis of the reconstructed image (for PSNR $\leq$ 26 dB). Regarding the JPEG 2000, this standard is always more robust than JPEG in term of the PSNR values.

For JPEG 2000, Figure 11.11 shows the reconstructed image after transmission with BER equal to $1.57.10^{-5}$, and a PSNR value of 39.73 dB, via the same Rayleigh channel.

All these tests show the vulnerability of the coders to transmission errors, and justify the introduction of a strategy for the protection of the information to be transmitted.

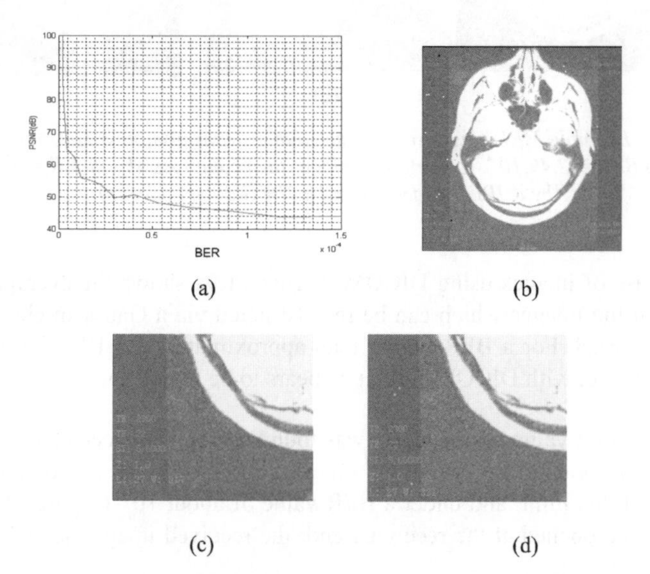

Figure 11.9. *Average behavior of DICOM (a) during transmissions by Gaussian channel of the MRI image shown in (b) with BER= $5.38.10^{-5}$. (c) and (d) are the zooms of the left lower part of the original image and image (b) respectively*

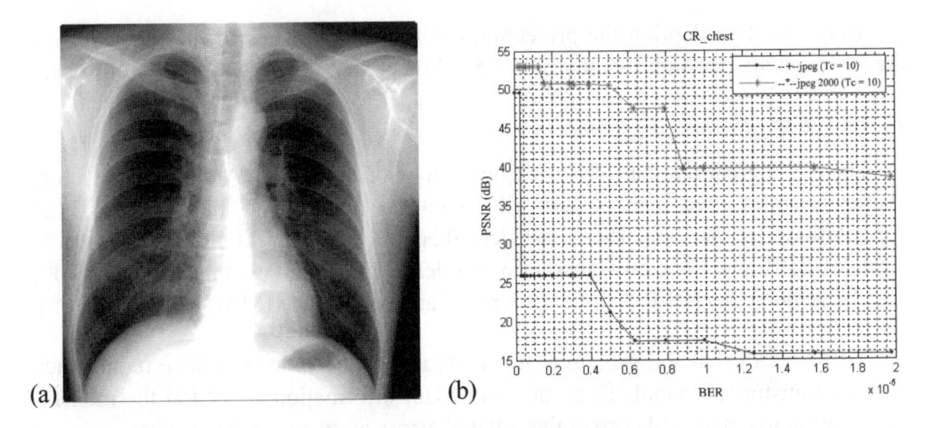

Figure 11.10. *(a) Original image of thoracic radiography compressed with JPEG and JPEG 2000 at a compression ratio CR = 10:1, and (b) transmitted in a Rayleigh channel: variation of the PSNR as a function of the BER*

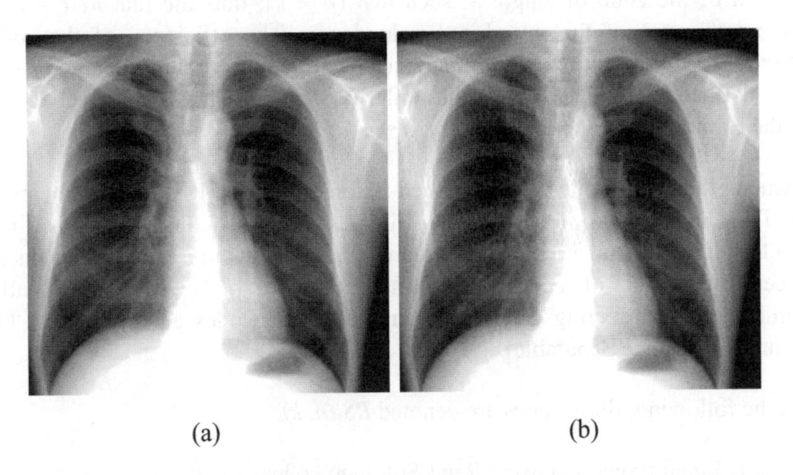

Figure 11.11. *(a) Compressed image with CR = 10:1 according to JPEG 2000; (b) reconstructed image via a Rayleigh channel with BER = $1.57.10^{-5}$*

11.4.3. *Presentation of some solutions and directions*

In transmission, the error correcting codes are generally used for the purpose of protection. In the case where the protection is restricted to one data flow, we can thus consider an equal protection. On the other hand, unequal protection can be considered as soon as the priorities are defined for each flow [ALB 96].

Moreover, by adapting the protection to the characteristics of the transmission, the BER can also be significantly reduced, while preserving an optimal useful flow.

11.4.3.1. *Use of error correcting codes*

There is a great number of Error Correcting Codes (ECCs). Among the different classical codes, we can specifically mention the linear block codes, the convolutional codes (e.g. [SOU 04] described the behavior of these 2 classes in terms of BER and SNR on Gaussian or Rayleigh channels), cyclic codes [DEB 00] and the turbo codes [BER 93]; for a more exhaustive list, see [MAC 04] or [LIN 05].

Let us recall that the principle of a correcting code is to introduce redundancy into the transmitted signal. Thus, this redundant information is used at the moment of reception to detect and correct the isolated errors or an entire data packet.

For each ECC, a certain number of characteristic parameters are defined. Therefore, if coding consists of associating with a given information word of length k, a word of the code of length n, such that ($n > k$); thus the rate $Red = (n-k)/k$ expresses the rate of the introduced redundancy. The efficiency of the code is characterized by the ratio $R = k/n$. The error correcting capacity, denoted t of the code is defined using a distance of dissimilarity, denoted d between two code words (e.g. the Hamming distance). Generally, $d \leq n-k+1$.

Among the cyclic codes, the RS (Reed-Solomon) codes [REE 60] are intensively used. They have the property of coding M-ary symbols. By this grouping of binary characters, the RS codes correct the arriving errors in burst mode. The distance between the words of codes is maximal, namely: $d = n-k+1$. This allows maximizing the correcting capacity. Therefore, these codes are optimal or MDS (Maximum-Distance Separable).

In the following, these codes are denoted *RS (n, k)*.

11.4.3.1.1. Equal protection using Reed-Solomon codes

We have chosen the classical codes *RS (255, k)*, which are well known for their use in spatial communications. Here $n = 255 = 2^8 - 1$. The source symbols are the coefficients of a Galois field polynomial of dimension 256 [RIZ 97], and t (with: $t = (255-k)/2$) is the capacity of correction allowed by the codes. Table 11.2 indicates, for both an *MRI* image and a thoracic radiography image (shown in Figure 11.8 and 11.10, respectively), the effect of adding a *RS (255, k)* code on the compressed image (size expressed in Kbytes) and on the other hand, its efficiency denoted R. We must note that the considered images are encapsulated in DICOM format.

Image type		***MRI***	***Radiography***
Initial volume		***129 Kbytes***	***378 Kbytes***
k = 245	Compressed volume with RS	135 Kbytes	394 Kbytes
	Efficiency R	95.56%	95.94%
k = 215	Compressed volume with RS	154 Kbytes	449 Kbytes
	Efficiency R	83.77%	84.19%
k = 185	Compressed volume with RS	179 Kbytes	522 Kbytes
	Efficiency R	72.07 %	72.41%

Table 11.2. *Effect of adding a RS (255, k) code*

If these correcting codes are applied to the whole *MRI* image (i.e. equal protection), a bounded (i.e. limited) BER is thus obtained, as shown in Table 11.3. This allows a perfect image reconstruction according to the Red redundancy rate. In addition, we consider here that the transmission is performed through a Gaussian channel.

That means, for example, that the *RS (255, 225)* code can correct the errors of the channel up to a BER of the order of 3.10^{-3}, for a data redundancy cost of 13.33%. Beyond that, the reception systematically includes errors. Consequently, for a BER higher than $1.8.10^{-2}$, no correction is possible whatever the rate of redundancy.

k	255	245	225	195	155
Red	0	4.08%	13.33%	30.77%	64.52%
R	1	95.56%	87.16%	76.33%	60.56%
BER limit	$9.46.10^{-7}$	$5.82.10^{-4}$	$3.2.10^{-3}$	$8.2.10^{-3}$	$1.73.10^{-3}$

Table 11.3. *Acceptable BER limits when adding RS (255, k) code on the DICOM MRI image of Figure 11.9*

We can also display the results using a limit curve. We thus obtain on the "colon" image shown in Figure 11.12a, the BER limit curve corresponding to an error-free reception with respect to the Red rate (Figure 11.12b).

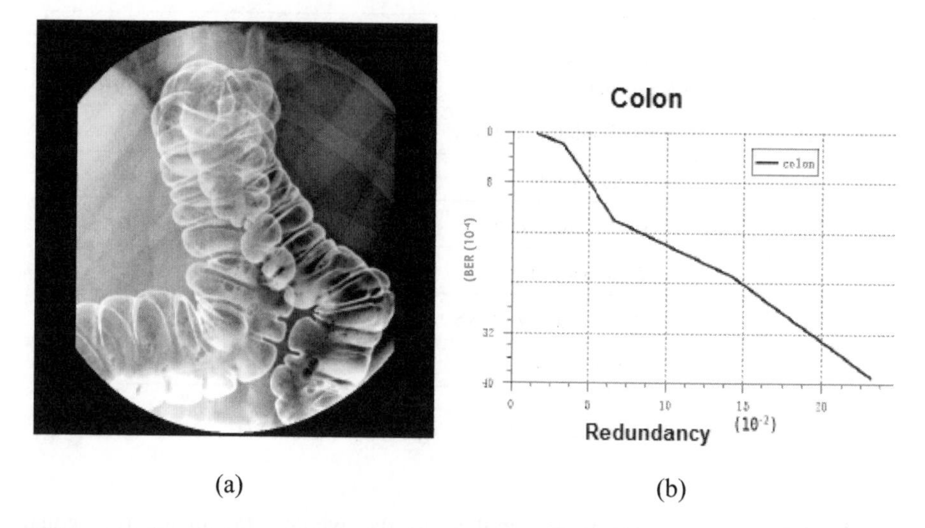

(a) (b)

Figure 11.12. *(a) Colon image; (b) curve of maximal BER with respect to the redundancy rate for an error-free reconstruction*

Finally, it should be noted that allowing a higher priority to the header of DICOM data, does not allow any additional improvement regarding the robustness related to the image.

11.4.3.1.2. Towards a content-oriented protection

The main limit of equal protection is that the reconstruction is "all or nothing". This leads to the idea of a content-oriented protection of the data where the reconstruction is more progressive. Moreover, it is clear that if we protect certain parts of the transmitted file, we may improve the performances in terms of global redundancy.

A first idea consists of using a fixed length binary encoder for high priority data. The results of simulation show a significant resilience to the errors, with a small efficiency loss. This supposes the use of scalable source encoders like the ones allowed by JPEG 2000, SPIHT, LAR (see Chapter 7) or some video encoders such as MPEG and H26x. Closer to the content, the Forward Error Correcting codes (FEC), which include a scalable protection of the source, have also been introduced. For example, Mohr *et al.* [MOH 00] use SPIHT source encoding, coupled with RS codes for a packet correction on both ATM transmission and MRI images. Likewise, signals with temporal constraints (e.g. video) can be coupled with RS codes having different correction capacities. This unequal protection allows a gain of expected quality at the reception, which is not negligible with respect to an equal protection.

The Mojette transform also handles this unequal protection, which is particularly well suited to overcoming the problem of packet loss experienced on some networks by significantly reducing the complexity.

11.4.3.2. *Unequal protection using the Mojette transform*

The Mojette transform [GUE 05] is a discrete and exact Radon transform. While its original use relates to tomographic reconstruction, it also presents good properties to ensure the integrity of data during a transmission on unreliable networks. In fact, it enables the easy representation of information by a set of redundant 1D projections. An example of this representation is given by Figure 11.13.

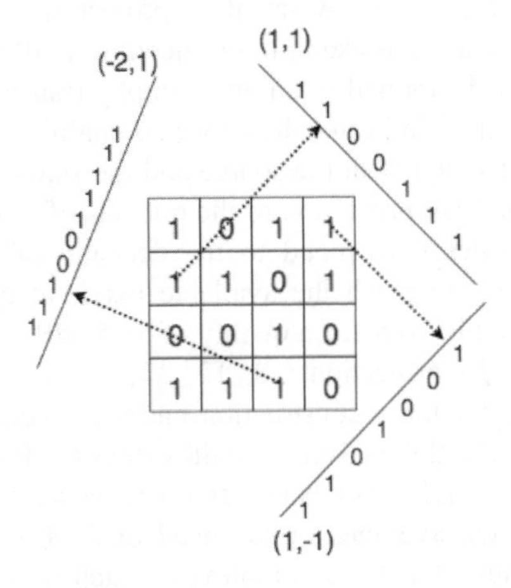

Figure 11.13. *Mojette transform of a 4x4 image. Projections (-2, 1), (1, 1) and (1, -1) are represented. Each element of projection results from the sum of the image elements in the direction of projection. In this example, the modulo 2 addition (XOR operator) is used*

These projections are regarded as many transport units (packets) whose loss or misordering does not disturb the reconstruction of the data at reception. From an encoding point of view, the 2D support of the information acts like a true geometrical buffer memory in the sense that the data flows using this support from which projections are calculated. By varying the number of projections necessary for the reconstruction of a geometrical buffer memory, it is possible to protect hierarchical information in an unequal way. Within this context, all projections are equivalent in terms of reconstruction. Progressive decoding of information is supported by an unequal number of received projections. For example, the low resolution of an image may require a low number of projections in order to be reconstructed (i.e. supporting a high rate of packets loss), whereas higher resolutions will require a greater number of projections. This mode of representation allows us to support variable rates of redundancy, depending on both the priority of the information and the status of network.

In this way, any source encoder having a scalable output representation can be protected. In [BAB 05], an unequal protection of lossless LAR encoder is carried out. In addition, an efficient comparison to the approaches proposed in section 11.4.3.1 has been performed. A number of projections are optimally allocated for each resolution of the encoder, depending on the quality of the image and of the characteristics of the transmission channel. The model of packet loss follows an exponential law here (i.e. the average value of the losses is given by parameter μ of the law). The progression of decoding depends on the number of projections received, in a deterministic way. A set of m_i projections out of N is sufficient to reconstruct the resolution i. An example of encoding is illustrated in Figure 11.14. The unequal protection is applied to an angiography transmitted through a channel where $\mu = 10\%$. The curve indicates the expected qualities in terms of PSNR at the reception taking into account both the source and the status of the channel. To each rate corresponds the optimal protection of the 6 levels of resolution delivered by the encoder. The singularities correspond to the decision of the system to send an additional flow in order to reach the available maximal quality according to the available rate. For example, to reach a quality of 38.51 dB, the 6 resolutions must be transmitted; for $N = 16$ projections: 8, 12, 13, 13, 15 and 16 projections are respectively necessary for the reconstruction under the conditions of transmission. For comparison purposes, the analysis rate-distortion for RS codes coupled to *LAR* is also given in Figure 11.14. For better comparison legibility, only the range of 0-5 bpp is displayed. On average, an overhead of 2.78% is recorded for *Mojette* encoding in this example, but for a complexity which is linear with respect to the number of pixels and the number of projections. Moreover, the increase in the data amount or the rate of loss, contributes to the reduction of the encoding cost.

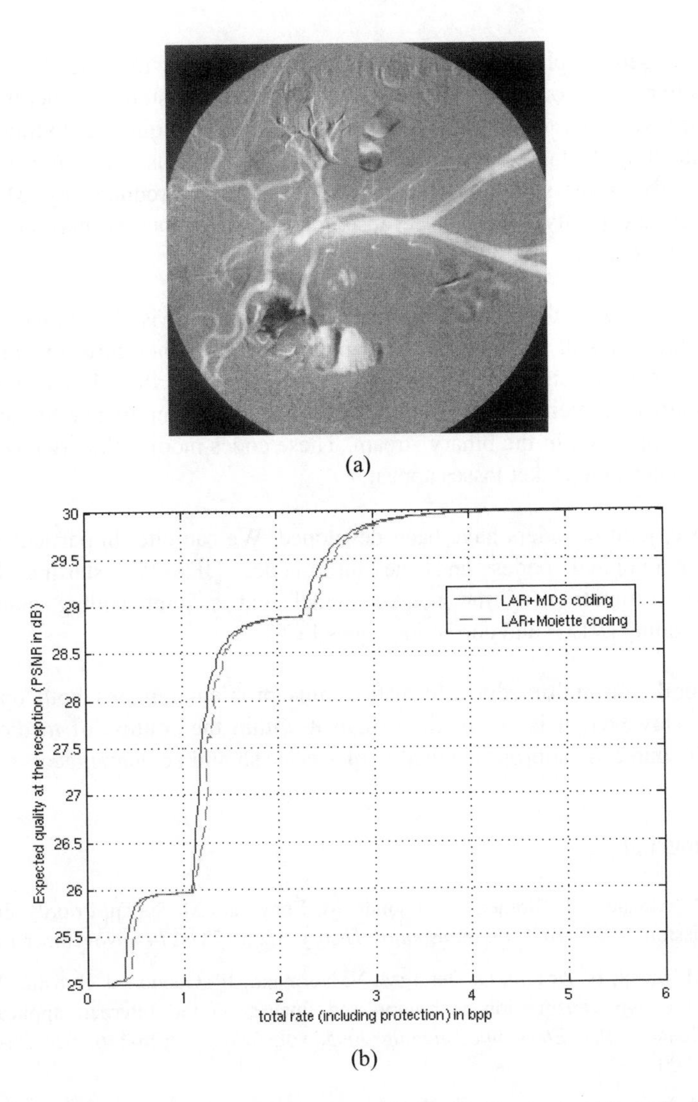

Figure 11.14. *Mojette unequal protection applied to LAR source coding of a 512(512 angiographic image: (a) preview of the raw angiographic image; (b) rate/quality analysis for two joint source-channel: LAR source coding with RS and Mojette coding*

11.5. Conclusion

The aim of this chapter was to highlight, on the one hand, the difficulties related to joint coding, dedicated specifically to the medical data in complex geographic environments and, on the other hand, to provide some useful prospects. The problem can be expressed simply: the more significant the compression rates become, the more sensitive the information provided by the encoder system is. Therefore, this makes it necessary not to separate the links constituting the source encoding and the channel encoding. In fact, some protection methods which use detecting/correcting codes can solve some situations, but require an increased redundancy. Moreover, their correcting capacity is limited because only one erroneous bit may involve the loss of the entire image.

Even if we know that the joint source-channel coding is the subject of very active research, it still remains far from being optimal. Therefore, unequal error protection codes are currently under development such as the Mojette transform which proposes an intelligent protection depending on the importance of the source information contained in the binary stream. These codes modify the organization of the source data when packet losses appear.

Likewise, joint decoders have been developed. We can cite, in particular, those built for convoluted codes and the turbo-codes [JEA 05], starting from a probabilistic estimation of the source model and a joint coding associating arithmetic coding (VLC) and correcting codes ECC.

In general, for multimedia information, this problem remains both open and promising. However, it is essential to solve it within the context of medical data reading. Of course, an appropriate quality of service should be guaranteed.

11.6. Bibliography

[ALB 96] A. Albanese, J. Blömer, J. Edmonds, M. Luby, and M. Sudan, Priority Encoding Transmission, *IEEE Trans. on Information Theory*, 42, p 1737-1744, November 1996.

[BAB 05] M. Babel, B. Parrein, O. Déforges, N. Normand, JP. Guédon, J. Ronsin, "Secured and progressive transmission of compressed images on the Internet: application to telemedicine", *SPIE Electronic Imaging 2005*, vol. 5670, p. 126-136, San Jose (CA), January 2005.

[BER 93] C. Berrou, A. Glavieux, P. Thitimajshima, "Near Shannon limit error correcting coding and decoding: turbo-codes", *IEEE Int. Conf. on Communication*, 2/3, p. 1064-1071, Geneva, Switzerland, 1993.

[BLA 98] S. Blake, D. Black, M. Carlson, E. Davies, Z. Wang, W. Weiss, An architecture for differentiated Services, IETF RFC 2475, available at http://tools.ietf.org/html/rfc2475.

[BRA 94] R. Braden, D. Clark, S. Shenker, Integrated services in the Internet architecture: an overview, IETF RFC 1633, available at http://tools.ietf.org/html/rfc1633.

[COM 06] P. Combeau, R. Vauzelle, L. Aveneau, "Efficient ray-tracing method for narrow and wide-band channel characterization in microcellular configurations", *IEE Microwave, Antennas and Propagation*, 153 (6), p. 502-509, December 2006.

[DEB 00] V. DeBrunner, L. DeBrunner, L. Wang, S. Radhakrishan, "Error control and concealment for image transmission", *IEEE Commun. Soc. Survey Tutorial*, 3 (1), p. 2-9, 2000.

[GOY 01] V. Goyal, "Multiple Description Coding: Compression Meets the Network", *IEEE Signal Processing Magazine*, 18, p. 74-93, September 2001.

[GUE 05] J.P. Guédon, N. Normand, "The Mojette Transform: the first ten years", *DGCI 05, Springer, Proceedings of DGCI*, p. 136-147, Poitiers, France, 2005.

[HAM 05] R. Hamzaoui, V. Stankovic, Zixiang Xiong, "Optimized error protection of scalable image bit streams [advances in joint source-channel coding for images]", *IEEE Signal Processing Magazine*, 22 (6), p. 91-107, November 2005.

[ISK 02] M. F. Iskander and Z. Yun, "Propagation prediction models for wireless communication systems", *IEEE Trans. on Microwave Theory and Techniques*, 50 (3), March 2002.

[JEA 05] M. Jeanne, J.C. Carlach, P. Siohan, "Joint source-channel decoding of variable length codes for convolutional codes and turbo codes", *IEEE Trans. On Communications*, 53 (1), p. 10-15, January 2005.

[LEE 93] C.Y. Lee, *Mobile Communications Design Fundamentals*, Wiley Series in telecommunication, 1993.

[LIN 05] S. Lin, D.J. Costello, *Error Control Coding: Fundamentals and Applications*, 2nd ed., Prentice Hall: Englewood Cliffs, NJ, USA, 2005.

[MAC 04] D.J.C. MacKay, *Information Theory, Inference and Learning Algorithms*, Cambridge University Press, version 7.0, November 2004.

[MOH 00] A.E. Mohr, E.A. Riskin, R.E. Ladner, "Unequal loss protection: graceful degradation of image quality over packet erasure channels through forward error correction", *Journal on Sel. Areas in Communication*, 18 (6), p. 819-828, 2000.

[PAR 92] D. Parsons, *The Mobile Radio Propagation Channel*, Wiley-Pentech Publication, 1992.

[PRO 00] J. G. Proakis, *Digital Communications*, 4th ed., McGraw-Hill, 2000.

[REE 60] I.S. Reed and G. Solomon, "Polynomial codes over certain finite fields", *Journal of the Society of Industrial and Applied Mathematics*, 8 (2), p. 300-304, 1960.

[RIZ 97] L.M. Rizzo, "Effective erasure codes for reliable computer communication protocols", *ACM Computer Communication Review*, 27, p. 24-36, April 1997.

[SOU 04] B. Souhard, Codage conjoint source canal: application à la transmission d'images fixes sur canal ionosphérique (in French), Thesis, Poitiers, France, 2004.

[TAU 03] D. Taubman, R. Prandolini, "Architecture, philosophy and performance of JPIP: internet protocol standard for JPEG2000", *International Symposium on Visual Communications and Image Processing (VCIP2003), SPIE*, vol. 5150, p. 649-663, July 2003.

[VAU 04] R. Vauzelle, Y. Pousset, F. Escarieu, "A sensitivity study for an indoor channel simulation", *Annals of Telecom*, 59 (5-6), p. 655-672, May 2004.

Conclusion

This book, dedicated to the compression of medical images and signals, is the result of the collective work of a number of physicians and scientific researchers from a dozen different French research teams. Working on the project has provided the stimulus for cross-laboratory exchanges and collaboration between researchers and various scientific communities. Indeed, a number of chapters have been co-written by researchers from different teams. In addition, this work goes hand in hand with the new policy implemented over the years in most scholarly societies on promoting medical and biological engineering and specifically the processing of medical signals and images.

Most specialists agree on the fact that it is essential and inevitable to proceed with the compression of medical data in order to guarantee a good Quality of Service in any PACS, wireless transmissions or even storage systems. We have understood in the course of our work that most compression techniques used on medical data today are of a rather "general application" in the sense that they can apply to medical images from a variety of sources. We thus believe that there is still room for improvement in this particular field before reaching an optimal performance for all encoders applied to medical images. Why not have specific encoders for ultrasound images or videos? Or a specific encoder for MRI images? Ideas are already bustling within the community, and many attempts at improvements have been undertaken. Systems must therefore adapt to the specificities of medical data, so that they can then become integrated into a DICOM standard to guarantee the durability of the image and its capacity to be shared and transmitted thereafter. Such solutions will not be found overnight, not necessarily for technical reasons but more as a result of financial, legal and organizational constraints. In the short term, we may begin by developing pre-processing and post-processing techniques that are either posterior or subsequent to already existing standards. Such changes should be made with the aim of improving the performance of compression procedures on medical data.

Other than the fact that compression of medical images is now accepted as necessary, the medical community have come to “tolerate” the use of irreversible methods. Nevertheless, the possible deterioration that results from these methods must be carefully controlled by objective and specifically targeted techniques of quality evaluation. Such techniques may therefore apply to monodimensional, two-dimensional or even three-dimensional data.

Finally, we must acknowledge the importance of security issues in any compression process; a problem that has been thoroughly addressed in the last two chapters of this book.

As referred to in the Preface, we have provided our community with the “MeDEISA” database, accessible on the Internet, so as to evaluate the performances of our own algorithms. This database is constantly evolving as it is intended to include images from the most recent acquisition procedures. Images from the MeDEISA have been used on many occasions in this work.

We truly hope that this work will serve as a useful guide to all students, engineers, professors, researchers and physicians for any further projects and research carried out in the field of medical data compression. Our aim was not to dictate the findings from one community but rather to gather the work from both the field of technology and that of medicine.

List of Authors

Elsa ANGELINI
TSI Department
Telecom ParisTech
Paris
France

Marie BABEL
IETR, CNRS
INSA–Rennes
France

Dominique BARBA
IRCCyN, CNRS
Ecole Polytechnique
University of Nantes
France

Atilla BASKURT
LIRIS, CNRS
INSA–Lyon
Villeurbanne
France

Christine CAVARO-MÉNARD
LISA
University of Angers
France

Joël CHABRIAIS
Centre hospitalier Henri Mondor
Aurillac
France

Gouenou COATRIEUX
LaTIM, INSERM
Telecom Bretagne
Brest
France

Olivier DEFORGES
IETR, CNRS
INSA–Rennes
France

Guy FRIJA
HEGP
Paris
France

Bernard GIBAUD
VisAGeS, IRISA, INSERM,
INRIA, CNRS
University of Rennes 1
France

Pierre JALLET
CHU
Angers
France

Christel LE BOZEC
SPIM, INSERM
HEGP
Paris
France

Patrick LE CALLET
IRCCyN, CNRS
Ecole Polytechnique
University of Nantes
France

Jean-Jacques LE JEUNE
Ingénierie de la Vectorisation Particulaire
INSERM
Angers
France

Khaled MAMOU
ARTEMIS Department
Telecom SudParis
Evry
France

Amine NAÏT-ALI
LISSI
University of Paris 12
Créteil
France

Christian OLIVIER
SIC
University of Poitiers
France

Azza OULED ZAID
SYSCOM
ISI, ENIT
Tunis
Tunisia

Benoît PARREIN
IRCCyN, CNRS
Ecole Polytechnique
University of Nantes
France

Françoise PRÊTEUX
ARTEMIS Department
Telecom SudParis
Evry
France

Rémy PROST
CREATIS, CNRS, INSERM
INSA–Lyon
Villeurbanne
France

William PUECH
LIRMM, CNRS
University of Montpellier 2
University of Nîmes
France

Jean-Yves TANGUY
CHU
Angers
France

Sébastien VALETTE
CREATIS, CNRS, INSERM
INSA–Lyon
Villeurbanne
France

Rodolphe VAUZELLE
SIC
University of Poitiers
France

Index